RUDDER, STICK and THROTTLE:

Research and Reminiscences on Flying in Nebraska

by

Robert E. Adwers

Making History
Omaha, Nebraska
1994

This is a saga about flying, flying machines and flyers everywhere, but in Omaha, Douglas County, and Nebraska in particular. The saga started with Orville and Wilbur Wright at Kitty Hawk, North Carolina, and probably won't end until it reaches outer space or beyond, or maybe even Heaven.

ILLUSTRATIONS

ACKNOWLEDGEMENTS

I would like to express my thanks to the following people and institutions who assisted me in my research:

Information and illustrations concerning the Baysdorfer brothers was kindly furnished by Betty Baysdorfer (Mrs. Don) Moucka, Judy Baysdorfer Foisey, Bonnie Baysdorfer Stavneak, Gerald Baysdorfer, and the late Hans Moeller.

Most of my research on Captain Frisbie's death was furnished by a wonderful lady, Lydia Bishop, long-time Norton, Kansas, resident and a retired Norton High School teacher.

Billie (Mrs. Halsey) Noyes provided information and photographs on early aeroplane flights at Waterloo, Nebraska.

A young librarian in Humboldt, Nebraska, Charles and Virginia Marburger, and Errold Gordon Bahl provided information on aviator Errold Grover Bahl's life. Frances Nordland Holmgren saw Errold Bahl and Charles Lindbergh land at Auburn, Nebraska. Errold Gordon Bahl also authorized me to use the contents of his mother's book about his father, "The Forgotten Pilot, Errold G. Bahl, Early Instructor and Friend of Charles A. Lindbergh."

Flora May "Brownie" Rimmerman Gow graciously shared information on her brother, Benjamin Rimmerman.

Libraries in Fremont, Auburn, Humboldt, Grand Island, Kearney, Broken Bow and Norfolk, Nebraska, provided assistance, as did the History Department at the W. Dale Clark Library in Omaha, Nebraska, especially Bob Potter and Bob Flood.

The libraries in Norton, Kansas; St. Joseph, Missouri; Sioux City, Des Moines, Ottumwa and Davenport, Iowa; Galesburg, Illinois; Terre Haute and Logansport, Indiana; Detroit, Michigan; and Rochester and Elmira, New York, were of great assistance.

The Historical Society of Douglas County, its library and archives, were a gold mine of resources. Roger Reeves (the director), Ann Haller (who knows where to look), John Brudney (who got me through the newspapers), Rita Jerins, and the greatest volunteer in the place, Monica Geier, provided a great deal of information.

The Nebraska State Historical Society in Lincoln, Nebraska, John Carter from their photo archives, and another great volunteer, Vince Goeres, were of great help.

The Western Heritage Museum in Omaha, Nebraska, and photography volunteers Vern Lenzen and Joe Masek were very helpful.

Marion Nelson, long-time Omaha pilot and flying-service operator, provided a lot of information.

Richard Joyce gave details on his experience during the raid on Tokyo. Old friends and World War II flyers George Seemann, Lee Seemann, Judson Hansen, Fred Wupper, Richard Lohse, William T. O'Hanlon, Lloyd Grimm, Owen Knutzen, and other friends who flew in the war shared their stories. Peter Lage (Navy), Richard Slaybaugh (infantry), and John Childe (artillery) shared their stories, too.

I am grateful for some great memories and conversations with my mom and dad, Nina and George Adwers. And I'll never forget those million hours of "hangar flying" with old pilots, young pilots, new pilots, good pilots, very few bad pilots and the old Red-tailed Hawk.

AUTHOR'S PROLOGUE

My mom told me I flapped my arms and ran as soon as I could walk and said, "Look, Mom! See bird! I can fly!"

I came by it honestly. Both my dad, George Clarence Adwers, and my mom, Nina Detrick Adwers, were flying enthusiasts. They saw some daredevil aeronaut fly his balloon at the Adams County Fair in Hastings, Nebraska, in 1902 when they were courting. They were at the Ak-Sar-Ben Fall Festival in 1906 when Charles K. Hamilton piloted the first airship to fly in Nebraska. Mom used to say, "Why, he flew his ship above Farnam Street right by the Omaha City Hall!" And indeed he had. They saw Glenn Curtiss fly the first aeroplane in Omaha in 1910, went twice in hot July weather to see his Flyers, and even took pictures with their brand-new $10 Kodak. They watched Lincoln Beachey do his famous loops at the Ak-Sar-Ben Fall Festival in October of 1914. My mom was carrying me then. I was born the next May, about a month-and-a-half after Beachey was killed in San Francisco Bay at the Panama-Pacific Exposition. I guess I was hooked on flying before I was born.

I remember the balloons flying at Fort Omaha. They were big and sort of scared me. The first aeroplane I saw was in the summer of 1920: It was a World War I de Havilland and was carrying airmail. She landed at the government field in Omaha, where the Ak-Sar-Ben race track is today. I was hooked! I remember frequently hounding my dad to take me out to the airfield in his 1919 Model T.

I still go out of the house or stop and look up whenever I hear a plane. There are a lot of airliners letting down for Eppley Airfield up there. When Offutt Air Force Base has a south "T-setting," my house is right on "base leg" for STRATCOM's (Strategic Command) longest runway. I get to see everything from peashooters to the B-1. Sometimes the Boeing 747 "Looking Glass" hangs right over my house. One day they had the Russian Foreign Secretary visiting, and SAC paraded our whole Air Force by—F-16s, B-52s, B-1s, KC-135s, DC-10s! Man, it was great! All kinds of fun flyers go by, too. One's going over right now—a Cessna 150, judging by the prop and engine sounds.

Last summer the Air Force Association to which I belong was allowed to look over the B-1 completely. They practically had to drag me out of the cockpit: I would give my right arm to drive her.

I still have my pilot's license and fly quite often. I really get the thrill of honest-to-God flying in a 1943 Stearman biplane tail-dragger that belongs to a young friend of mine, Robert R. Alf. I even loop or roll it when I'm feeling my oats.

But the birds, aren't they something? A hawk soars, riding the thermals, and a blackbird, pumping furiously, dives at him. Robins fly into trees and never hit a branch. All of them roar in, drop their gear and flaps, and (boom!) land on a twig, fence, weed or post. Hummingbirds, geese, ducks, pheasants, quail, sparrows, wrens and those shiny blue-black swallows—oh, how I always have envied the birds! They flap their wings and *fly*. I had to settle for the next best thing: I learned how to fly an airplane a long time ago. . . .

"Look, Mom! See bird! I can fly!"

Robert E. Adwers
Elkhorn, Nebraska
July 1994

INTRODUCTION

Birds fly; almost all do. A few animals and reptiles sort of fly, even some fish. Clouds float in the air. Leaves glide; so do snow flakes. Cottonwood seeds soar to the whims of the winds, as do hats, paper, dust, skirts and sometimes houses. Kites do a kind of flying. Some people hang in gliders, sail, soar and even sky dive. Athletes sometimes do exciting and dramatic flying. But only the birds fly with their bodies, their muscles, their feathers, their know-how. Okay, bugs and insects fly, but not the way birds do.

Darius Green and his Flying Machine

2

Archeology and history confirm that people have always dreamed of flying, from Daedalus and Icarus to Darius Green and his flying machine.

"Darius Green and his Flying Machine"

"The birds can fly an' why can't I?
Must we give in," says he with a grin,
"That the bluebird an' phoebe
Are smarter'n we be?
Jest fold our hands an' see the swaller
An' blackbird an' catbird beat us holler.
Doos the little chatterin', sassy wren,
No bigger'n my thumb, know more than men?
        Jest show me that!
        Ur prov't the bat!

Hez got more brains than's in my hat,
An' I'll back down, an' not till then!"
He argued further: "Nur I can't see
What's the use of wings to a bumblebee,
Fur to git a livin' with, more'n to me—
        Ain't my business
        Important's his'n is?
        That Icarus
        Made a perty muss—
        Him an' his daddy, Daedalus.
They might a' knowed wings made o' wax
Wouldn't stand sunheat an' hard whacks.
        I'll make mine o' luther,
        Ur suthin' ur other." . . .

As a demon is hurled by an angel's spear,
Heels over head, to his proper sphere—
Heels over head, and head over heels,
Dizzily down the abyss he wheels—
So fell Darius. Upon his crown,
In the midst of the barnyard, he came down.
In a wonderful whirl of tangled strings,
        broken braces and broken springs,

Broken tail and broken wings.
Shooting stars and various things;
Barnyard litter of straw and chaff,
And much that wasn't so sweet by half.
Away with a bellow fled the calf,
And what was that? Did the gosling laugh?

'Tis a merry roar from the old barn door.
And he hears the voice of Jotham crying,
"Say, D'rius, how do you like flyin'?"
Slowly, ruefully, where he lay,
Darius turned and looked that way.
As he stanched his sorrowful nose with his cuff.
"Wal, I like flyin' well enough,"
He said; "but the' ain't such a thunderin' sight
O' fun in 't when ye come to light."[1]

There were flying carpets, flying gods, flying cupids and flying angels. In the fifteenth century, Leonardo da Vinci designed helicopters and parachutes. Seventeenth-century French philosopher Bernard de Fontenelle doubted "there shall never be a Communication between the Moon and the Earth." He noted that

> several have found the secret of fastening Wings, which bear them up in the Air. . . ; I cannot say indeed they have yet made an Eagles Flight, or that it doth not cost now and then a Leg or an Arm to one of these new Birds.[2]

How true this was! The philosopher concluded, "The Art of Flying is but newly invented, it will improve by degrees, and in time grow perfect; then we may fly as far as the Moon."[3] Had Fontenelle been around, he could have written the same thing in 1903. Wouldn't he

---

[1]John Townsend Trowbridge, "Darius Green and his Flying Machine," in *Omaha Evening World-Herald,* October 22, 1907.

[2]Bernard de Fontenelle, *A Plurality of Worlds,* trans. by John Glanville (1686), as quoted in Alvin M. Josephy, Jr., ed., *The American Heritage History of Flight* (New York: American Heritage Publishing Co., Inc., 1962), p. 36.

[3]*Ibid.*

have been amazed if he had witnessed Neil Armstrong's "one giant leap for mankind" in 1969?

Is there anyone today who hasn't lain on a grassy slope and watched a Red-tailed Hawk fly lazy-eights against the blue backdrop of space? Is there a boy, a man, a girl, a woman who hasn't thought, "If only I could fly with you on your great open stage?"

Two brothers from Dayton, Ohio, were the first people to truly fly—to live their dream in a flying machine, the *Wright Flyer.* Orville and Wilbur Wright didn't become the first to fly by luck alone: They worked long and hard at it. They studied, designed, tried and erred. The pair read all they could find about flying and made inquiries to the Smithsonian Institution. They studied the glider flights and data of Otto Lilienthal, a German who made the first successful airborne flight, and they consulted with Octave Chanute, who was a pretty good aeronautical engineer. The Wrights observed the birds, made models, and even made a wind tunnel to test wing design. They designed, built and flew gliders. Next, Wilbur flew one while lying prone, guiding it by twisting and shifting his body weight. The work the Wright brothers began in 1898 was well documented: Their notes, photos, drawings and diaries are fantastic; their research and investigation was unbelievably detailed.

The two were ready to build a powered flying machine by 1902. After consulting with the U. S. Weather Bureau for a location with steady winds, open spaces and little population, they picked Kitty Hawk, North Carolina. There, the would-be aviators built a shop, a hangar and living quarters. They spent most of the summers of 1901 and 1902 experimenting at Kitty Hawk, returned to Dayton to build a plane and engine in their bicycle shop, and then traveled back to Kitty Hawk in November 1903. What they were about to do would change the world forever.

Orville and Wilbur Wright accomplished the world's first powered, sustained, steered flight. To their credit, they did it on their own—no government funds, no consultants, no subsidies—just plain old perseverance, ingenuity, talent and desire. It was a real American free-enterprise project. When the Wright brothers flew their *Flyer* over the sands of Kitty Hawk on December 17, 1903, they joined the company of great American inventors like Robert Fulton, Cyrus McCormick, Eli Whitney, Thomas Edison, Harvey Firestone, and Henry Ford.

BOOK ONE:

A New Bird in the Sky

The Wright Flyer, December 17, 1903

CHAPTER ONE:
THE DAY THE WORLD CHANGED

The world changed forever in twelve seconds and about one hundred feet at 10:35 a.m., Thursday, December 17, 1903. Only seven people knew it: John T. Daniels, W. S. Dough, A. D. Etheridge, W. C. Brinkley, Johnny Moore, and Wilbur and Orville Wright. They were there when it happened. Wilbur won the toss to take the controls, but his first attempt to fly failed. Orville then

#orm No. 168.

THE WESTERN UNION TELEGRAPH COMPANY.

—— INCORPORATED ——
23,000 OFFICES IN AMERICA.   CABLE SERVICE TO ALL THE WORLD.

This Company TRANSMITS and DELIVERS messages only o . conditions limiting its liability, which have been assented to by the sender of the following message. Errors can be guarded against only by repeating a message back to the sending station for comparison. and the Company will not hold itself liable for errors or delays in transmission or delivery of Unrepeated Messages, beyond the amount of tolls paid thereon, nor in any case where the claim is not presented in writing within sixty days after the message is filed with the Company for transmission.
This is an UNREPEATED MESSAGE, and is delivered by request of the sender, under the conditions named above.
ROBERT C. CLOWRY, President and General Manager.

RECEIVED at

176 C KA CS 33 Paid.   Via Norfolk   Va

Kitty   Hawk N C Dec 17

Bishop M Wright

       7 Hawthorne St

Success four flights thursday morning   all against twenty one mile

wind started from Level with engine power alone   average speed

through air thirty one miles longest 57 seconds inform Press

home Christmas .   Orevelle Wright   525P

Orville Wright's telegram

took flight. It was a raw, windy morning at Kitty Hawk, near Nags Head, North Carolina, and, when Orville flew, the world and all things living in it, on it and above it changed.

Hardly anyone got excited about those happenings at Kitty Hawk. Western Union even garbled Orville's historic message to his father, Bp. Milton Wright, in Dayton, Ohio, by misspelling Orville's name and giving the wrong time for the day's longest flight.

Nor did Orville's "inform the press" inspire newspaper editors to epic editorials, stories or extra editions. In fact, the few stories printed were fragmentary and misleading. To be fair to the nation's press, editors were skeptical; they were weary and wary of claims that had proven false and of flying machines that couldn't fly. Just nine days earlier, the august scientist and Chairman of the Smithsonian Institution, Professor Samuel Pierpont Langley, had suffered a second failure in an attempt to fly his huge machine above the Potomac River. Langley's claims and subsequent actions later made the Wrights furious.

How did Omaha, Douglas County, and Nebraska take the astounding news of the Wrights's flight? To tell the truth, very few knew about it or gave a darn! Edward Rosewater, editor of the *Omaha Daily Bee*, was so little impressed that his story entitled "Is A Real Flying Machine" (shown at right) didn't appear until two days after the flight, on December 19, 1903. It was buried on page nine, between advertisements for "Montana Diamonds" and a cure for "Special Diseases of Men." Other headlines on the page read "Finds Wife Dead on Floor" in Des Moines, Iowa, "Livery Drivers on Strike" in Chicago, two drunks "Shows [*sic*] Blair a Lively Time," and (holy cow!) "People Contract Anthrax" near Auburn, Nebraska.

The *World-Herald,* not to be scooped, printed the identical story the same day, placing it on the editorial page with the

IS A REAL FLYING MACHINE

Ohioan's Invention Rises, Sails and Lights as Desired.

ENGINE AND PROPELLERS THE FORCE

In First Test Machine Soars Through Space for Three Miles and Then Descends to Selected Landing Place.

NORFOLK, Va., Dec. 18.—A successful trial of a flying machine was made yesterday near Kitty Hawk, N. C., by Wilbur and Orville Wright of Dayton, O.

The machine flew three miles in the face of a wind blowing at the registered velocity of twenty-one miles an hour and then gracefully descended to earth at the spot selected by the man navigating the car as a suitable landing place.

The machine has no balloon force, but gets its force from propellers worked by a small engine.

Preparatory to its flight the machine was placed upon a platform near Kitty Hawk. This platform was built on a high sand hill headline: "Successful Trial of Airship." Other stories on the page that day included: "Egg Famine in Lincoln" and "Carlisle Indians Play Today."

Credit should be given where credit is due, though: Both the *Bee* and the *World-Herald* did report something about the day the Wright brothers flew. They both simply reported that a successful trial of a flying machine was made on December 17 at Kitty Hawk, North Carolina. Whoever wrote the story reflected the public's general indifference.

Later, the Wrights tried to set the record straight. On January 6, 1904, in Dayton, they gave a statement to the Associated Press. The *Bee* printed it in on January 7, 1904. The *World-Herald* also printed the release verbatim that day. It placed the release on page three, next to the continuation of the big story about a train wreck near Topeka, Kansas, that took twenty-eight lives, and a story reporting that Frank Natejicke, a farmer near Schuyler, Nebraska, was killed instantly by Union Pacific Train No. 2 as it sped sixty miles per hour past the Schuyler depot. The *Bee's* story appeared on page eight along with the Kansas train wreck, but with no mention of Natejicke.

Flying through the air in a machine wasn't big news to Nebraska folks in 1903 and 1904. It wasn't very interesting to most people in the United States, Great Britain, France, Germany, Russia, Italy, Africa, or South America or to several million Chinese, Japanese, Indians or Arabs. Most folks in Omaha, Douglas County and Nebraska didn't get worked up about the fact that *the world had changed,* except maybe for a few boys, a few tomboy girls, Omaha's Baysdorfer brothers (Otto, Gustav A. and Charles), some brothers up in Holt County, Nebraska, named Savidge, maybe a Minnesota kid named Lindbergh, and a few others scattered around the country.

CHAPTER TWO:
LEARNING HOW TO FLY

Humans get along very well on land. They scoot, hop, walk, trot, jog, run, jump, slide and perform some great feats. They also do well in water. Almost everyone can swim: A few swim fast and far; some swim with precision and grace; and others just look great in swimming suits. But none of us ever have been able to flap our arms, kick our legs, bend our bodies, and fly through the air like a bird.

Although we are unequipped by nature to fly, some of us always have yearned to do so. We simply had to figure out how it was done. The world's best flyers just do it naturally: When mom and dad think their fledglings are ready to fly, they push them out of the nest, and away they go. They aren't always completely successful, but the young birds do fly, and in a short time they become wonderful flying machines.

The First Flights at Kitty Hawk

The Wrights's four flights on December 17, 1903, were truly miraculous, since the Ohioans didn't instinctively know how to fly like the birds. The first humans to achieve controlled, sustained and powered flight, they soon discovered that the *Wright Flyer* went where it wanted to go or where the wind took it. Each flight ended when the machine decided to hit the sand.

Fortunately, the plane never fell from a great height, nor was it going fast. Still, it was damaged on each flight. On the first, the engine crank lever broke, and the rudder skid cracked. The left wing tip was damaged on the third flight. On the fourth flight—the longest at eight hundred feet—the plane landed so hard the front rudder frame was damaged extensively, so the brothers decided to call it a day. Then, while they were carrying the plane back to the shop with help from the Kitty Hawk rescue station crew, a sudden gust of wind turned it over and really broke it up.

What a day it was: Four flights, and the Wrights were still alive! In fact, they had suffered no injuries. Knowing they were going to have to learn how to really fly their machine, the pair returned to Dayton.

Flight and Aircraft Theory

Why were they successful? Well, first the Wrights designed and built a wing that would create lift when pulled through the air. Second, they designed and built a gasoline engine that was light and powerful enough to drive a propeller that would lift itself, the aeroplane and the person in it. They knew they needed to steer the machine, so they put elevators out in front for up and down ("pitch"), rudders for right and left ("yaw"), and warped the wing tips right and left for banking ("roll"). That's why it flew! Sounds simple enough, but everything had to coordinate precisely.

1904 and 1905

Back in Dayton, the Wrights had the only aeroplane in the world that really flew. Still, they had to work on flying, so they located a one-hundred-acre meadow that would make a suitable landing field. It was about eight miles from town and was called Huffman Prairie. Then, they built plane number two, which had many improvements over the *Flyer.* The brothers stored and saved the original *Flyer*—more will be said about it later. They also made a new starting track with a weight-and-derrick system for launching. All total, the Wrights made 105 flights in 1904, the longest lasting five minutes eight seconds and covering a three-mile course. Most importantly, they *learned how to fly in a circle,* which required complete control of the aeroplane.

Amos I. Root, publisher of the very popular magazine *Gleanings in Bee Culture,* happened to be at Huffman Prairie on the day in 1904 on which Wilbur made history's first full-circle flight in a powered machine. In the first published eyewitness report of a Wright brothers' aeroplane flight, Root explained that

> God in His great mercy has permitted me to be, at least somewhat, instrumental in ushering in and introducing to the great wide world an invention that may outrank electric cars, the automobiles, and all other methods of travel. . . . It was my privilege on the 20th day of September, 1904, to see the first successful trip of an airship, without a balloon to sustain it, that the world has ever made, that is, to turn the corners and come back to the starting-point . . . [and] there was not

another machine equal to such a task . . . *on the face of the earth. . . .*[4]

Physicists claim that, theoretically, it's impossible for a bumblebee to fly. The bumblebee doesn't know that, and now Root knew humans could fly too.

The Wright Brothers Learn to Fly

Wilbur Wright wrote, "During all the flights we had made up to this time we had kept close to the ground, in order that, in case we met any new and mysterious phenomenon, we could make a safe landing."[5] Mainly, they stayed alive and learned.

At the end of 1905 they had made many improvements in the engine, propellers, controls and structural strength of their newest *Flyer,* and on its forty-sixth flight she stayed airborne thirty-eight minutes three seconds. This represented thirty circles around the field and some twenty-four miles! Wilbur exclaimed, "Our 1905 improvements have given such results as to justify the assertion that flying has been transferred from the realm of scientific problems to that of useful arts."[6] The Wright brothers now were ready to place their flying machine on the world market.

Epilogue

Still, the world wasn't impressed. A few French, English and Americans had tried to fly since Kitty Hawk, but none had succeeded. In Nebraska, boys flew kites, jumped off haystacks with an umbrella, or held out their arms and ran, pretending they could fly. Every now and then a daring aeronaut would thrill folks at a county fair by dramatically parachuting to the ground from his hot

---

[4]Amos I. Root, *Gleanings in Bee Culture* (January 1, 1905), in *American Heritage History of Flight,* p. 152.

[5]Wilbur Wright, "Summary of Experiments of 1904," in Lynanne Wescott and Paula Degen, *Wind and Sand: The Story of the Wright Brothers at Kitty Hawk* (New York: Harry N. Abrams, Inc., 1983), p. 150.

[6]Wilbur Wright, in Wescott and Degen, *Wind and Sand,* p. 150.

air balloon. No one as yet had flown an aeroplane over Omaha or any other spot in Nebraska.

That four-wheeled marvel, the horseless carriage, was winning acceptance everywhere, but with the automobile you almost always had four wheels on the ground. People could live with the automobile. In Omaha there were twenty-seven auto dealers selling thirty-one makes of cars. Yet many things were to pass before the aeroplane became a reality.

Flight evolved from balloons to airships to aeroplanes. People first flew with success in lighter-than-air balloons. On June 4, 1783, the French brothers Joseph-Michael and Jacques-Etienne Montgolfier recorded the first balloon flight when their unmanned hot-air balloon flew to a height of over three thousand feet, floated majestically for more than ten minutes, then dropped lightly to earth over a mile downwind. An eyewitness to the flight, physics professor Jacques Alexandre César Charles, was an expert on the laws governing gases. He suggested that they fill the balloon with hydrogen, so it would really fly.

The Montgolfiers did just that, even though hydrogen was difficult to produce in large volumes. They filled another unmanned balloon with hydrogen, and this one soared. Benjamin Franklin, our ambassador to France, was in the crowd and marveled at the flight. Someone turned to Franklin and muttered, "What good is it?" Old Ben smiled and replied, "My friend, what good is any new born baby?"[7] Next, the Montgolfiers built a bigger hot-air balloon and sent up a sheep, a rooster, and a duck in it. All three animals survived the landing and thus became the first real aeronauts.

On November 20, 1783, Jean François Pilâtre de Rozier (no relation to Mike Rozier) and François Laurent, Marquis d'Arlandes, made a five-mile untethered flight of more than fifteen minutes on the outskirts of Paris. The first men to see the earth while flying through the air, they were given that great name, "aeronauts."

Balloons, beautiful as they were, weren't as interesting to Americans as they were to Europeans. At the start of the twentieth century, balloons were used to entertain people at fairs, expositions and carnivals in the United States. True, they had been used earlier in the Civil War for artillery spotting and observation. They even flew cross-country, but aeronauts in them had to go the way the wind wanted, so they weren't considered practical. Nonetheless, balloons proved vital to the development of flying, for almost all those who became aeroplane 'drivers' began their careers in a balloon.

---

[7]Josephy, *American Heritage History of Flight*, p. 41.

Balloons didn't reach Nebraska until the late 1890s. Perhaps the first flight was made by a traveling aeronaut calling himself "Professor La Wing" (or some other fancy name) to entertain local gentlemen, ladies and youth with his daring balloon ascension.

The first recorded balloon flight in Omaha was made at the sixth annual Ak-Sar-Ben Carnival on July 27, 1900, when the *Omaha Daily News* arranged for the famous Murphy brothers to fly a balloon race.

As usual in Nebraska, the wind and weather canceled the first two days' flights. Then, finally, the weather cooperated, and Prof. Sam "King" Murphy guided his wonderful balloon up between the buildings surrounding the carnival grounds, 17th to 20th Streets, Dodge to Farnam. He floated eastward over the city. The crowd, estimated by city officials to be at least six thousand people, cheered and stood in awe as Murphy soared overhead, describing the wonderful view he had of Omaha.

After drifting for more than thirty minutes, Murphy landed

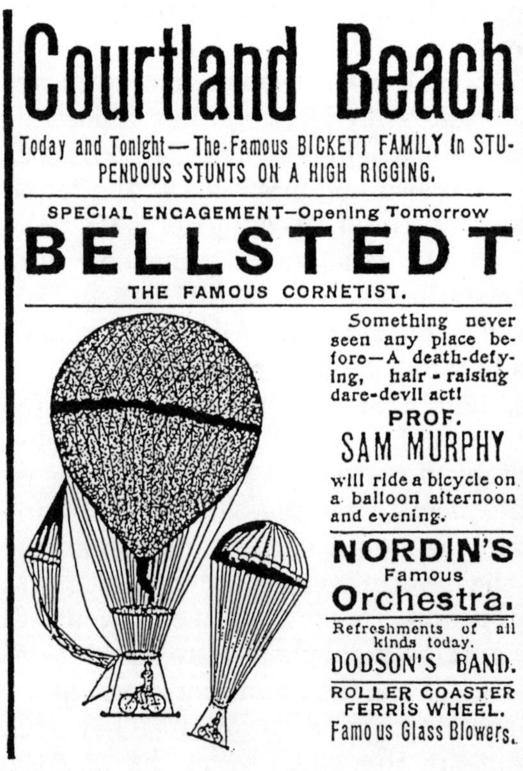

by the Union Pacific Missouri River bridge, just east of the Metz Brewery. A crowd of young men and boys enthusiastically followed his flight to assist his landing. One of the young men was Charles Baysdorfer, about whom more will be said later. Murphy was welcomed like the hero he was. The trophy room at the Brewery was turned into the "throne room," and there Murphy majestically held court and received crowds of admirers.

He flew at the Ak-Sar-Ben Fall Festival, Lake Manawa Park, Krug Park, and county fairs by the dozens. He could be found at the Nebraska State Fair or anywhere there were paying spectators. Murphy finished his Omaha career in 1904 at the famous Courtland Beach on beautiful Carter Lake, near what is Eppley Airfield today. Then "King" Murphy disappeared, balloon and all, like the Wizard of Oz, never to be seen again.

Another early Nebraska balloonist, John Waldorf Hall, came to Omaha in 1902. That year's city directory lists him as an aeronaut who resided at 617 South 17th Street. His next-door neighbor was Charles Baysdorfer, who roomed at 609 South 17th

Street and soon would become involved in air flight. Hall was the resident aeronaut at Krug Park. Owned by Omaha's Fred Krug Brewing Company and located at 52nd Street and Military Avenue, Krug Park was the city's top amusement park in the early 1900s. Gallagher Park is there now.

Hall thrilled thousands with his spectacular flights. On special occasions, when someone paid his additional fee of $100, he would suspend a cannon under two balloons and send it up in the air to over one thousand feet. There, with a spectacular explosion, the aeronaut would be fired from the cannon's mouth, dramatically would prolong opening his parachute to the last moment, and then float to *terra firma*. Wow, what a sight! Those early aeronauts were real risk-takers. Although many believed Harry LaThoma, who lived in Omaha in 1911, was Hall, LaThoma always denied it.

Otto, Gus and Charles Baysdorfer were among the most important of Nebraska's early aircraft enthusiasts. Beginning in 1903, Charles Baysdorfer flew balloons, and he and his brothers built and flew airships beginning in 1906. But all of this is covered in more detail below, in Book Two.

The last-known Nebraska aeronaut of the early 1900s was Harry Hensley, a painter who lived at 34th and Curtis Streets. He gave it up in 1904, not any too soon it appears from this *World-Herald* story:

> Harry Hensley . . . is strangely affected by making parachute leaps, and this morning he was taken to the police station in the ambulance and was cared for by the police surgeons. . . .
>
> His illness is a mysterious affection. He is a victim of a sinking heaviness at the pit of his stomach, as though he were falling. It is the same sensation that he experiences when he has "cut loose" from the balloon and falls some 150 feet before the parachute opens. Everyone has a similar feeling when going down in a fast elevator. . . .
>
> . . . The anxiety he undergoes lest at some time the parachute should fail to open is so continually upon his mind that nearly every night he dreams that he is falling. . . .
>
> Often he says the same sensation comes to him while he is walking down the street. He will grow dizzy and seem to be shooting downward at a terrific speed. . . . It was while he was in one of those attacks that he was taken to the police station this morning. . . .
>
> It is thought that he may be compelled to give up his business.[8]

Balloonists remained popular in Nebraska well into the twentieth century. I remember in 1928 one of them flew his hot-air balloon from the center of Krug Park. Dressed in gymnastic tights, he ascended sitting on a trapeze bar and performed gymnastics as he

---

[8]*Evening World-Herald,* August 17, 1904.

rose. His parachute was in a long tube resembling part of the trapeze suspension; thus he appeared to be hanging by his knees with no parachute at all. Then, at about six or seven hundred feet, almost directly overhead, he hung by his knees, and suddenly fell! He fell quite far before his chute pulled out of the tube and opened. Women screamed, a half-dozen fainted, strong men paled, and kids were suddenly quiet. As a young spectator, my heart skipped a few beats too, as he landed where Benson High School's football field is today.

CHAPTER FOUR:
AIRSHIPS

A sufficiently large balloon filled with hot air, hydrogen or helium can lift into the air an attached frame containing an engine-driven propeller, a rudder and a person to drive it. In the summer of 1898, Alberto Santos-Dumont, a Brazilian living in France, built and flew the first successful lighter-than-air ship. From this invention derived all airships: Zeppelins, dirigibles and blimps. Santos-Dumont was a darn good aeronaut, but he tired of having the wind determine his flight plan. He built an engine with a propeller, made a frame with a rudder attached, hooked it to a balloon, and flew the thing. He entertained the French with many flights, both good and bad, and won ten thousand francs for being the first to fly from St. Cloud to the Eiffel Tower and back in one hour.

The first airship flight in the United States was made at the St. Louis World's Fair in 1904. The fair promoters planned a balloon race to Washington, D.C. Although a bunch of balloons entered, only two got into the air, as the weather was so inclement. They both came down near St. Charles, Missouri, a few miles northwest of the fairgrounds. Santos-Dumont planned to be in the race with his new airship, but his invention mysteriously was damaged beyond repair while being unloaded at the fairgrounds. Santos-Dumont was justifiably distraught, but absolved his rival aeronauts of duplicity. After hearing of the *Wright Flyer,* he was anxious to return to France to build an aeroplane.

Santos-Dumont's competition was led by "Captain" Thomas S. Baldwin, who had brought his airship, the *California Arrow,* to St. Louis. Fair promoters said that Captain Baldwin was formerly of the U.S. Army, but he actually was a balloonist and parachutist from upper New York State who billed himself as "Captain." As an aeronaut, Baldwin could indulge in creative billing. The previous year he had flown a balloon and jumped at the Montgomery County Fair at Red Oak, Iowa. George Yager of Red Oak, later of Omaha, knew him.

Baldwin's contribution to flying was in teaming up with Roy Knabenshue, then considered the best balloonist in America, and Glenn Curtiss, a young motorcycle racer from New York. Curtiss, a genius at building engines for his cycles, held the world speed record for motorcycles—a mile in 56.666 seconds. Also on Baldwin's team was Charles K. Hamilton, Knabenshue's best student.

This team built the *California Arrow* in Los Angeles and brought it to St. Louis for the fair's great race. Piloted by Knabenshue, the *Arrow* was one of the two that went up in the race on July 20, 1904. It was the first American airship ever to fly. The crowd of thousands went wild over this flying balloon, and soon the *Arrow* and its crew were flying exhibitions around the country.

Gould C. Dietz, an Omaha lumber-coal dealer and real estate man, was Chairman of the 1906 Ak-Sar-Ben Festival. While on a business trip to New York City, Dietz witnessed one of Knabenshue's flights. He was hooked and began a lifetime's involvement with aviation in Nebraska. Omaha has never really recognized the tremendous part Dietz played in Nebraska's aviation history. Even though he wasn't a pilot, Dietz could almost be considered the "father" of aviation around here. From Hamilton's airship (1906) to Baysdorfer's (1907) to Curtiss's airplane (1910) to Beachey's loops (1914) to

Hamilton's 1906 Omaha airship flight

the Air Mail Service (1920) to the Aero Congress (1921) to the Omaha Municipal Airport (1925 to 1950) to the Omaha Air Races (1931) to his death in 1960, Dietz was there, flying with the Red-tailed Hawk. I think a plaque should be put up at Eppley Airfield, honoring him.

Dietz signed Knabenshue's airship to perform at the Ak-Sar-Ben Festival. According to their contract, Charles K. Hamilton would bring the ship to Omaha for $3,000 and attempt to make a successful flight each "pleasant day" of the fair. Hamilton was to receive an additional $500 each time he succeeded in keeping the ship floating above the carnival grounds for fifteen minutes. Not bad pay for flying in 1906!

Unfortunately, two carnival days didn't meet the definition of a "pleasant day." Everyone was champing at the bit. Finally on Friday, September 28, 1906, the weather cleared, and Hamilton flew his wonderful airship. The crowds went wild as Hamilton guided his machine over downtown Omaha. Louis R. Bostwick, Omaha's premier photographer, documented this first flight of an airship in Nebraska.

My dad and mom saw Hamilton fly up Farnam Street by City Hall! Then the aeronaut flew the great ship back to the carnival grounds, between 17th and 20th Streets and Dodge to Farnam, where his ground crew lowered the ship to earth and tied it down. Charles Baysdorfer and his brother Gus were part of that crew. Charles Hamilton became the toast of Omaha, honored by all, and he got more ink in the local papers than all the world events of the day. This was Omaha's first view of a successful airship aeronaut.

Having worked with Hamilton's crew during preparations for this first successful airship flight in Nebraska, Omaha's Baysdorfer brothers began building their own airship in 1906. They first flew it in Omaha in August 1907, but more about this will be said in Book Two below.

Early in 1909, the U.S. Army Signal Corps completed the installation of a balloon plant at Fort Omaha, and aviation activity commenced at that location. It was in April that Capt. Charles F. Chandler piloted the first balloon at that site, *Signal Corps No. 12*. Lt. J. Ware accompanied him as a passenger. The successful flight ended about ten miles northeast of the Fort near Crescent, Iowa.

Later in 1909, the *Signal Corps Dirigible No. 1* was brought to Fort Omaha by Lieutenants Benjamin D. Foulois, Bamberger and Winters. "Captain" Thomas S. Baldwin had built the ship and Glenn Curtiss the engine, and it had been flown with some success in St. Louis. Lieutenant Foulois, the third Army Signal Corps officer to be trained by the Wrights at Fort Myer, Virginia, in the first Army aeroplane, later became a major general and head of the Army Air Corps. He is credited with the development in 1935 of the most powerful bomber in the world, the XB-17, which became the B-17 *Flying Fortress* of World War II fame.

Now that we're up to date on balloons and airships in Omaha and Nebraska, we'll return to aeroplanes.

The aeroplane evolved slowly. There were several reasons for this, and foremost was the Wrights themselves. They took all of 1904 and most of 1905 to learn to fly and then to improve their machine. Then, feeling strongly that, as the first to build and fly a successful aeroplane, they were entitled to a profit, they kept their knowledge to themselves and patented their wing-rolling system.

The second reason underlying the slow development of aeroplanes was communications. In the early years of the century, information spread only via word-of-mouth, letters, newspapers, periodicals, wireless and a very few telephones. This meant months, or years, before information got around the world.

In the third place, most of Europe and the United States already had what were considered effective transportation systems: Networks of railroads covered the land; steam-driven ships crossed the seas; and travel by air wasn't deemed necessary. In addition, that young giant—the automobile—gave individuals the freedom to discover the open road.

The greatest impediment to flight, though, was that the atmosphere above the earth is hostile to humankind. It doesn't have enough oxygen to keep people alive above eighteen thousand feet. It's got weather problems—unbearable heat and cold, winds, rain, hail, snow, fog, clouds and violent storms. This meant that the aeroplane wasn't as easy to adapt to as was the horse, or the donkey, camel, elephant, cart, wagon, buggy, train or automobile.

The Wrights stopped flying in October 1905. For the next two-and-a-half years, the pair tried to sell their aeroplane, first to the U.S. government, then to the governments of France and Great Britain. In 1907 they offered their aeroplane to several European governments for $250,000. They had no takers. The French had always been leaders in aviation, and, when France showed some interest, the Wrights stored the aeroplane they had brought across the Atlantic in LeHavre, France. Then they returned home to Dayton.

One reason for the lack of interest in Europe and the United States was that there were now imitators. On October 23, 1906, about three years after the Wrights's flight, Alberto Santos-Dumont flew some eighty feet in an aeroplane he had built. On July 4, 1908, that young motorcycle racer and balloonist Glenn Curtiss successfully flew his *June Bug* before several hundred people at Hammondsport, New York. Then in Great Britain, an American, S. F. Cody (no relation to Col. William F. "Buffalo Bill" Cody), built *British Army Aeroplane No. 1* and flew it on October 16, 1908.

Curtiss's *June Bug* started an historic conflict with the Wrights. The Wrights had patented their wing-warping system for roll or bank control in 1906; then in 1907 a Frenchman named Robert Esnault-Pelterie built roll controls he called ailerons into a glider and wrote a paper about roll control. Curtiss also knew that roll control was essential for successful flight, and he built wing-tip ailerons for banking his *June Bug* in 1908. The Wrights thought Curtiss's ailerons infringed on their patents, and, after a long, bitter fight in the courts, the Wrights won.

Ailerons were here to stay. All aeroplanes flying in 1908 except two, Blériots and Antoinettes, were almost exact copies of the *Wright Flyer*. These planes were the Voisins, Farmans, Curtisses and Codys. All provided some kind of bank control; a few had wheels instead of skids; most were biplane pushers, with the pilot sitting in front; and the engines usually were built by the plane's designers. Many were planning and building aeroplanes, and not a few were killing themselves in their crazy contraptions.

Finally, in February 1908, the U.S. Army accepted the Wrights's bid of $25,000 to build an aeroplane for military use. It would have to carry two men, have fuel for 125 miles, and be able to fly at least ten miles nonstop at an average speed of forty miles per hour. A month later, they sold aeroplane patent rights in France to a syndicate of Parisians for $100,000. Suddenly, the *Flyer* had come to life.

Since they hadn't flown in over two years, Orville and Wilbur Wright went back to Kitty Hawk and flew their modified 1905 machine. The pilot sat up in this version, but the Wrights retained their previous launcher-and-skids landing gear. Then they went to work modifying, improving and building the first U.S. Army aeroplane.

By midsummer, *U.S. Army No. 1* was almost ready to fly, so Wilbur sailed to France to assemble and fly the ship they had stored in LeHavre in 1907. There he thrilled Europe with record-setting flights. Louis Blériot considered the Wright machine superior to any European machine, including his own.

Orville's flights in the new two-seater, *U.S. Army No. 1,* at Fort Myer early in September were spectacular. He broke record after record as he effortlessly circled the parade ground, banking and turning, up and down. The new ship far exceeded the tough specifications the U.S. Army required, and Orville flew it with complete control. The Army, Washington officials, and politicians were all dumbfounded—and it's a rare thing when politicians are speechless!

Unfortunately, on September 17, the last day of testing *U.S. Army No. 1* in two-man flights, Lt. Thomas O. Selfridge, the first Army passenger to ride with Orville, also became the Army's first fallen aviator. The flight seemed picture-perfect; the ship flew like a bird; and then a prop broke! This triggered a succession of mechanical failures. Orville tried to guide the plane to the ground, but it wouldn't fly. The crash killed Lieutenant Selfridge and injured Orville so severely that his recovery took a long time.

Nonetheless, the Army was pleased with Orville's flights and was sold on the plane's performance, so Wilbur returned from France and immediately started building a plane to be delivered early in 1909. The stage was set, and the curtain would soon rise on aviation.

By 1909, aeroplanes were being flown and built everywhere—in the U.S., Great Britain, Canada, the Austro-Hungarian Empire, Sweden, Russia and even Turkey. Louis Blériot flew across the English Channel from France to England in his latest monoplane. Wilbur Wright thrilled a million New Yorkers with a flight up the Hudson River from Governors Island to Grant's Tomb. All over the world, newspapers offered huge cash prizes for every aspect of flight—speed, endurance, city-to-city, altitude, fastest take-off, and for crossing the English Channel. One even offered £10,000 for the first flight to Mars and back in a week. Americans like Glenn Curtiss, Glenn L. Martin, Grover Loening and Donald Douglas joined the Wrights in the aeroplane business.

CARRYING BALLAST.

From the Washington Sunday Star.

Finally, on June 11, 1909, almost six years after the flight at Kitty Hawk, President William Howard Taft decorated the Wright brothers with a special medal honoring them for having first successfully flown an aeroplane. A week later the Wrights's home town of Dayton honored them in a daylong celebration with a parade, reception and fireworks. The festivities ended with a pyrotechnical display on the riverfront, complete with the brothers' eighty-foot-high portraits entwined in an American flag.

The year 1909 also saw the first international air meet, held August 22-29 at Reims, France. The world's great aviators assembled there with their aeroplanes: Voisons, Blériots, Farmans, French-owned Wrights and Glenn Curtiss's new *Golden Flyer*. There were thirty-eight aeroplanes in all. The Wrights, however, were missing at Reims. Orville was recovering from the crash at Fort Myer, and Wilbur was building the new Army ship. The meet was a week long flying spectacular that ended with Curtiss nosing out Blériot by six seconds in the twenty-kilometer race for the Gordon Bennett Trophy. Curtiss won it with a total time of fifteen minutes and fifty seconds! Just recently, on March 12, 1990, a retiring Air Force SR-71 spy plane flew from California to Washington, D.C., in sixty-two minutes—at slightly over three thousand miles per hour. It would have taken the SR-71 a mere forty-four seconds to fly Curtiss's twenty-kilometer course!

One month after Reims, an international exposition of aerial locomotion opened in Paris with 318 exhibitors. More than 100,000 people attended the show. In Berlin, Germany, "The World's Foremost Aviators" flew in a meet held October 1. The first aviation meet in Great Britain was held at Doncaster, England, from October 15 to 23, and the first one in the United States was scheduled in Los Angeles for the week of January 10, 1910.

Six years after Kitty Hawk and the triumph of the *Wright Flyer*, aviators—those daring, dashing, death-defying men—had become the toast of America. People were wild to see them fly. They were treated like heroes, wined-and-dined gods of aviation. The world, its cities, towns and girls were theirs. They lived fast and recklessly.

The Wrights and Curtiss started flying schools, mainly to put on flying exhibitions across the country, and the schools were booked far beyond capacity. Their graduates became the first barnstormers. People were willing to pay good money to see their exhibitions, so aviators, birdmen, flyers, aeronauts and pilots began staging flying shows all over the United States. The planes were dismantled, crated and shipped by train from town to town.

Often, these aviators' time was short. Aircraft were fragile, the engines disastrously untrustworthy, the weather treacherous, and very few pilots really knew how to fly. Aeroplanes didn't fly very long or far: A ten-minute flight was a record, and ten miles cross-country was darn near an impossibility. Mortality was high, so anyone who survived his first flight became an aviator.

But they kept at it. Some survived, and the public was enthralled.

How did Omahans and other Nebraskans take this increasing aviation activity? Well, first, the newspapers started covering aeroplane disasters—crashes and deaths. Soon they were reporting successful events, speculating on the future of the new bird and covering flights, especially those of the Wrights and Curtiss. By

An Up-to-date Elopement

1910 a handful of Omahans and Nebraskans even had begun toying with aeroplanes.

34

According to Frederick Pries, who lived two miles north of Florence, Nebraska, at Pries Lake, all aeroplanes were copies of his own boyhood invention. After interviewing Pries in 1910, a *World-Herald* reporter said Pries claimed:

The Aeroplane Is an Imitation, and His Old Invention Has the Flying Principle. . . .

Fifty-one years ago, Frederick Pries, a 14-year-old boy living in Denmark, constructed a flying machine . . . with wings to be propelled by the occupant. In this he used to travel twenty or thirty feet.

Thirty years ago [around 1880], Frederick Pries . . . constructed another flying machine . . . and in this Mr. Pries says he flew as much as 700 feet at a time, making three turns.[9]

If Pries's story is true, he should be considered the world's first true aviator.

Another Nebraskan, "Professor" U. Sorensen of Berwyn in Custer County, was an aeronaut, really a balloon driver. He built what he called an airship, although it really was a glider that looked very much like an engineless *Wright Flyer.* On June 13, 1909, Sorensen tied his ship to a huge balloon, ascended, climbed into the

Prof. Sorensen's Flying Machine just before the ascension.

9*Omaha World-Herald,* October 30, 1910.

ship, cut the balloon loose, and away he went. The *Custer County Chief* described how Sorensen quickly found he had no control over the glider as it plummeted to the ground and smashed to pieces. The birdman hit the ground sitting up and was knocked unconscious. Altogether, he came out of the disaster shaken and with only a few bruises.

John and Matt Savidge, two other Nebraska flyers, saw the flight and confirmed much of the newspaper account. They said the Professor was saved by the glider spiraling down like a maple seed. No matter, Sorensen was still the first Nebraskan who flew a glider and survived.

A few months after Sorensen's fiasco, Omahans Otto, Gus and Charles Baysdorfer started building an actual aeroplane. The details of their work will be covered in Book Two below.

Late in 1909 Gould Dietz (who had been associated with the 1906 Ak-Sar-Ben Fall Festival at which Charles Hamilton flew the airship), James J. Deright (a very successful Omaha auto dealer) and John M. Guild (Commissioner of the Commercial Club) formed the Aero Club of Nebraska. They applied for a charter from the Aero Club of America, which had been chartered by the Aéro Club of France. By the way, the Aéro Club of France is still in existence today. The Aero Club of Nebraska soon acquired a large membership and was very successful in promoting a statewide interest in aviation.

The First Flight in Nebraska

On June 8, 1910, a group of Aero Club of Nebraska members formed the Midwest Aviation Meet Company at a meeting of the Commercial Club. Their goal was to finance an aviation meet to be held July 23-27 under the auspices of the state Aero Club.

The new company was capitalized at $100 per share for a total of $10,000. The officers were: President, James Deright; Vice President, Thomas R. Kimball (a prominent Omaha architect); Secretary/Manager, Clarke Powell (Omaha's first auto dealer); and Treasurer, Gould Dietz.

This group authorized Powell to secure either the Wrights or Curtiss to fly in Omaha. A real stem-winder, Powell had begun selling Cadillac and Packard cars at 15th and Davenport Streets in 1902. Then he built a garage at 21st and Farnam and added Oldsmobiles to his sales. He also opened the first auto parts store in Nebraska. Having just sold his auto business to Deright, Powell set to work organizing the Midwest Aviation Meet Company and the meet itself.

Powell first contacted the Wrights in Dayton. They were completely booked. In fact, the two brothers were to fly in Nebraska at the State Fair in Lincoln that September. Every Nebraskan knows about the rivalry between Lincoln and Omaha, and Powell just couldn't let Lincoln win. He went after Curtiss and caught up with him flying a show in Minneapolis-St. Paul, Minnesota.

Curtiss said he was completely booked. He said he would be flying in Sioux City, Iowa, on July 23 and in a meet in Atlantic City, New Jersey, on July 14. Powell was discouraged but not defeated, so he offered Curtiss $10,000 to fly in Omaha for the five-day meet and said the city would treat them like kings. He added that the Wrights were booked for the Nebraska State Fair in September, so, if Curtiss flew in Omaha in July, he would become the first to fly an aeroplane in Nebraska. This convinced Curtiss—to beat the Wrights was worth anything. He shook hands with Powell, and they boarded the night train for Omaha. This was June 16, 1910.

Curtiss needed to find a field to use for the Air Meet. The next day, Curtiss, Powell, Kimball, Deright and Mogy Bernstein drove around town in Deright's new Locomobile touring car, looking for a site. The committee favored Creighton College's athletic field, but Curtiss wasn't thrilled with it. He said it might do, but he knew J. C. Mars (one of the Curtiss Flyers) wouldn't fly from it. So, off they drove to the Benson Gun Club, near 45th & Military. Curtiss was pleased. It was open, didn't have many houses, and was a little hilly, but Curtiss said he would have Mars look at it when he was done at Sioux City.

Next they drove over to Chief Meteorologist L. A. Welsh's house on 37th Street to get information on wind and humidity. Curtiss then left by train for New York. He was going to attempt to break the world distance flight record the following week, flying from Albany to New York City nonstop in his brand new plane, the *Hudson Flyer* (and he succeeded). He said the signed contract for the Omaha Meet would be sent as soon as he got back to New York.

Everybody was elated. The great Glenn Curtiss himself was going to fly in Omaha. Powell raised $16,000 from the business community to underwrite the meet and started planning grandstands and all the facilities such an event required. Omaha got excited—an aeroplane was going to fly over town!

A week went by, and no contract came from Curtiss. Powell wired the Curtiss headquarters at Hammondsport, New York, but received no reply.

Finally, on Sunday, July 3, Curtiss confirmed the July 23 through 27 dates and said he himself would arrive with his *Hudson Flyer*. Powell and company had feared that another city with more money might get Curtiss. Breathing a sigh of relief, they arranged with the Jesuit Fathers to get the meet site ready at Creighton Field. While this site was easily accessible, it was a poor place to take off and land planes. No one on the Committee had seen a plane fly, and no one knew the requirements for an airfield.

Then, James C. "Bud" Mars, probably the best Curtiss exhibition pilot, who had just flown in Sioux City, arrived in Omaha to inspect the site. He rejected Creighton Field as totally unsuitable. Powell took him out to the 45th & Military site, which Mars enthusiastically approved. Powell had to locate the landowners, who turned out to be Frank W. Carmichael and Erastus A. Benson, after whom the town of Benson was named. They had just platted the area from Military Avenue to Maple Street and from 45th Street to 47th Street as Omaha's newest and best subdivision and named it "Clairmont." Approaching them about using their land for the meet, Powell wisely noted it would draw a lot of people to the area. Benson made a deal with Powell: These aeroplanes and grandstands also could help sell real estate. . . . Served by both the North 45th Institute and Benson streetcars, the site was far more accessible than had been thought.

Finally, work started building stands for five thousand and fixing an area for the planes, fuel, oil and other equipment. A

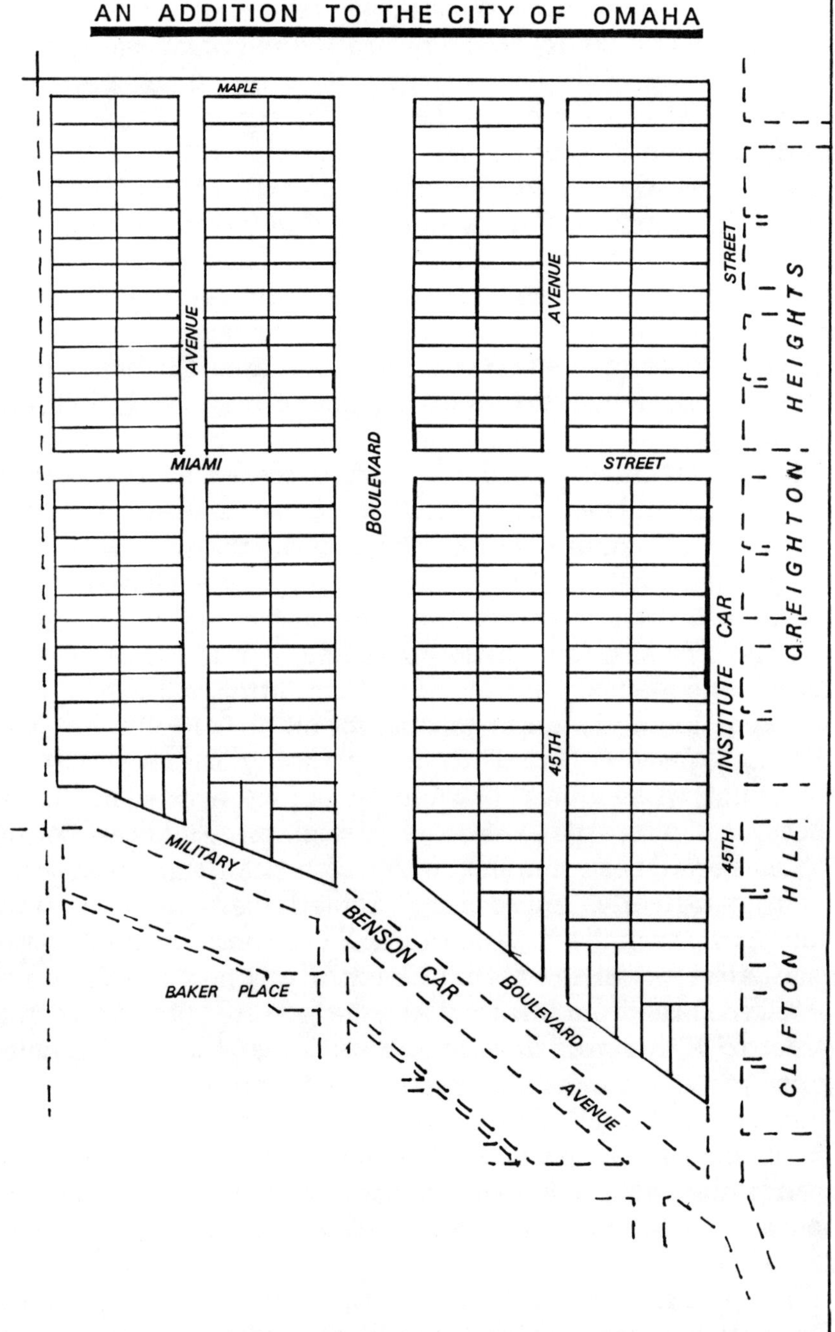

CLAIRMONT.
AN ADDITION TO THE CITY OF OMAHA
MAPLE
AVENUE
AVENUE
STREET
CREIGHTON HEIGHTS
MIAMI
STREET
BOULEVARD
45TH
INSTITUTE CAR
45TH
CLIFTON HILL
MILITARY
BENSON CAR
BOULEVARD
BAKER PLACE
AVENUE

contract was let to Lee McGreer to build the stands. Three thirty-by-thirty-foot tents to house the aeroplanes and six thousand feet of canvas for fencing was ordered from the Omaha Tent and Awning Company. Both stands and fence had to be up four days before the meet.

Powell expertly promoted the coming air show with the help of Bernstein, Edgar H. Allen, Harry A. Tukey and Guild. The Signal Corps from Fort Omaha planned an exhibit of government balloons and hoped to fly the new Baldwin Airship *Signal Corps Dirigible No. 1.* Hot-air balloon flights were to be held by aeronaut John Waldorf Hall and Prof. Thomas "Daredevil" Andrews. A model-aeroplane-flying contest for boys was to be held in conjunction with the Meet. The Union Pacific, Burlington, Northwestern, Rock Island and Illinois Central Railroads all scheduled excursion rates to Omaha. Promotional news stories and advertisements appeared in the *World-Herald, Bee, Omaha Daily News,* and in all the neighboring town papers as far away as Lincoln. Tickets were put on sale at many locations. In short, people were getting excited.

Finally, on Thursday, July 21, Elmer Robinson (Curtiss's chief mechanic), Lyman Bouten (his assistant) and a crew of mechanics arrived at the 13th and Webster Street Station of the Northwestern Railroad. Along with them were two aeroplanes packed in a half-dozen crates. The new *Hudson Flyer* came in on the Rock Island. George Johnson, Omaha's premier heavy hauler, achieved a "first" when he hauled the aeroplane crates to the meet field in his new 1910 Reo truck. It took four trips to move all the crates. Johnson noted he never would have made it in time using just his horses.

Crews worked expertly and furiously to assemble all three planes by Friday evening. They were ready to test the engines and double-check everything by Saturday morning. Many neighborhood boys and men came to watch, including all three Baysdorfer brothers, who by helping during the entire week were learning about aeroplanes—at least Curtiss aeroplanes.

Frank A. Tillman, the Curtiss manager and advance man, really was pleased with the field, stands, enclosure tents, promotion and ticket prices. He summed it up by saying that Powell had planned the meet beautifully and had made better arrangements for the Flyers than St. Paul, Chicago, St. Louis or Minneapolis.

Then Curtiss, Mars, Eugene Ely and John A. McCurdy arrived on the Rock Island night train from Chicago at seven o'clock

Saturday morning. Ely's appearance was a surprise. Only a week ago he had been battered and bruised in a crash in Toronto. He had even been reported killed! McCurdy, a young twenty-three-year-old Curtiss student, was brought in, in case he should be needed.

Mars and Ely brought their wives with them. The ladies were a great addition to the flying group and caught everyone's attention with their charm, wit and beauty. They were met by Aero Club members and lodged at the Paxton Hotel. After lunch at the Omaha Club, the ladies were driven to the aviation field in autos furnished by the Omaha Automobile Dealers Association, which Powell had founded in 1903 and which still exists today.

The aviators were here. The aeroplanes were here. Mayor James C. "Cowboy Jim" Dahlman proclaimed Saturday afternoon July 23 an official holiday. The stage was set seven years after the Wright Brothers flew for the aeroplane to fly in Nebraska!

Day One

The people started arriving a little after noon on July 23, and by 2:30 the stands were full. Nobody was turned away. There were over a hundred automobiles parked inside the grounds and an estimated seventy-five more outside on Military Avenue, 45th Street and the edge of the field. The air was electric, hot (ninety-four degrees) and windy, and the refreshment stands were full. Hawkers sold programs, balloons and knickknacks. The gentlemen of the press were all there to cover the event, which was properly photographed by Bostwick and his assistant, Homer O. Frohardt.

The Army balloon flew above the grounds; three Curtiss aeroplanes were poised near the southeast corner of the field; and the aviators and mechanics were bustling around the mechanical birds. Dignitaries and Aero Club members conferred with solemn dignity. Police Sgt. Henry C. Cook was in charge of the traffic detail, and Chief John J. Donahue, under fire to be removed from office by a legislative committee, promised that his mounted police patrol would be there the next day.

Curtiss, an Episcopalian, asked Dean George A. Beecher, the popular rector of Trinity Cathedral and a member of the Aero Club, to bless the planes. As Old Glory flew briskly in the southwest wind, at exactly 3 p.m. Lt. William N. Haskell, in charge of the Army troops, dropped his sword. The cannon boomed; Dan

Mid-West Aviation Meet

FIVE DAYS, COMMENCING JULY 23

Grounds 45th and Military Avenue   **Hours 3 to 6 P. M**

THE MOST FAMOUS AVIATORS IN THIS COUNTRY TODAY WILL ASSIST

MR. GLENN H. CURTISS

The world's greatest aviator, in making this the greatest aeronautical event in this country since the International Meet at Los Angeles.

This is positively the only western meet in which MR. GLENN H. CURTISS himself will take part. He will use the same aeroplane in which he made his famous flight from Albany to New York.

J. C. MARS whose name is now known all over the civilized world, will give daily exhibitions of his daring "glide" in his Curtiss aeroplane. From a height hundreds of feet above the earth, he shuts off his motor and plunges like a bullet to the earth. It is a feat rarely attempted by any aviator.

This is one of the few opportunities that have been offered in this country to view every type of craft that flies, viz: SPHERICAL BALLOONS, DIRIGIBLE BALLOONS, and HEAVIER THAN AIR MACHINES.

Mr. Horace Wild, the most daring and famous operator of the Dirigible Balloon in the world today, will give daily exhibitions. His balloon is modeled after the famous Zeppelin Dirigible.

EUGENE ELY one of the most intrepid and venturesome of the American aviators, will also take part in a special machine which is now being constructed in the Curtis factory. It is the latest production of Curtiss' genius.

GOVERNMENT EXHIBIT

The Signal Corps from Fort Omaha will give daily exhibitions of the uses of the captive and dirigible balloons in modern warfare. This is one of the few opportunities which has been offered to the people of the Mid-West to witness a government exhibition of this kind.

It will be practically impossible to furnish tickets to the big crowds at the grounds to advantage, and the management has therefore placed tickets for sale at the following points:

Crissey's Pharmacy ... 24th and Lake	Barnes Drug Company ... 40th and Dodge	Rome Hotel ... 16th and Jackson	Moritz Meyer Cigar Store ... 1314 Farnam
Saratoga Drug Company ... 24th and Ames	Beaton Drug Company ... 15th and Farnam	Her Grand ... 16th and Howard	Henshaw Hotel ... 1509 Farnam
Red Cross Pharmacy ... 17th and Cuming	J. H. Merchant ... 16th and Howard	Courtney & Co ... 17th and Douglas	Murray Hotel ... 14th and Harney
Walnut Hill Pharmacy ... 40th and Cuming	Merchants Hotel ... 1508 Farnam Street	O. D. Kiplinger ... 13th and Farnam	Loyal Hotel ... 16th and Capitol Avenue
Bemis Park Pharmacy ... 33d and Cuming	J. L. Brandeis ... 16th and Douglas Street	Hayden Bros ... 16th and Dodge	L. C. Gibson ... 501 N. 24th Street, South Omaha
J. H. Schmidt Drug Store ... 24th and Cuming	Paxton Hotel ... 14th and Farnam	Sherman & McConnell ... 16th and Dodge	Fisher, McGill Co ... 24th and N Sts, South Omaha

J. A. Clark Drug Co ... Cor. Broadway and Main Sts., Co. Bluffs

PRICES FOR THE MEET   -   -   -   Adults 50c, Children 25c, Grand Stand $1.00, Automobiles $1.00

Desdunes' band struck up "Something Doing Every Minute," and over seven thousand people gave a mighty cheer! Cy Amons, dubbed the "Jocose Ballyhoo Boy" by the *World-Herald* and *Bee* reporters, was the announcer. With the help of his tremendous lungs, he immediately got the huge crowd's attention and announced that the wind was too severe and the humidity was too slight for the flyers' safety and that there would be a short delay.

The crowd took the announcement good naturedly. There was much to see: three aeroplanes, the flyers, Edna (Mrs. James C.)

Curtiss's 1910 aeroplane flight over Omaha

Mars and Mrs. Eugene Ely in their New York fashions, and a dozen small boys flying their kites just outside the fence. A woman fainted, overcome by the heat and the exhausting strain of standing up in the crowd. Gus A. Renze, Omaha's great artificer and Ak-Sar-Ben impresario, took in everything, no doubt getting ideas for a Den Show or Electric Parade.

The aviators knew better than to fly, but finally Curtiss climbed into the seat of the *Hudson Flyer,* the biggest plane. A mechanic turned the propeller, and the eighty-horsepower, eight-cylinder engine barked to life. The people behind his plane found out what "prop wash" is as it blew hats, hair, skirts, and dust and dirt all over. Curtiss signaled to four mechanics to pull the chocks, and the big bird started moving, and then it lifted off the ground about where Fontenelle Boulevard is today. It took off toward the northwest—*the first aeroplane ever to fly in Nebraska!* The house on the left in the photo above is where the Military Theatre is today. Note the Benson street car on Military Avenue.

As the ship rose into the air, the huge crowd fell almost silent, watching the plane take a few dips before gaining altitude. Curtiss made a big flat turn toward the east, dipped and bobbed, turned west and came back over the stands at about two hundred feet. As Curtiss headed west toward Krug Park, there was a mighty ovation from 7,700 throats.

As flights go, it was not a pretty sight. The wind was

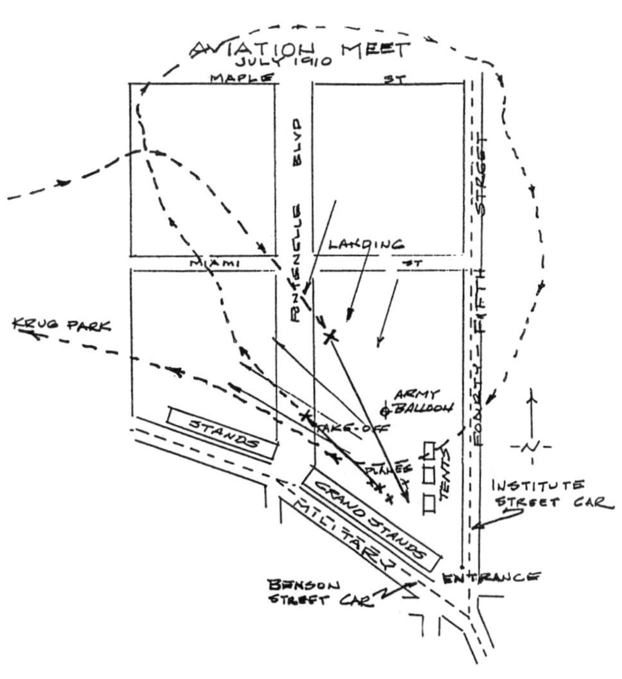

Route of first airplane flight in Nebraska, made by Curtiss, July 23, 1910, 3:15 pm.

frightful. Although Mars was fearful, he started his engine. Ely said he shouldn't be flying, but the crowd loved it! Mars started across the field, then his sixty-horsepower plane rose into the air, bobbed and weaved, came down and bounced up into the air again. Mars wisely shut the engine down and drove (now called *taxiing*) the ship back to the tents to wait until the wind calmed down.

Meanwhile, Curtiss had flown almost a mile and a half, as far as Krug Park. As he sailed over the park, the Saengerbund folks (a

German singing group) were having a picnic, which quickly broke up as they scrambled to see the flying machine. With the wind worsening, Curtiss landed nearby in Kehren's wheat field, where Benson High School and Monroe Junior High Schools are today. Scores of people joined the Saengerbund folks, and two dozen automobiles rushed into the recently-threshed field. As Curtiss waited for a lull in the wind, Kehren came puffing up the hill from his house, looking none too happy, but really just wanting to see the aeroplane. Curtiss misinterpreted Kehren's expression and decided the wind was the lesser of the two evils. He hit the throttle, flew back to the aviation field, drove to the tents and got out to the roaring cheers of the 7,700 fans inside and the 1,000 outside. He announced there would be no more flying until the wind died down. It was only 4:00 p.m. Dietz suggested that the aeronauts fly their balloons, but they claimed the wind was too strong. Dietz roared, "Well, we have a contract. You'll either go up or out, and I don't care much which."[10] John Waldorf Hall and Prof. Thomas "Daredevil" Andrews fired up their balloons and flew into the sky perched on their trapeze bars. At about four hundred feet, they both pulled their chute-release hooks and dropped. Hall's parachute opened immediately, and he floated gracefully to the ground, to the delight of the crowd. Andrews thrilled the people when his chute stuck. He knew he could ride the balloon down as it deflated, but the crowd didn't. When he disappeared beneath the trees east of the field, the crowd assumed he was in trouble. They were greatly relieved, giving him a great cheer, when he walked back onto the field. After this, the crowd was somewhat more patient. Soon the heat (then 105 degrees), the wind and the waiting took their toll, and the crowd started to disperse.

Shortly after six that evening the wind died down. The machines were started, and Curtiss made a magnificent flight, circling the field four times. Mars then flew expertly, thrilling the remaining two thousand fans with his famous dives. He would start about three hundred feet up and dive until it seemed he would hit the ground. Even Ely, despite his bruises, wanted to practice to impress his father, who was on his way from Davenport, Iowa, to see him fly for the first time. He made several flights in a forty-horsepower ship. The show ended with Curtiss and Mars flying

---

[10]*Omaha World-Herald,* July 24, 1910.

around the field together, Curtiss a hundred feet above Mars. The first aeroplane flights over Nebraska were a smashing success.

That evening, the aviators were guests at a large reception and dinner at the Paxton Hotel, attended by Omaha's elite. The flyers, Edna Mars and Mrs. Ely had captured everyone's affection. Even Curtiss, a quiet man in contrast to his dramatic flights, seemed to enjoy the evening. Mars was the favorite—friendly, handsome and gregarious. Ely, a congenial Iowan, and McCurdy, an upstate New York boy, both seemed starry-eyed about the whole affair.

Day Two

Sunday, July 24 dawned clear and bright. By noon the temperature was eighty-one degrees, the humidity twenty-nine percent and the wind out of the north at only ten miles per hour. When church was over, people started to arrive at the field. Many had picnic baskets. Kids were there in droves. The streetcars were running nose to back. Automobiles were everywhere. Sergeant Cook said that there must have been two hundred autos there. Powell later reported that $310 was taken in at the auto gate.

Displeased, Dietz commented that there had to be three thousand dead-heads outside the fence and three hundred autos (his estimate may have been a little high because there were only 639 registered automobiles in Douglas County at the time). The *Bee* reported that there were seven thousand people watching from beyond the fence. Clarke G. Powell commented that it was a shame. The *World-Herald* gave the freeloaders a going-over as well:

> It seems peculiar that when Omaha men put up money to bring this grand event to this city, there should be Omaha business men who stand for hours on outlying hills rather than spend a dollar to boost the project.[11]

Not to be outdone, Council Bluffs residents climbed to the top of Fairmont Park to reach a splendid viewpoint from which they could observe the aeroplanes circling over Omaha. Many had opera and field glasses. The *Bee* commented that, if the success of the flights continued, reserved seats could be sold in the park and at

---

[11]*Omaha World-Herald,* July 25, 1910.

James C. "Bud" Mars at the Mid-West Aviation Meet

Point Lookout where President Lincoln, looking across the river to Omaha, decided to start building the railroad to the Pacific Ocean.

But this day belonged to aeroplanes, not to trains. By 3:00 p.m., there were almost ten thousand people in the stands and on the field. Best of all, my father and mother were there. Much of what follows will be my folks' story. Dad was the chief engineer of the Bennett Company, which owned substantial property in downtown Omaha. They lived at 3712 North 23rd Street and drove out to the aviation field in the company car, a Brush Run-about. Both Ward M. Burgess and Fred A. Nash, owners of the company, were members of the Aero Club and stockholders in the Mid-West Aviation Meet Company, so I bet Dad had a complimentary ticket. Being somewhat of a mechanical genius, Dad met Curtiss and the others and spent the entire day with the aeroplanes. Mom brought her brand new Kodak, which she had just bought for ten dollars from the Burgess-Nash Company at 16th and Harney Streets.

Curtiss had the *Hudson Flyer* parked near the fence; spectators were allowed to come close, and his mechanics answered questions. Mars, Ely and McCurdy talked to everyone. Edna Mars, almost as popular as her husband, always watched her husband's flights, and that day their son, nine-year-old Thomas Baldwin Mars, came too. Young Mars attended military school in Lincoln,

Nebraska, and had been named after none other than "Captain" Thomas Baldwin, with whom Mars had flown balloons and dirigibles before he took up aeroplanes.

At 3:30 p.m. the "Jocose Ballyhoo Boy" announced that flying would start in fifteen minutes. The darn wind had come up, and it was gusty. Nonetheless, according to the *World-Herald,*

> Mars, who has been rightly termed a "dare-devil," started out, bucking the jerky north wind. He attempted to turn to the west when he reached the limit of the field, but couldn't negotiate the maneuver in the teeth of the gale. Alighting, he turned his plane around and started back. Ely was just

*Photo of Curtiss and Flyer taken
by my mom at the Mid-West Aviation Meet.*

> rising from the earth, but came down to give Mars the right of way.

> The "daredevil" came hurtling toward the grandstand about fifty feet from the ground. Making a quick swerve to the east he found he had gathered too much speed with the wind, and as he jammed his ground brake into the turf to avoid running into the flock of autos outside of the ropes, he broke the brake and a guy wire. . . . Ely clattered forth and

endeavored to make the turn in the gale, but met with a diminutive disaster breaking a rib and a stay.[12]

Next, Curtiss tried to fly his craft in the wind. He was able to negotiate the turn, but, when forced to land, he broke a rib and a wheel. Knowing the birdmen were trying to perform under extremely adverse circumstances, the crowds gave them a hearty cheer.

At this stage the Signal Corps boys began to inflate the big government balloon, which distracted the crowd while the three biplanes were being repaired. My dad, who helped, always said they were super mechanics and great improvisors. Curtiss was talking with a group of admirers when Robinson told him the repairs were made. The great aviator bowed to his audience and asked them to excuse him while he flew a little.

Then the wind calmed considerably; the crowd buzzed, expecting something rather frisky, and they got it.

A grimy mechanician twirled the sturdy propeller and ducked out of danger, while the terrific crackling, spluttering and roaring of the huge motor drowned all other sounds. The propeller disappeared as if by magic and a half dozen assistants clung desperately to the tugging plane. The slender, retiring figure of Curtiss stood for a moment trying the mechanism, then he calmly adjusted his cap, stepped into the tiny seat, looked about, nodded his head and stormed northward across the field.

In exactly six seconds he left the ground and mounted at an astonishing angle until he was 100 feet in the air. For six minutes the breathless crowd was treated to aerial antics, which would have been witchery not so many years ago. Four times he described wide circles about the field, and in these trips he dipped, lunged and careened like one possessed, until the last ounce of reserve on the part of the audience was gone and they sent up to him a mighty roar of appreciation. When he alighted, as softly as a floating feather, a throng of privileged persons rushed to the plane and wrung the

---

[12]*Ibid.*

master's hand as he stepped from the seat with his omnipresent plaintive smile.[13]

The crowd had just settled down from Curtiss's flight when daredevils Hall and Andrews bobbed up serenely in their hot air balloons, nimbly clambered about on the trapezes, and dangled from their parachutes. Around a thousand feet, both cut loose and descended together, casually talking back and forth to the delight of the people. The aeronauts landed at the southwest corner of the field. Their balloons descended about a block west of the field, and a score of boys and young men ran to retrieve them.

Suddenly, far to the northeast, at Fort Omaha, the airship could be seen ascending and heading toward the park. Everyone was gazing at the approaching ship when it gradually descended.

Presently, the "Jocose Ballyhoo Boy"

Nina Detrick Adwers's photo of crowd watching Curtiss in flight

announced that Lieutenant Haskell, the airship's commander, had telephoned that the propeller had broken, and the trip would be postponed until Monday. The ship's engine had been built by Curtiss, and the ship itself had been manufactured for the Signal Corps by none other than "Captain" Thomas Baldwin.

Ely's plane was ready, and he hoped to fly well for his visiting father. His plane, the oldest of the three and with only a forty-horsepower engine, was missing strokes as he took off, so he couldn't gain much altitude. As he turned to the east, the ship hit the

---

[13] *Ibid.*

ground with its right wing, smashing several ribs and ending flying for the day. Fortunately, Ely wasn't hurt, but he was very disappointed that his dad hadn't seen a good flight.

That day Mars directed the work on his plane's magneto and helped Curtiss's crew. This man with the astronomical name had a big place in the hearts of his fellow aviators and workmen. Even so, when reporters asked young Thomas Mars who was the best aviator, he responded, "Curtiss," with a Marsian grin.

Meanwhile, the "Jocose Ballyhoo Boy" was welcoming a large group that had driven down from Fremont, and special trains had brought enthusiasts from as far away as Lincoln, Grand Island, Nebraska, Sioux City, Iowa, and Des Moines. Charles W. Martin, an Omaha insurance man and realtor, busily was taking pictures of the day's flights with his inevitable motion picture camera. This was the uncle of today's well-known Omaha and Nebraska historian Charles Martin. The latter Martin told me that as a boy he saw the movies of Curtiss flying; unfortunately, the film had a nitrate base and deteriorated long ago.

Then the terrific racket of Curtiss's powerful motor again thrilled the crowd. The famous aviator adjusted his seat, waved his hand and ripped off to the northwest at tremendous speed. He seemed to leave the ground almost instantly when a tiny switch of the forward control shot the *Flyer* into the air as though propelled from a cannon. Testing the wind and his machine by various dives and lurches, he sped around the field four times, rapidly gaining height. He then swept by the stands at about three feet above the ground, heading straight west, and drove steadily toward the red setting sun. Farther and farther he flew until he faded into nothing and vanished from sight over a distant grove, leaving only the empty summer sky.

Mars knew Curtiss was off on one of his joy flights, so he fired up his sixty-horsepower machine and thrilled the crowd with his famous dives. He flew by the stands like a roller coaster before landing so softly it would not have cracked an egg.

Now, I can tell you what Curtiss was doing. The night before at the Paxton, John W. Redick (insurance man, son of the Honorable Judge William A. Redick, and member of the Aero Club and the Omaha Country Club) spoke with Curtiss about flying over the Country Club, which was then located south of Krug Park. Benson, who owned the land the Meet was being held upon, also told Curtiss that Kehren (of the wheat field) really wanted to see the aeroplane and asked the aviator if he would land on the wheat field again. So, when the wind started behaving, he decided to satisfy their requests. Spectators at Krug Park almost dislocated their necks as he circled four times over the Country Club. The golfers cheered; the folks on the verandas cheered. It was an unforgettable sight, as Redick told me many times. Curtiss then dragged Kehren's wheat field, turned, and landed soft as a bird. Crowds immediately gathered. Kehren got to see the plane up close, then Curtiss waved the gawkers back, hit the throttle, and flew back to the aviation field.

Meanwhile, Mars had just finished his flight. Out of the west, Curtiss's *Flyer* swooped over the field at a tremendous speed, banked to the north almost to the Deaf & Dumb Institute, came back, and landed like a feather.

It was a little after six in the evening. Some of the crowd had left, but flying conditions were almost perfect. Curtiss decided to show them some flying. He and Mars took off side by side. The *Flyer* gained altitude fastest, so Curtiss flew at about two hundred feet and Mars at one hundred. They circled the field together. Mars could turn inside the *Flyer*, but Curtiss would zoom by on the straight-away. They made four flights around the field before landing together, much to the relief of the audience, most of whom thought they surely would kill themselves.

Thousands of Nebraskans had seen their first aeroplane and some magnificent flying. They went home tired, hot, windblown. My mom and dad never forgot it. Thank goodness they were there to tell me about that wonderful day.

Day Three

In spite of terrible weather on the third day of the Omaha Aviation Meet, July 25, a record was set. A high wind, with frequent gusts over thirty miles per hour, raced over the aviation field all afternoon and evening. The committee delayed opening until 4:00 p.m., and, since it was a Monday with most people at work, only 5,500 people were on hand.

Mars was the only one to bring his plane on the field that day. With the favorable high wind and the long field, he had decided to try to break the world record for the shortest takeoff run.

Two weeks earlier at Atlantic City, Curtiss in his *Flyer* had made the world's shortest takeoff in eighty-seven feet.

The *World-Herald* had put up a silver cup and a prize of $75 to anyone breaking a record, so it became quite an affair. The Aero Club appointed a panel of judges for the event—Clarke G. Powell, James J. Deright, Lieutenant Haskell, Charles Baysdorfer and J. W. Canham, a carpenter from David City, Nebraska, who was picked from the crowd. Canham had been studying aviation and was building a monoplane in his shop. It ran twenty-eight feet across and thirty-one feet long and was made mostly of steel tubing with some bamboo and spruce. He hoped to fly to Omaha sometime during that fall, but there is no record that Canham made it off the ground.

Equipped with a one hundred foot steel chain, the judges moved to the north end of the field to set the takeoff course. The "Jocose Ballyhoo Boy" announced to the crowd that everything was ready. Mars fired up his sixty-horsepower engine and taxied out to the starting point. The wind was blowing fiercely from the southeast. The aviator knew that the danger in such an attempt came from tilting the front control up too fast, for a high wind would flip the plane straight up and over on its back. As a matter of fact, Mars had done just that once before and, luckily, walked away with only bumps and bruises.

The "Jocose Ballyhoo Boy," being a circus announcer, kept the crowd on needles and pins with his spiel as Mars took off. The *World-Herald* later reported that Mars took an enormous chance by flying in what was almost a gale. He headed the plane into the wind, and, after a very short run, the plane took to the air. Mars flew only a short distance at about twenty feet off the ground and then landed. The distance of his run before take-off was measured at fifty-three feet four inches. That beat Curtiss's previous record by thirty-four feet. Mars said he had won the cup, even though he couldn't fly. Curtiss was delighted with Mars's new record, but he announced there would be no more flying until the wind diminished.

However, there were still plenty of thrills. *U.S. Army Captive Balloon No. 12*, which seemed to be securely anchored, broke loose and rolled to the west in a particularly heavy gust of wind. At that point, Pvt. Otto Mais seized the cord to release the hydrogen gas. He held on tightly, and, as the balloon rose, he was pulled into the air. When the gas finally was released, Mais was so caught up in the rope that he couldn't let go. For a moment it appeared Mais might be injured, or even killed, but fortunately the rope broke. He fell only ten feet and suffered a small injury to his hand. The balloon was recovered soon afterwards and taken back to Fort Omaha, where it was repaired for the next day's events.

It was now 6 p.m. Powell ordered "wind checks" given to the crowd, and that ended the third day. The YMCA and the Boys Aero Club did get to visit with McCurdy and Boulton. They were shown the aeroplanes and had their questions answered. McCurdy told them that they would be the future flyers and suggested that they build a glider first. As you will soon see, they did just that.

That night, the aviators were treated like kings, as Powell had promised. They were the guests of honor at the Ak-Sar-Ben Den Show, *A Ride on Halley's Comet,* performed at the Den, located at 20th and Grace Streets. In those days Ak-Sar-Ben Den Shows were for men only, and their contents were kept secret. At the show's end, the crowd roared for a speech from Curtiss, who deferred to Mars. Mars entertained them with his wit and thanked them and Omaha for their gracious hospitality.

The women had dinner at the Omaha Club, and Ms. Burgess and Nash had arranged for James Clayton and Ella McLeigh to sing the latest Broadway numbers. Clayton and McLeigh were playing at the Boyd Theatre at 15th and Harney. The Boyd was part of the Bennett properties. That Monday evening it rained heavily, making the farmers happy and cooling the air for next day's flying.

Day Four

On July 26, day four of the Omaha Aviation Meet, the complaint of the aviators was that the humidity was too slight. No humidity! They found that their propellers had no push behind them, or so wrote a *World-Herald* reporter on July 26.

According to aviation theory, a plane's propeller doesn't push or pull as such. The propeller screws through the air just like a wood screw. The pitch of the propeller determines how far it goes with each revolution—a coarse thread goes farther than a fine thread. As it spins, the prop pulls or pushes the wing through the air to create lift. Our atmosphere has density; a square inch of it weighs 14.7 pounds at sea level. Temperature and humidity make it more or less dense, and the more dense the air, the better the prop works.

54

Omaha is one thousand feet above sea level, so the air is less dense than at the Atlantic or Pacific Oceans. This fact, combined with a high temperature and a low relative humidity, caused the so-called plight of the aviators on day four. All the planes had "fixed-pitch" props, so they did have a more difficult time flying.

The newspaper headline that day read, "Humidity Too Slight For The Intrepid Flyers."[14] Nonetheless, Curtiss and Mars made the best flights of the meet that day and held the four thousand spectators spellbound during the entire sizzling session. Mars and Ely made short up-and-down flights into the brisk northwest wind. Curtiss then roared off the ground in the big *Flyer,* seemed to go straight up and then turned west toward the Country Club. "He's down!" groaned the populace as he sank behind the clump of trees at the horizon. Soon a mighty cheer went up, as he skimmed back again and tore over the grandstands at over seventy miles per hour. Four times he circled the field, then made his usual gentle and genteel landing, scarcely startling the dust when he reached the ground.

Eugene Ely, just before takeoff

Then *U.S. Army Captive Balloon No. 9,* which substituted for the runaway *No. 12,* appeared over the northern trees. It had been walked up from Fort Omaha. It was immediately sent up to the heavens where it stayed. Curtiss, taking advantage of a chance for a dramatic spectacle, took off once more and circled the balloon five times at around 150 feet. It was a beautiful sight.

---

[14]*Omaha World-Herald,* July 26, 1910.

Aviator Mars, who was very popular with the spectators, took off in the four-cylinder plane and did some fancy flying. He circled the balloon four times at about seventy-five feet, climbing and diving as he went around. This roused the people, and they rewarded him with applause. Curtiss then flew over the grandstands and dropped numbered cards to the audience as he sped over them. The cards were produced by the Ryan Jewelry Company, and whoever picked up number 1313 was to receive a $20 diamond ring. Curtiss made four complete circles of the field, dropping cards each time he flew over the stands. The young boys and girls had a ball shagging the cards, fighting, jumping, running and looking for the lucky 1313. Finally, the "Jocose Ballyhoo Boy" announced the winner was W. Mortinson, a young man from Harlan, Iowa, and a student at the Ames Agriculture School of Iowa (now Iowa State University). He came to Omaha to see the meet, evidently to his great benefit.

Next, Curtiss judged the model-aeroplane contest. There were eighteen entries from boys under fourteen. The boys launched their models from the tonneau of Deright's Locomobile. Some even flew a little. Finally Curtiss, after consultation with Mars, Ely and McCurdy, picked the winners. The first-place prize of $20 went to Frank Engstrom of Omaha, the $10 second-place prize to Wayne Moore of Geneva, Nebraska, and third place belonged to Kenneth Norton from Omaha. Hall and Andrews made their usual great balloon flights and landed almost in the stands with their parachutes.

The day was very memorable for Clarke Powell. Lieutenant Haskell allowed him to go up in the basket of the balloon. Powell said the view was magnificent, and when Curtiss and Mars flew around the bag it was thrilling, though a little scary. When the soldiers started to crank down the balloon, however, something broke, and Powell was stranded for about an hour. His many good friends encouraged him with catcalls and gems such as, "How's the weather up there?" He finally got down, and the crowd gave him a great hand. The fourth day ended with four thousand very happy spectators.

That night a big reception and dinner honored the aviators at the Country Club. They were all made honorary members of Ak-Sar-Ben. Edna Mars and Mrs. Ely were presented beautiful pearl fans. Powell received praise for his great organizational talent. The committee was given a standing round of applause. Gould Dietz said

56

that he expected to know soon if the Aero Club of Nebraska would receive a charter from the Aero Club of America. Lt. Col. William A. Glassford, Commander of Fort Omaha, and Lieutenant Haskell attended in full uniform, and Omaha's "400" were there. Curtiss modestly thanked everyone and said that they had never been treated so hospitably. He commented, "I am sorry that the weather conditions were not ideal, for then we would have done even more."[15] Mars starred as usual: When he asked for champagne, he poured some in his World Record Silver Cup and toasted the prairie people of Omaha for inviting them to fly.

Day Five

On July 27, the fifth and last day of the meet, the crowd saw some darn good flying! The only record broken on that day was the temperature.

With a sweltering, melting, sizzling leap the mercury in Mr. Welch's [Omaha's meteorologist] pet thermometer aviated some itself. The weather man simply cut all the guy ropes and let 'er go until it looked like the top of the glass would be the limit. When the little register of the deeds of hot and cold air up in the weather mixing shop on government building quit sky galloping and put the soft pedal on the altitude at four o'clock yesterday afternoon, the 104 notch was reached.[16]

Actual flying didn't begin until 6:00 p.m. Curtiss made three spectacular flights, and Mars surpassed all his previous performances. He made four circles around the government balloon and had everybody on their feet. On the fourth circle, he swept down from about three hundred feet, looking like he was going to fly right through the hydrogen-filled bag. Then he banked around it, with his wings almost perpendicular to the ground; he came so close that the balloon seemed to follow his flying machine! Then he zoomed back over the stands, circled to the west, headed back south and attempted to land. He met the ground with a resounding thud that placed his already battered machine in the hospital for good.

---

[15]*Omaha World-Herald,* July 28, 1910.

[16]*Ibid.*

The crowd's cheers probably were heard all over Omaha—Mars could have been elected mayor or maybe even governor! The closing of the first aeroplane flights in Nebraska couldn't have been described better than in this account from the *World-Herald:*

It was a faithful crowd of some 4,000 souls which fought it out under the baking, boring, scorching sun of yesterday and waited until nearly 8 o'clock to have a last glimpse of Curtiss and Mars in their inspiring conquest of the

Clarke Powell (left) and Glenn Curtiss at the 1910 Mid-West Aviation Meet

vast aerial seas. As usual, they were well rewarded, for as the red sun made its evening bow before hurdling the hazy horizon, a speeding cross passed over its face and the distant crackling of his throbbing motor told a multitude of Omaha admirers that Glenn Curtiss was on the wing the last time at the Mid-West meet. With a graceful swoop, his fragile sky craft dipped to the earth as if in a farewell curtsey, and a

58

long reluctant sigh swept over the field as the throng slowly sought the gates.[17]

The *Bee* added,

Flying has beyond a vestige of all doubt been demonstrated, and Omaha now knows how it is done. That certain other things have come into the general education, marking the new science as a most difficult one, only further impresses those who have seen the flying.[18]

Thus, the first aeroplane meet in Nebraska came to an end.

Clarke Powell unanimously was given credit for the meet's success. He announced that all the expenses and guarantees had been met, and there possibly would be receipts beyond this mark. He quickly acknowledged the forty men and firms who put up $100 apiece to bring off the meet: Thomas R. Kimball, Gould Dietz, J. L. Brandeis & Sons, William H. McCord, Ward Burgess, David Cole, C. H. Hayward, F. D. Haller, George W. Peck, Charles H. Pickens, Herman Peters, J. J. Deright, Dr. William J. Bradbury, Frank J. Taggart, Martin Brothers & Company, Daniel Baum, Jr., Thomas H. Matters, Arthur H. Fetters, Updike Grain Company, Fred Metz, Victor B. Caldwell, E. M. Fairfield, Edgar H. Allen, William M. Glass, Archie W. Carpenter, Byrne-Hammer Dry Goods Company, Wright & Wilhelmy Company, Omaha Rubber Company, Orchard-Wilhelm Company, F. W. Wead, Hastings and Hayden Thompson, the Thompson Belden Company, Harry Tukey, Luther L. Kountze, J. DeForest Richards, Miller Stewart and Beaton Carpet and Furniture Company, Henshaw Hotel, Thomas Kilpatrick & Company, Fred A. Nash, Charles F. Weller and the Rees Printing Company. Powell had contributed as well.

Afterwards, the *Bee* concluded, "Omaha's aviation meet has really put it on the aeronautical map and the flights of Curtiss here have been reported all over the country."[19]

---

[17]*Ibid.*

[18]*Omaha Bee,* July 28, 1910.

[19]*Omaha Bee,* July 29, 1910.

What Became of Those Intrepid Flyers?

Glenn Curtiss became one of the most successful aeroplane designers and builders in aviation. His company built the first flying boats for our Navy, the famous JN-4D Jenny, many other aircraft and also the OX-5 engine. He was involved with several aviation companies, but by 1920 he was on his way out and became a Florida land developer. He died in 1930 at the age of fifty-two. Curtiss and the Wrights were fiercely competitive, and their successor companies continued the rivalry even after these men were no longer associated with their companies. Before he died, Curtiss admitted that much honor and credit was due to the Wrights because of their achievement in flight.

Ely, about to land
on the USS Pennsylvania

Eugene Ely became one of Curtiss's best flyers, despite miserable luck in Omaha. On November 14, 1910, four months after he flew in Omaha, Ely flew a Curtiss plane off the deck of the USS *Birmingham* in Hampton Roads, Virginia—the first flight ever off the deck of a ship. Two months later, on January 18, 1911, he made the first trip in the other direction, flying out from the Presidio grounds overlooking San Francisco Bay to land on the quarterdeck of the USS *Pennsylvania.* The aviator was killed on October 19, 1911, while flying at the Georgia State Fair in Macon, Georgia. As usual, the field was too short and too crowded; there were too many trees; and the wind was from the wrong direction. His Curtiss stalled out as he made a turn, too low and too slow, trying to avoid trees and a barn. He was the 101st aviator to be killed since that December day at Kitty Hawk, but Eugene Ely flew with the old Red-tailed Hawk. He was the first "Top Gun."

Mars had a long career. In 1911 while flying a Curtiss Exhibition in Japan, he gave Crown Prince Hirohito of Japan his first plane ride. Much later he commented that the world would have been better off if the plane carrying him and Hirohito had crashed. In a meet at Erie, Pennsylvania, he was reported killed. Seriously injured in the crash, he did survive to serve in the Army in World War I as a pilot instructor. After the war, he built one of the first airports in the United States in Westchester County, New York. Mars quit flying in the 1920s and went into the gas-engineering business. He died in Los Angeles in 1944. His wife Edna and son Thomas, who so long ago were the hits of the Mid-West Aviation Meet, survived him.

After the Mid-West Aviation Meet, others soon flew the Nebraska skies. The Baysdorfers were among them. Close behind them was M. J. Cole. Cole designed an unusual aeroplane while recovering from a fall from a grain elevator in Grand Island. According to newspaper accounts, Cole invented a machine unlike any other plane—a triplane with a new steering device. Its two-cylinder, air-cooled engine developed twenty horsepower and drove a seven-foot propeller that in no way resembled those developed by the Wrights or Curtiss. According to W. F. Maloney, the aviator's manager, Cole was invited to fly his bird at the Ak-Sar-Ben Festival, and he had many bookings in Oklahoma. The inventor seems to have gotten his plane in the air once at Courtland Beach amusement park near Omaha. After that, Cole, his aeroplane and Maloney dropped out of sight; perhaps they went to Oklahoma and became Sooners.

Flying the Nebraska State Fair

Then Arch Hoxsey flew during the Nebraska State Fair. The Fair opened Monday, September 5, 1910, to the largest crowd in history. Hoxsey, with his Wright biplane, was apparently the reason.

> At 1:15 the engine was set in motion, with Aviator Hoxsey in his seat. After trying out the machinery for a minute or two he put on the power. The aeroplane skilled along the rail for almost its entire distance, near the end taking a sharp upward curve and proceeded to mount rapidly in the teeth of the wind. At the height of several hundred feet he turned to the right and went flying with the wind down to the east end of the track, when he swerved sharply

downward. At a short distance from the earth he shut off the power, the machine righted itself and floated slowly down to the earth, its momentum carrying it a few yards along the ground.

Cheers greeted the flight's ending. From almost the beginning the airship was in full sight of everybody on the grounds, and its dippings and dashings, in perfect imitation of the birds was followed enthusiastically by the thousands of eyes fixed upon it.[20]

Almost nineteen thousand more Nebraskans had seen their first aeroplane flight.

The next day, Hoxsey gave the crowd a real thrill. After a beautiful flight circling the center field of the track four times, waving to the crowd and drawing a tremendous cheer from the thousands, he came in to land. A gust of wind tipped up a wing, and he crashed into one of the large horse barns on the northwest corner of the track. It wrecked the plane, but miraculously the aviator crawled out. He was scratched, bruised, battered, and a bit woozy, but he walked away from the landing (and any landing a pilot can walk away from is a good one, especially in a 1910 aeroplane). Dr. A. I. McKinnon, who examined Hoxsey, said he was bruised, strained, sprained, and skinned up, but would be all right the next day.

Hoxsey went to his room at the Lindell Hotel in Lincoln to recover. Incidentally, Hoxsey was one of the Wrights's best pilots. In fact, Orville said many times that he was the best, but the Wright Aeroplane was out of business for any more flying at this Fair. President Rudge of the Fair Association wired Dayton immediately asking for another machine and a man. He might as well have asked for the moon! There was no way the Wrights could get a plane there until at least a week later. It's hard to believe now, when planes fly anywhere in the United States in a couple of hours, but not in 1910. Chairman Hendershot of the Board of Managers said,

The aeroplane flights were our big card this year. . . . When we were negotiating for flying machine flights, we had an offer from the Curtiss people for just half the money the Wrights wanted. The Wright figure was $10,000 for twenty

---

[20]*Lincoln Evening News,* September 5, 1910.

flights. We chose the latter because we believed, from what we could learn that their's [*sic*] was the more successful machine and that the people would have the best.[21]

I don't know how the Fair Board came out with the Wrights, but Hoxsey did make five great flights at the 1910 State Fair. The aviator was killed two months later in Los Angeles, the day after he set a world altitude record of 11,474 feet.

Meanwhile, in the spring of 1910 Joseph W. Miller, educational director of the Omaha Y.M.C.A., had organized the Boys Aero Club of Omaha. Sgt. Clarence Adams of Richfield was signed on as instructor. Adams had been the post electrician at Fort Omaha and was building a biplane at his shop in Richfield, Nebraska. The first boys in the club were Harry A. Sackett, Harry Coesfeld, Louis Wade, Art Frencer, Fridolf Engsrom, Hugo Heyn, and Roy Whitmore. Under Adams's direction, they built a biplane glider. That fall, on October 15, they attempted to fly it on the Thompson Farm, near Richfield. Whitmore was chosen to be the pilot. A neighbor pulled the glider with his car until it lifted, immediately reared up, did a nose-dive, smacked the ground, and flipped over on its back. Whitmore miraculously walked away, but the glider was a wreck, and parents of the young members of the Boys Aero Club put an end to more flights. Adams kept working on his aeroplane, though, and every now and then the *World-Herald* or *Omaha Daily News* would run a story on him, but there's no record of his ever flying a machine.

---

[21]*Lincoln Evening News,* September 6, 1910.

Many early adventurers were killed flying, and unfortunately aviators still are dying today. Aeroplanes crashed in 1910 for the same reasons they crash today: pilot error, faulty design and construction, structural failure, engine failure, instrument failure, other mechanical failure, weather, or other unknown failures. But the biggest reason of all is that *the aeroplane quits flying—stalls!* The wing, or airfoil, is shaped so that it will create lift when it moves through the air. It takes so much power, or speed, or both, to move it fast enough to create enough lift to make it fly. The amount of power or speed, or both, varies with the attitude of the wing. Straight and level flight, let us say, takes 1 Power-Speed. To climb takes 1+ Power-Speed. To turn or bank take 1+ Power-Speed. To dive or go down takes 1- Power-Speed.

Aeroplanes in 1910 were flimsy at best. They had to be light because there just were no good lightweight engines available with enough power to fly a plane. So, the engine ran at full power, all the time, except when landing. Whenever the pilot or the wind or the air or whatever made it climb too steeply, turn or bank, it would quit flying, or *stall,* and the only way the pilot could get it flying again was to dive, to get the nose down. To regain flying speed in the dive, the aeroplane had to be high enough above the ground not to crash.

Lack of power plagued flying and aeroplanes from day one at Kitty Hawk. I grew up flying internal-combustion reciprocating engines. I flew some pretty sophisticated engines in World War II, and even then the cry was, "More Power!" This problem finally was solved when the Germans and British built the first jet engine. Now we fly 747s, B-52s, DC-10s, and B-1Bs that weigh over two hundred tons at speeds faster than sound. The *Wright Flyer* weighed under six hundred pounds with Orville in it.

By 1910 aeroplanes were flying all over the world, and some very exciting things were happening in the air over Nebraska.

BOOK TWO
THE BAYSDORFER BROTHERS:

The Wright Brothers of the Prairie

CHAPTER ONE:
THE BAYSDORFER INVENTORS

In the summer of 1887, Robert and Katherine Bayersdorfer moved from Davenport, Iowa, to Omaha. Robert had become fed up with civilization and moved to the West. He was a cigarmaker by trade. Omaha was something of a cigar-making center at that time. Some fine, long-leaf tobacco was grown locally, and still is, down by Saint Joseph, Missouri, and Omaha had a string of cigar lofts. There was a big loft in Deshler, Nebraska. Greg Hornbostle's foundry is in the building now. It was started just after the broom factory and before the manure-spreader factory there.

The Bayersdorfers had a big family: Emmy, Ella, Matty, Annie, Otto, Gustav ("Gus") and Charles. The children, by the way, soon changed their family name to "Baysdorfer." The family lived at 1520 South 13th Street. Robert went to work for the West & Fritscher Cigar Company at 14th and Capitol Streets. Otto, the eldest Bayersdorfer boy, always had been very mechanical and worked for an electrician after he finished grammar school in Davenport. He had built a very fine demonstration lightning rod model, as well as a model electric streetcar. Although only sixteen, Otto was a kind of "whiz kid" with electricity. Gus and Charles graduated from the eighth grade at the old Pacific School in Omaha. The three brothers worked in print shops, a box factory and a number of different shops. Otto and Gus even worked for Burkley's Printing Company. They were super mechanics who could make things run and who could sometimes revamp machines to improve their performance. They had the touch.

In 1892 Otto Baysdorfer started a bicycle shop at 14th and Dodge Streets. Charles and Gus soon were working with him. They were good, and business prospered. Their business grew until they had a really fine machine shop. This sounds much like the beginning of the Wrights's career, except the Baysdorfers's dad was just plain Robert, while the Wrights's was a bishop.

Otto got the automobile bug in 1897, so he decided to build one. He designed and built a lightweight, water-cooled, one-cylinder, three-horsepower gasoline engine with a "special mixer" for combining the gas and air. The engine was so compact that it fit under the seat. It was fitted with Otto's new "mummer," which smothered the exhaust sound and made the car nearly noiseless.

Most automobiles were very heavy, but Otto's two-seater weighed only 350 pounds. He even had a differential drive using belts to apply power independently to each of the rear wheels.

In the summer of 1898, the Baysdorfers gave Omaha its third automobile and the first built in Nebraska. They proudly called it the *Ottomobile,* and it moved over the streets every bit as well as Emil Brandeis's $5,000 electric. The only other automobile to be seen in Omaha at this time was gasoline-powered and had been exhibited on tour in the Midwest by a Chicago company. The *Ottomobile* attracted crowds wherever it went. It could even climb the steep Dodge Street hill from 16th to 20th Streets with Gus and

Otto and Gus Baysdorfer driving the Ottomobile
on 40th Street near Joslyn Castle in Omaha

Otto in it. (At that time, Dodge Hill was about forty feet higher at 20th Street than it is today, and the door to Saint Mary Magdalene Church at 19th Street was originally where the windows are now —some twenty feet above today's street level.) Otto and his wife could make the nearly four-mile trip from their home at 711 North 16th Street to Forest Lawn Cemetery in twenty minutes on three-cents worth of gasoline. Otto thought his car capable of going about twenty-five miles per hour, but there was no safe place to reach such a reckless speed without scaring people and horses. Moreover, a recently passed city ordinance restricted horseless carriages to a maximum speed of ten miles per hour. Life was slower in 1898.

The *Ottomobile* did have one little drawback—it didn't have a radiator. When the engine-cooling water heated up, the seats got hot, necessitating a stop to let both the passenger and engine cool off—small enough price to pay for such a great Ottomobile! Later, when Otto and Gus were asked why they had not gone into the auto-making business, they said they didn't have enough capital; there were other things to do; and besides, they were busier than heck.

Otto and Gus were inventive mechanics operating a lucrative shop that kept them busy throughout the year. They were both married, but brother Charles was single, a little footloose, and free to drive and run machines. He talked Otto and Gus into building him a motorcycle, which they did. Charles then got to drive the first motorcycle in Nebraska in 1899. It was a good machine, and Charles became an expert rider. The engine was the Baysdorfer brothers' first attempt at building an air-cooled engine. Less than one horsepower, it ran the bike really well. Best of all, the rider didn't have to stop to let the engine cool off.

Otto and Gus built several more gasoline engines for motorboats. They kept making improvements until their reputation brought them all the business they could handle. As more new cars appeared in Omaha, their shop became one of the busiest in town, and they grew to be experts on ignition and electrical systems.

CHAPTER TWO:
BAYSDORFERS AND BALLOONS

The aeronaut "King" Murphy had hooked Charles on flying in 1900, when Murphy flew his hot air balloon at the Ak-Sar-Ben Fall Festival. Charles helped in any way Murphy would let him. Charles spent more time with the balloon than he did at the shop. He never got to fly, but he learned a lot about balloons.

As noted earlier, John Waldorf Hall, an aeronaut, came to Omaha in 1902 and happened to move next door to Charles. That did it! Charles and Hall became friends, then Charles became his assistant and learned about ballooning. He still worked for his brother Otto, but in the spring of 1903 he went with Hall touring the country making balloon flights at fairs, shows, and amusement parks.   After several months and many flights, Waldorf said Charles was ready, and in August he soloed as "The Great Sir Charles" at the Decatur County Fair in Leon, Iowa. Charles made a great flight, and his parachute drop was a gem.

Charles came back to work for Otto after the ballooning season was over. In the spring of 1904, he was off again with Waldorf, who had many bookings back east in Illinois, Indiana and Ohio. He met some of the greatest balloonists in the country: Roy Knabenshue, Lincoln Beachey, Charles K. Hamilton and Tom Baldwin, who had an idea for making an airship—by adding a motor to the balloon.

Charles became an excellent aeronaut. His fondest memory was an event that happened in Cincinnati. He decided to give the crowd gathered at a big amusement park a real thrill, so he had a chest-pack parachute in addition to the regular hanging chute. At about fifteen hundred feet he released the regular chute, which opened in good shape; suddenly he fell off the trapeze bar, free-falling. To the crowd it was obvious the result was going to be tragic. He opened the chest-chute in plenty of time to land safely. What an uproar! Women fainted; people were hysterical; and it scared the heck out of almost everyone. The cops weren't amused and said that if he did it again he would land in jail. Charles had become a true, slightly dingy aeronaut.

Somewhere along the way, Charles caught typhoid fever and returned to Omaha a sick man. Otto and his wife nursed him back to health. This took most of the winter, then Charles went back to work in his brothers' prosperous shop. Otto was making lightning-rod demonstration kits for several lightning-rod companies, including Montgomery Ward. He also manufactured a portable X-ray machine and sold about a hundred of these to doctors. Charles didn't barnstorm with J. Waldorf in 1905. He was recuperating, and Otto and Gus needed him in the shop.

Gus and Charles started a two-car taxi cab line with rebuilt wrecks—a Brush Runabout and a Pope-Toledo Sedan. Baysdorfer Auto Taxi Livery was the first of its kind in Omaha. Charles drove, and Gus kept the taxis running. The shop was now located at 1620 Capitol Avenue. They proudly showed off their new letterhead and had a telephone connected. Since it was about seventy-five feet from

Otto Baysdorfer's business letterhead

the back of the shop to the office, Otto and Gus built the first radio, or wireless, in Omaha. With it, they could now ring a bell in the shop from the office, without any wires!

Next, they invented a gasoline lamp with a very bright light. The brightness was caused by the pressure on the gas to the burner from a water reservoir floating on top of the gasoline. It worked beautifully and sold well. Before Otto could make an even better lamp, the electric-dynamo arrived to provide electric lighting.

Charles's talk about balloons gradually got his brothers interested in flying. One day, he told them that he had once spoken with a man at an Indiana fairground who had seen the Wrights fly their aeroplane in Dayton. At that time in 1904, the Wrights had the only flying aeroplane in the world. This story roused Otto and Gus's interest in aviation. The seed was planted.

CHAPTER THREE:
THE AIRSHIP EPISODE

Charles Hamilton arrived in Omaha for the Ak-Sar-Ben Fall Festival on the Rock Island from Des Moines, Iowa, on September 25, 1906. He brought the *Knabenshue Airship,* and the Baysdorfer brothers met him and offered their help, which he was pleased to accept. He remembered Charles from two years prior in Indianapolis. Charles and Gus spent the entire week helping Hamilton's two mechanics. Whenever he could, Otto came over to the airship tent, located on the Kings Highway on the carnival grounds at 17th Street between Dodge and Farnam. The ship was assembled and kept in the tent. First, the forty-foot long spruce frame was assembled. The five-horsepower, two-cylinder, air-cooled engine was installed in the frame. The engine drove a seven-foot ash propeller on the front of the frame. A four-foot by four-foot rudder for steering was on the back. The pilot stood on the frame. In level flight, he stood in the center. When he wanted to go up, he moved to the back, and to go down he moved to the front. The frame was fastened to a four-foot mesh rope netting completely surrounding the bag. The bag was sixty feet long and twenty feet in diameter. It was made of varnished Japanese silk. All this was assembled by Tuesday night, and the bag was scheduled to be filled Wednesday morning.

In the early 1900s, hydrogen was very difficult and expensive to make. To produce enough hydrogen to fill an airship's bag, Hamilton's hydrogen generator required a ton-and-a-half of iron filings and scrap, 1,500 pounds of sulfuric acid and enough hot water to fill a dozen large bathtubs. All this took more than twelve hours to manufacture. Filling a bag with hydrogen also is quite a dangerous maneuver. As the lightest of all elements, hydrogen would appear to be the gas of choice when filling a balloon. It can, however, slip through a hole that isn't there. It is most difficult to keep contained, especially in a varnished silk bag. It is also the most flammable of all gases and will ignite with a lone spark. This is what happened on the fateful night of May 6, 1937, at Lakehurst, New Jersey: Germany's *Hindenburg*—the last of the great airships filled with hydrogen—exploded into flame and was destroyed completely in thirty-two seconds. Thus ended air travel in the giant dirigibles.

The weather on the Wednesday scheduled for take-off was terrible—rain with a gale-force wind. The bag couldn't be filled. Thursday was a beautiful day with a wild wind. Once again, no flying. This turned out for the best, though. The clutch controlling the propeller had not been working properly, so Otto had time to take it back to the shop and rebuild it. Hamilton said it was now the best clutch he had ever seen.

As Friday's weather forecast was promising, the hydrogen generator was fired up Thursday evening. Would-be spectators grumbled, asking where the airship was. The biggest crowd of the week, almost seven thousand, were in attendance at 4:00 p.m. when the great airship was walked out of the tent by the Baysdorfers and the rest of the crew. Hamilton climbed into the frame, ordered the engine started, and then signaled to release the ropes. The *Bee* reported that the airship flew like a huge bird towards 18th & Douglas Street. Hamilton showed that he had complete control over the ship. He flew over the carnival grounds, toward City Hall and the telephone building, and returned six minutes later to his starting point. Hamilton made three more flights during the carnival, and everyone thought it was spectacular. On September 28, 1906, Omaha, Douglas County and Nebraska saw the first airship ever to fly here. This was the flight Mom and Dad saw.

After this, Otto and Gus began studying and learning everything they could about airships. Charles also was interested, but not so much as his brothers. Soon, Otto and Gus decided to build their own.

The Baysdorfers didn't leave behind much information about the many things they invented or built. Otto's son, the late Gerald Baysdorfer, worked with his father for many years, but said that he never talked much about those days. Gus's daughter Betty Baysdorfer (Mrs. Don) Moucka said about all she knew was that her dad had built an airship and aeroplane. These two descendants have furnished important information (such as family history, newspaper clippings, printed materials, photographs and, of course, their memories) and have been invaluable consultants for this story.

Gerald Baysdorfer said that his father would grab any piece of paper to make sketches and notes for his inventions. Unlike the Wrights, who kept meticulous records, data, and drawings, Otto Baysdorfer never made formal drawings. Fortunately, Bostwick and Frohardt took photographs of the airship, but no Bostwick or

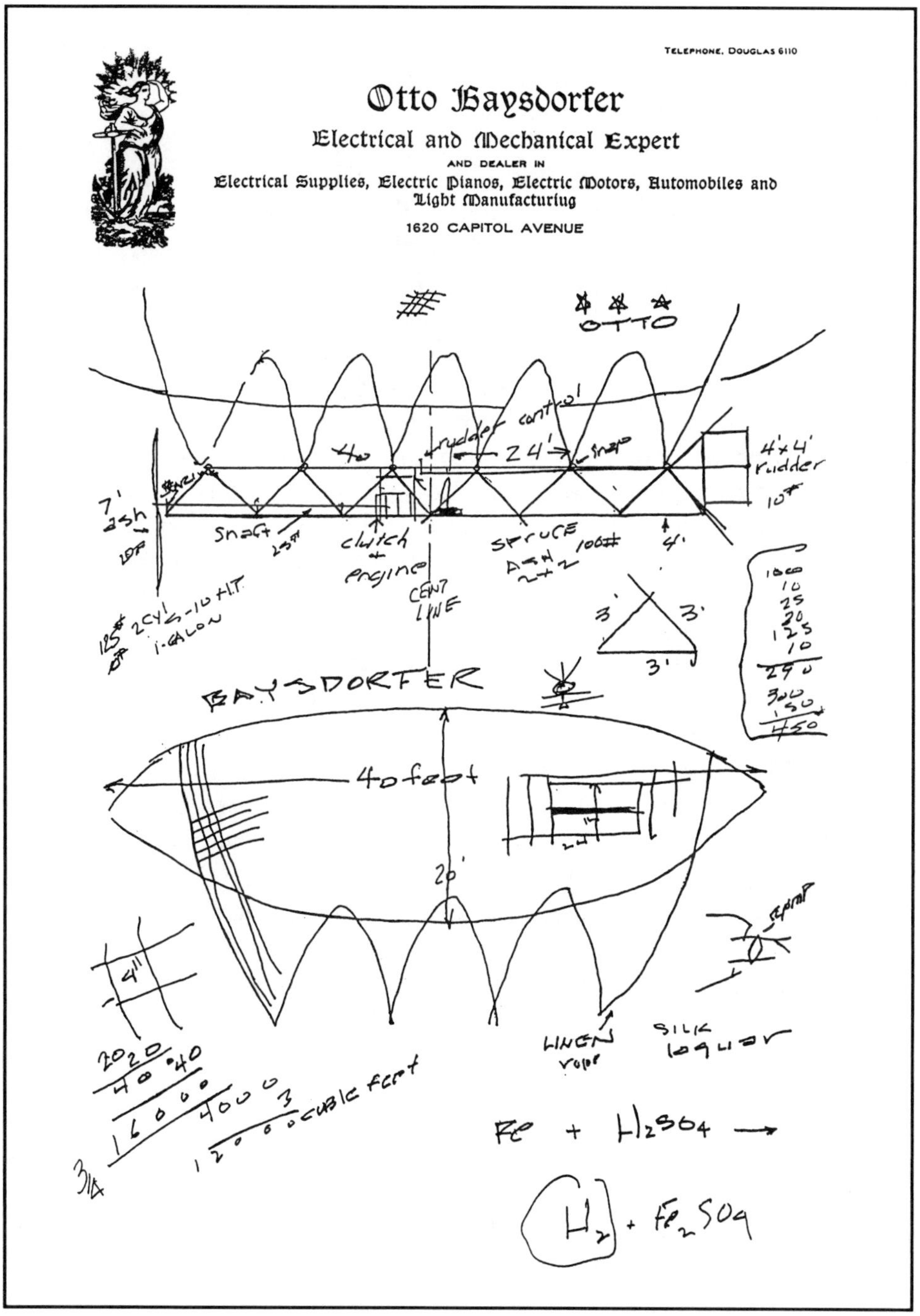

Plans for the Baysdorfer airship,
drawn by the author in the manner
of Otto Baysdorfer

Frohardt photograph nor Baysdorfer drawing of their aeroplane has ever been found.

When Otto and Gus felt they had enough information to design their own airship, they began with the engine first. With twenty or thirty engines already built in their shop, they weren't starting from scratch. Otto worked mainly on the engine, and Gus and Charles worked on the frame. With the promise of flying their ship past the City Hall, they talked the city fathers into letting them build the bag in the City Market Building at 14th and Capitol. It was an ideal place—long and wide with a high ceiling.

By New Year's, 1908, the Baysdorfers had the frame ready to assemble and were almost ready to test the new engine. They made remarkable progress despite the fact that they had to make a living with their shop and repair business and still work and pay for the airship. They had no big investors nor subsidies—sound familiar?

Otto and Gus were doing electrical work for the Boyd Theatre at 15th and Harney. Omaha's leading theatrical house was always looking for novel entertainment, so Otto built a moving picture projector. Gus talked his friend Homer Frohardt, photographic assistant to Louis Bostwick, into producing a film featuring a flying pigeon. The three-minute epic was sixty feet long and had to be re-threaded each time it was shown. The Boyd Theatre thought it was great and booked it. They had Parker Haight, a young fellow from Waterloo, Nebraska, who worked at the Boyd, help them. It was a sensation and ran for days, until the film wore out. My dad saw the film. Otto always claimed he had the first movie in Omaha, and so far no one has proven him wrong.

Back to the airship. The bag was fifty-two feet long and eighteen feet in diameter. It was of Japanese silk. The silk was cut into twelve-inch-wide strips dipped in linseed oil. Otto and Gus's wives and a number of their friends hemmed and sewed the one-foot panels of varying lengths together in long rolls. The individual rolls were brought from the ladies' homes to the Market House, sewn together on the big floor and finally joined to form an eighteen-foot-diameter tube. The ends of the tube were sealed, and an umbilical cord was sewn in the bottom. Air was pumped into the bag to inflate it. The entire surface was given an exterior coat of varnish. After two days of drying, the interior was varnished. When dry, the bag was tested for leaks. A four-inch mesh netting, made of one-quarter-inch cotton rope, was put over the entire bag. The netting

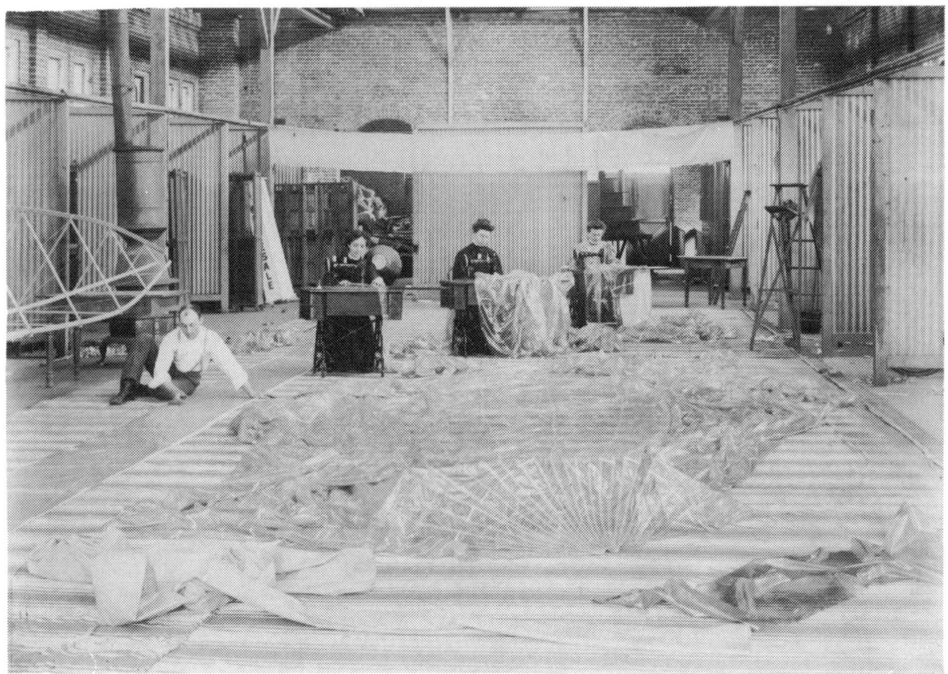

Assembling the Baysdorfer airship

had stringers hanging from it, so the frame could be fastened. The women wove the rope netting together. Self-motivated and self-supporting, the creative Baysdorfers were Nebraska's pioneers in aviation.

The frame was forty feet long, made of two-inch-square laminated spruce. It was triangular in shape, with the isosceles triangles bolted an equal distance apart on three long stringers. Piano wire was used for stiffeners. The two-cylinder air-cooled engine developed eight horsepower and weighed only seventy-five pounds, including the two-gallon gasoline tank. The engine was almost an exact copy of the Curtiss engine in Hamilton's ship, plus several of Otto's ideas—his gas-air mixer, his own insulated spark plug, his friction-belt clutch, his "mummer," or muffler, and more cooling fins. It was made in the Baysdorfer shop. The engine drove the six-foot paddle-type ash propellers on the front and back. Having two props was Otto's idea to give the craft greater stability in flight. This was about the only bad idea Otto had. The props were fastened to the engine, located a little aft of the center of the frame, by a long drive shaft. A three-by-four-foot rudder was on the back. It was controlled with a lever rod. The frame was fastened to the bag stringers with harness snaps to eye bolts fixed to the bottom of the triangles of the frame.

The airship, when assembled, weighed just under three hundred pounds. According to the calculations of young Dr. Herbert A. Senter, the new chemistry professor at Omaha High School, the bag would hold almost ten thousand cubic feet of hydrogen—enough lift to float 475 pounds. As Charles weighed no more than 145

Otto and Charles Baysdorfer inside their airship bag

pounds and had flying experience, he was to be the pilot.

Now came the really tough part—generating the hydrogen. Hydrogen was produced by pouring sulfuric acid over iron filings and adding very hot water. Sulfuric acid raises havoc with almost anything it touches. The Baysdorfers needed almost two hundred gallons of it, as well as over one-thousand pounds of iron and one hundred gallons of hot water. Hydrogen burns furiously. In their box-making days, they had built pickle vats for the Haarmann

Vinegar & Pickle Company, so they got four used vats from them. Each vat had a tight cover with a three-inch hose coming out the top. The hoses connected into a collector box, connected to a fan pump that pumped the hydrogen into the bag with enough pressure to inflate it. Sounds pretty simple, but they found it took fourteen anxiety-ridden hours to fill the bag. Altogether the materials cost $175.

Now that the Baysdorfer Brothers were building an airship, they received lots of help and advice—some bad, some good. John Waldorf Hall was very helpful. He knew a great deal about balloons and how to build bags. Parker Haight not only helped with the ship, he also drove the taxi for Charles whenever he could. Their machinist friends at the Union Pacific shops saved filings from their machines and probably machined a thing or two. The city fathers dropped by to see the progress. Ak-Sar-Ben grew interested and wanted them to fly at the Fall Festival. Gould C. Dietz also became enthusiastic about the project. There was quite a bit of community interest, and Frohardt saw to it that he and his boss took pictures.

The ship was completed by summertime. All they needed was a place to fly. Dietz suggested using Dietz Field, at 30th & Sprague Streets, and the brothers agreed. It was a good site: It was open, and the only building nearby was Druid Hill School. A tent big enough to hold the ship was put up on the field. The balloon was carted out from the Market House by horse teams, and then they spent a week reassembling it. As of yet, the bag had never been filled with hydrogen.

Then another good thing happened. George E. Yager (who was born in Red Oak, Iowa, but grew up in Omaha) had been over to Clarinda, Iowa, for the Fourth of July. Baldwin and Curtiss had flown their airship, the *California Arrow*, at Clarinda, and Yager was fascinated. He managed to become a helper for the two flyers, assisting them in running the hydrogen generator. When he got back to Omaha, he introduced himself to the Baysdorfers and offered his help. He and Charles soon became friends, and the two of them set up and ran the hydrogen generator.

On Friday, August 16, the airship was completely assembled and ready to fill. The generator was charged with iron filings; the sulfuric acid was poured in; and, finally, the boiling water was added. It started making hydrogen. By noon on Saturday, the bag was full, and the ship was tugging at the restraining ropes. A big crowd gathered. When they walked it out from the tent, the people cheered in loud approval. Charles climbed onto the frame, straddling it like a bike. The engine was started. After checking everything, Charles signaled to let her go. The airship rose majestically into the air, slowly turning to the right. He moved back, and the ship's nose went up, but it still was turning. Fortunately, there was no wind. He moved forward, and the nose came down, but the ship still was turning.

They said later that they all knew at the same time what had gone wrong: The two props were making it autogiro, that is, were turning the airship in circles. Charles did the right thing. He shut the engine off. The revolving stopped; Charley moved forward; the ship came down; and the ground crew grabbed the ropes. He landed about two blocks south at 30th and Manderson. They walked the ship back to the tent, and Otto and Gus started to work removing the back propeller. The crowd thought it was great, and it did fly! The Baysdorfers vowed to have it rebuilt by the next day, for they had no idea how long the bag would hold hydrogen. At $175 a shot, they had no desire to fill it again.

Sunday afternoon the airship was ready to fly again, and my dad was there to watch. He walked up to Dietz Field with Douglas County Commissioner Tom Falconer and J. O. Hiddleston, who were friends and neighbors. The weather was perfect, so out she came. The engine and new prop were tested. Charles climbed aboard and said let her go. Up she went. Charles moved back. Up went the nose, and this time there was no turning. He leveled off at about two hundred feet, then made a slow turn to the east and flew above Sprague Street, then a turn south to about Pratt Street, then a 180-degree turn back north. She flew like a bird, and Charles had her under complete control. He did another 180-degree turn to the left and flew south down to Evans Street, then another 180 degrees back to the field. He moved forward—the nose started down. He cut the engine, and she settled down. The crew grabbed the ropes. The Baysdorfer airship had made a perfect flight—the first airship made in Nebraska and flown by a Nebraskan! Charles was just the right man for the job. All his ballooning and parachute jumping had paid off. No seat, no cabin, no cockpit, no safety belt, just straddling that frame—Charles moved back and forth, hanging on with hands and feet.

The Baysdorfers knew they had to redesign the rudder after taking off the rear prop, and they couldn't do it before the next weekend. By then so much hydrogen would have leaked out that the ship wouldn't lift off, so they disassembled the bag and frame and took it all back to the Market House, which had been moved over to 10th and 11th Streets, Jackson to Howard Streets, where Omaha's Old Market is today. They rebuilt the rudder and added an elevator to help control up and down.

Dietz and other Ak-Sar-Ben people had watched the flight and struck a deal to have the airship flown at the Fall Festival the last week of September. Whatever they were paid, it probably didn't come close to Hamilton's take the previous year.

George Yager was personable and charming and seems to have been an early "hot dog." A hack politician, Yager belonged to the "Square Seven" and was a staunch supporter of the Mayor Dahlman administration, which gave him the concessions in Riverview Park. He knew Baldwin and Curtiss, said he was an expert on airships, certainly knew how to run the hydrogen generator, and made several good suggestions about the rudder. Although Yager knew little about flying airships, the *World-Herald* story on him dated November 24, 1929, portrayed him as having designed, built, and flown the Baysdorfer Airship. Despite it all, Charles and Yager got along well; the pair later formed a company to build and fly airships.

Otto and Gus decided that they had better get back to work at the shop, so the airship was more or less turned over to Charles and Yager. It was decided not to fly it until the Festival, only a little over a month away. Otto figured he had spent $5,000 and a lot of time building the ship.

Otto Baysdorfer was one-in-a-million. His kind is rare in any age, but seem few and far between today—not much education by today's standards, usually just grade school, sometimes high school, but probably a very high I.Q. and running on all eight cylinders all the time. He was a mechanical whiz, who could take a project from conception to completion. The airship was a success, so Otto's interests turned elsewhere. He had an idea about an electric band to go with his electric piano. Gus was probably a lot like Otto. Charles's talents lay in another direction: He was cut out to be an aeronaut, an aviator and a good fisherman.

84

Back to the airship. There was a lot of talk about Halley's Comet due in 1910 after a wait of seventy-six years, so Charles and Yager named the ship the *Comet.* Incidentally, Halley's Comet was just a dandy in 1910. The comet was so bright that almost everyone got to see it, unlike its measly appearance in 1987.

The Ak-Sar-Ben Fall Festival opened Friday, September 27, 1907, and one of the big attractions was to be the flight of the *Comet.* Over six-thousand spectators were on hand when the *Comet* was launched. Charles was at the controls. He signaled the ground crew to cast off the lines. The *Comet* majestically rose about thirty feet before a gust of wind blew it into nearby telephone poles and wires. There it stuck, tearing a hole in the bag. Charles sacrificed his dignity to slide down one of the landing ropes. They had actually tried to fly off 17th Street, between Dodge and Douglas, with buildings, poles and wires everywhere. Charles told the crowd that he'd fix the ship and be back the next day, but he'd get the ship a hundred feet in the air before turning her loose.

They didn't get the bag repaired until Tuesday. By then it had lost so much hydrogen, they had to make more, and this required firing up the generator. This took another day. She was finally ready to fly Thursday, and fly she did, pretty spectacularly. According to the *World-Herald,*

> It was with many protests from those who knew many things about airship sailing that Baysdorfer prepared for his second flight from the carnival grounds in Omaha Thursday afternoon. Horace B. Wild[e], who is, perhaps, one of the greatest aerial navigators in the world, stoutly protested against the procedure. He advised his friend by all means not to attempt this flight. He stated that the upper air currents would carry the ship away and beyond control.[1]

Baysdorfer knew all this, but, whatever the cost, the public had to be satisfied. For days, the huge crowds of carnival-goers had waited for a successful flight, and it was up to him to demonstrate the possibilities of flight with their marvelous airship. The ascent was made at 6:30 p.m. Thursday evening. At an angle of nearly forty-five degrees, the airship rose under Baysdorfer's guidance, shot over the City Hall, and was off on what looked like a pretty flight.

---

[1] *Omaha Evening World-Herald,* October 4, 1907.

The surface breeze had died down, and everything was working beautifully. An admiring cheer rose up from the audience as the ship sailed gracefully up over Omaha's highest steeples some five hundred feet above the city.

Then things started to go wrong. The ship encountered a strong overhead current, but few of the spectators realized anything was amiss. The fruitless efforts at tacking made by Baysdorfer appeared to be just pretty maneuvers to the crowd below. Realizing he was up against a wind which far exceeded the fourteen-mile-per-hour speed of his ship, Baysdorfer sailed with the wind southward.

Charles was able to keep the *Comet* on even keel, but he kept drifting south, even with a full throttle. Then another disaster struck: The throttle rod came loose, and the engine only would idle. The *Comet* started drifting out of control. Charles cut a shroud line with his pocket knife, somehow tied it to the throttle, and was able to get the engine back to full on. When he reached Gibson, the Missouri River became more ominous, so he thought he'd better bring her down. He moved forward, and she started down. The closer to the ground he got, the more control he had. Finally, near where the South Omaha Bridge is today, he saw a fisherman and called to him to grab the landing rope.

"Vat you tink I am, here to tie up storks? Catch holdt of yer own rope." With that the fisherman disappeared into his cabin.

The Shaliham brothers, who had a fishing camp below the mouth of the Mud Creek sewer, happened to be out in their boats as the airship was passing. The first of the brothers missed the rope, but the second, Frank Shalihan [*sic*], succeeded in catching hold of it and in a short time had it towed to shore.[2]

Then the brothers slowly pulled the *Comet* to the ground, where it was tied down.

Meanwhile, in Gould Dietz's big Locomobile, Otto, Gus and Horace Wilde arrived in South Omaha to help secure the *Comet* for the night. Charles told the Shalihams that he would be back in the

---

[2]*Ibid.*

morning. Dietz, being the man he was, gave the Shalihams a five-dollar bill to make this one of their better fishing days.

How did Horace Wilde get into the act? George Yager, who had met Wilde at Red Oak, Iowa, called him on the "Bell" in Chicago and asked him to come to Omaha to see what he thought of the *Comet*. Wilde was impressed and thought Charles, Yager and the *Comet* could be worked in with his airship. Friday morning, all the Baysdorfers, Yager, Wilde and a few governors of Ak-Sar-Ben arrived at the Shalihams's fishing cabin. It was decided that Wilde—an experienced aeronaut—would fly the *Comet* back to the Kings Highway. Charles protested, but Otto and Gus agreed, so Wilde took off at about 1:00 p.m.

The flight started well. Wilde took her up to about a thousand feet, made several turns, and headed for downtown Omaha. Thousands watched the airship as it floated gracefully through the air. She was right above 9th and Hickory when the prop shaft snapped; something tore a hole in the bag; and Wilde had his hands full calming a pretty wild airship. He got her under control and skillfully guided the *Comet* towards the Saint Joseph Hospital grounds. Wilde made it, but she was coming down fast, and he hit with a great thud and broke quite a few things, including his left leg. Later, while the Saint Joseph doctors were patching him up, Wilde said that he could have landed on the table if the skylight had been open.

Thus ended the flying of the *Comet* for the 1907 Festival. Really, it was a pretty good show. They flew, and people enjoyed the spectacle. Unfortunately, the Baysdorfers weren't paid a thing because none of the flights started and ended at the Kings Highway. In any case, the *Comet* was a good airship, and Charles and Yager, with Otto and Gus's blessings, were planning to take her on tour.

Wilde invited them to bring the *Comet* to his tent at the International Aeronautic Contests to be held in St. Louis beginning October 21. They worked around the clock to make repairs and improvements, crated her up, and shipped her by freight on October 15. All of them went, and it was a truly memorable experience. They were in the company of the world's best aeronauts. Balloons from England, Germany, France, Spain, Italy and the United States were entered in the James Gordon Bennett Cup Race. The starters for the American teams were Lt. Frank P. Lahm of the U.S. Army (who won in 1906), J. C. McCoy (a New York millionaire), and A. R.

Hawley (a Wall Street broker). The German team won, flying almost a thousand miles and landing in Maine. The Baysdorfer boys were competing with the best airships and flyers in the world.

Lincoln Beachey won the airship race, flying the one-and-a-half mile course in four minutes and forty seconds. Jack Dallas, a teammate of Beachey's, was second with six minutes and ten seconds. "Captain" Tom Baldwin came in third with seven minutes and five seconds. The fourth competitor, the *Comet*, was entered by Charles Baysdorfer of Omaha, Nebraska, and flown by Horace Wilde (according to the newspaper). Actually Charles flew it, since Wilde's leg was still in splints. Although Charles made the turn at just under three minutes, which placed him in contention, he had to drop out when the engine quit a little over a mile out. The winner's prize was $2,000.

The Baysdorfer brothers came home from St. Louis with fire in their eyes. They knew the *Comet* was as good as any airship there, and Otto planned to spend the winter making her the best.

Here were three brothers from Omaha who had built what was probably the fourth or fifth airship ever in the United States, and it was almost as good as those of the most famous aeronauts in the country. Most people, then as now, didn't know where Nebraska was, and thought of the "Wild West" when Omaha was mentioned. They also had a great time with "Baysdorfer who?" But the *Comet* made them take notice.

Otto and Gus had to devote their full time to the shop. Business was great, and they began advertising they would pay $100 to any customer who brought them a generator they couldn't fix. They went into a partnership with Louis Flescher, an Omaha investor, and built and sold Flescher Motorcycles. They built the engine, farming out some of the machine work to their friend Paul Melchor, who had a shop over on 13th Street. They were busy, so they turned most of the rebuilding of the *Comet* over to Charles and Yager, who formed Baysdorfer and Yager Airship Manufacturing, located at 1620 Capitol Avenue.

Charles, under the guidance of his brothers, did most of the work on the ship. The engine was rebuilt, and the frame was shortened. The bag was repaired, and a pointed nose and tail, like Beachey's in St. Louis, was added. Yager spent most of his time contacting fair boards, amusement parks and various groups in Iowa, Illinois and Missouri about booking the *Comet*, and he had considerable success. Meanwhile, both Charles and Yager continued to drive their taxis, and Charles chauffeured for several Gold Coast families on special occasions.

In May 1908, Baysdorfer and Yager Airship Manufacturing unveiled the new *Comet* with a great flight from the vacant lots on 13th Street between Cass and Webster Streets. Charles flew her for thirty minutes over the Union Pacific Railroad shops, the wholesale district and downtown. It was one of those beautiful May afternoons, perfect for flying. Bostwick and Frohardt took several pictures of the flight.

1911 postcard showing the Baysdorfer airship in flight

They had their first booking on Memorial Day at Creston, Iowa. The Creston show was not too impressive. The weather was cloudy and very rainy, and all the ceremonies were canceled. Fortunately, Charles had decided not to fill the bag, but they had the expense of getting the ship and equipment to Creston. After much deliberation, the committee gave the flyers $25 for expenses. It was a beginning.

Their next booking was the Fourth of July celebration in Oskaloosa, Iowa. The weather was good; the crowd was good; and Charles made a great flight. Everybody was happy, and the committee paid them $500. The *Comet* finally had made some money, and the company was on its way—they hoped.

They flew at Electric Park in Waterloo, Iowa, on August 2. The *Waterloo Reporter* noted that the airship lifted off at 6:45 p.m. and ascended to several hundred feet. Baysdorfer and the *Comet* then flew for half an hour before landing in a nearby oat field. Charles had run out of gas and made a forced landing. No damage was done, but it was dark before Gus (who had gone with them to Waterloo) and Yager got there. There was no more flying that day. That caused a hassle with the park people, who wanted to reduce payment to $350 instead of the $700 that had been agreed on. That was one of the perils of barnstorming with an airship in 1908.

The rest of the summer went about the same way. When the weather was good, and everything went right, they flew and got paid. Charles did all the flying. Sometimes Gus and Otto would go to wherever they were when the ship needed repairs or to help them out. It was a precarious business, not only risking life and limb, but being flush or broke. So Charles, with Otto and Gus's blessings, sold the *Comet* to Yager. The price was rumored to be $5,000.

Charles came back to Omaha and the shop. Yager persuaded Horace Wilde to fly the *Comet* for $200 per week. They flew it all over—St. Louis, Kansas City, Louisville, Indianapolis and elsewhere. Wilde finally wrecked it in Minneapolis at the Minnesota State Fair in September 1909.

Charles Baysdorfer had joined an exclusive group of men. He was an aeronaut. He knew how to fly a balloon, and he had graduated to airships. There was one more place for him to go—aeroplanes—and to become an aviator. Charles couldn't stop until he had flown an aeroplane.

CHAPTER FOUR:
THE BAYSDORFER AEROPLANE

The Baysdorfers had wonderful memories of building a great airship, which was now part of the history of the Wild Blue Yonder. Then one day in Tony's Barber Shop, Gus saw a picture and a drawing of the Wrights's new Army Flyer in a week-old copy of the *Chicago Tribune*. He could hardly wait to get to the shop to show it to Otto. That did it! They decided to build an aeroplane!

It was 1909, the year the world discovered the aeroplane. After the great meet at Reims, France, in August 1909, newspaper stories, photographs and drawings of aeroplanes were printed in all the newspapers. *Scientific American* magazine published a detailed story on Curtiss's *June Bug*. Otto and Gus wore out their copy reading and studying the aeroplane. Soon Otto's sketches started to appear. They decided they would buy an engine and propeller—the only things they thought they couldn't build. Wheels were no problem: They could get them from their bicycle suppliers. The design of the first Baysdorfer airplane started taking shape.

By New Year's Day 1910, they were ready to build. First, they found a place in which to build the machine—next door to the shop in the back of Fletcher's Cigar Store at 1622 Capitol Avenue. This structure had been a cigar-making loft, so it had a high ceiling and an open room about thirty-by-fifty feet. The cigar store occupied a small part in front, and there was a barber shop in the other storefront. Originally, M. B. Fletcher had come from Waterloo, Nebraska, where he had a pool hall and barber shop.

The first thing they had to have was an engine and propeller. Charles had met a fellow named Grey while flying the *Comet* in Peoria, and he said he was going to build airship engines in his Muncie, Indiana, shop. He was contacted, and, sure enough, he was building engines. He could build and deliver one in about six months. He called it the Grey Eagle. As he was anxious to have one in an aeroplane, he gave them a bargain price of $1,375 for the four-cylinder, fifty-horsepower, water-cooled, 1,250 rpm Eagle. It weighed only 145 pounds. Grey said he knew a propeller manufacturer by the name of Sensnich who had designed a prop for his engine that could be bought for $100. Later, Sensnich became one of the principal makers of wood props. They made a deal, and

Otto sent Grey a $100 down-payment and received a drawing of the engine.

Now all they had to achieve was the aeronautical engineering that goes into designing and building an aeroplane, without ever actually having seen an aeroplane. Fortunately, the *Scientific American* story was detailed enough so that it could serve as their bible.

The wings were built first, then the airframe—the framework to which the wings, control surfaces, landing wheels, engine, fuel tank, cooling system and pilot's seat are fastened. The wings had no

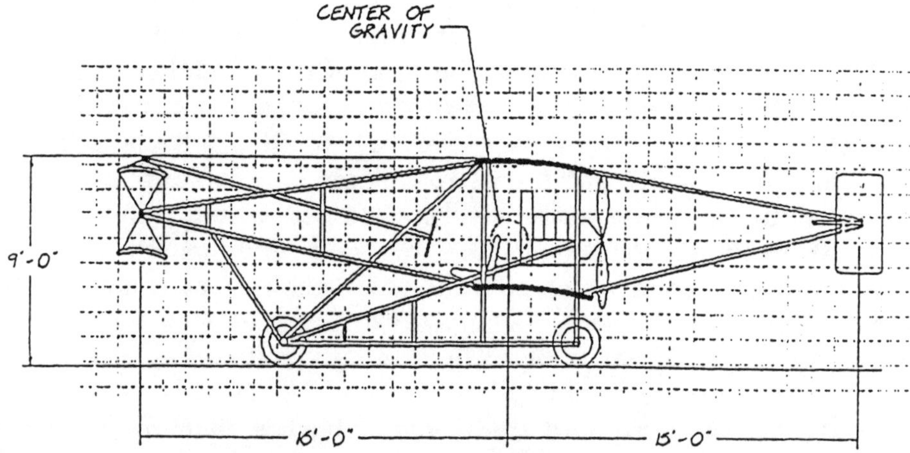

Elevation of the Baysdorfers' first aeroplane,
drawn by Robert E. Adwers

spar. The ribs were laminated ash, one-inch wide, two-inches high and four-and-a-half feet long. They were curved to a nine-foot radius by steaming before gluing. The struts were also one-inch ash, five feet long, tapered from three inches at the center to one inch at the ends. The leading and trailing edges were one-inch oak, half-rounded. The ribs were fastened to the edges with wood screws. The wings were internally strutted with piano wire and turnbuckles. Top and bottom wings were made in two sections that were bolted together, so they could be dismantled and crated. They were covered with the same materials used on the *Comet's* bag—silk, linseed oil and varnish.

All the control surfaces were made like the wings, except that the framework was not so thick. The part of the airframe that contained the engine, pilot's seat and landing gear, and to which the wings were connected, was made of one- and two-inch oak. The engine mounts were one-inch steel tubing. The front elevator and rear rudder controls were connected to the center section with bamboo tubes and rounded ash stringers. All fittings were made of thin gauge steel, and all wires were three-sixteenths-inch steel. The wheels were fourteen-inch heavy bicycle rims.

Everything was fastened together with nuts and bolts, screws, tape, fish glue and the grace of God. Before assembling all this together, somebody (it was probably Otto) had to do some weight-and-balance-figuring to get everything located in the right place. Evidently they did, because the aeroplane flew. These three brothers from Omaha, Nebraska, built an aeroplane in 1910. The aeroplane had no name, so I have given it one—the *Red-tailed Hawk*.

I have made the drawing on the left and listed the specifications after considerable study of the many newspaper stories, photographs, and accounts collected in my research in Omaha and Waterloo, Nebraska; Des Moines, Ottumwa and Davenport, Iowa; Galesburg, Illinois; Terre Haute and Logansport, Indiana; Buffalo and Elmira, New York; and from my conversations with Gerald Baysdorfer and Betty Baysdorfer (Mrs. Donald) Moucka and eyewitnesses of the first flights, Anna Otte Boettger of Irvington, Nebraska, and the late George Frost of Elkhorn, Nebraska.

Otto and Gus did most of the work on the *Hawk*, assisted by Charles and their friend Parker Haight. Charles was now chauffeuring for the Metz and Nash families, who lived west of 36th Street in the Harney and Farnam Gold Coast area, and Parker still worked at the Boyd Theater. Charles soon became well-acquainted with J. J. Deright, who sold Locomobiles and Stoddard-Daytons, the autos the Nash's and Metz's owned. He was the same J. J. Deright who helped organize the Aero Club of Nebraska and was the President of the Mid-West Aviation Meet Company. Deright and Hyland B. Noyes, Chief Engineer of the Omaha Street Railway Company, had just purchased Clarke Powell's Auto Parts Store at 2154 Farnam Streets (the same Clarke Powell who so successfully ran the Mid-West Meet). Noyes was the son of Senator Isaac Noyes, prominent citizen, politician, substantial landowner and farmer at Waterloo, Nebraska, twenty miles west of Omaha.

The Grey engine arrived in May. It was not quite as good as they had hoped. At 198 pounds, it weighed almost fifty pounds more than specified, and one cylinder leaked in the water jacket. Otto and Gus soon had the leak fixed and made other "Otto-improvements" on the ignition system and carburetor. They also rebuilt their engine mounts. The engine never developed much over forty horsepower. The prop came with the engine—a real plus. The airframe was completed after changing the mounts, then they started assembly of all the components. They hoped to have her ready to fly soon.

Then, fortunately, they got some first-hand experience with aeroplanes. Curtiss was going to fly in Omaha in July, so Deright arranged for them to assist Curtiss's mechanics, as we have already read. Otto and Gus learned a lot and made some important changes to the *Hawk*. Not many aeroplane builders had Glenn Curtiss himself check over their ship and make suggestions for improvements. They were shown how to balance the ship when they got her together in one piece. They even learned how to build the crates in which to ship the *Hawk*.

Charles hung on every word the pilots had to say. Charles was a balloon driver, an aeronaut, as was Mars. They had a lot in common, so Mars took him under his wing and gave him a flying course. Pure luck played a great part in their success, not only in getting help from one of the world's greatest aviators and aeroplane builders, but in receiving a course on how to fly as well. Years later, while helping build Martin B-26B bombers during World War II at the Fort Crook plant (where STRATCOM is today), Gus Baysdorfer said that Glenn Curtiss himself helped them build the *Hawk*.

When the Curtiss meet was over, the brothers went back to the shop and started making changes on the *Hawk*. They were unable to finish the improvements until October, for they had to stop from time to time to make a living.

Meanwhile, the dealers on automobile row became very interested in the project. They gave encouragement and advice (and, I think, some capital) to the Baysdorfers, and the shop continued to get a lot of business from them. Noyes invited them to make the first flight at his family's farm two-and-one-half miles north of Waterloo, on Douglas County Road 84 today. They had a hayfield almost a mile long north and south and a half-mile wide. The corn had been picked in a field west of the hay. This was a perfect place to make the first flight of the first aeroplane built in Nebraska and to be flown by a novice Nebraska pilot. They quickly accepted Noyes's offer. The Omaha Automobile Dealers Association arranged with the Scott Tent and Awning Company to put up a tent large enough to cover the *Hawk* at the Noyes's hay meadow. Allen C. Scott (the company's owner) later was credited with developing the basic concept of the modern parachute with a pilot and a main chute in 1918. The plane was packed into three crates. Another crate was filled with tools, parts and miscellaneous equipment.

On Friday, November 18, the *Hawk* and the tent were hauled to the Union Pacific freight depot—the same building that houses today's computerized railroad system control—and loaded for the trip to Waterloo. It would arrive on local freight *No. 27* at 2:00 p.m., Friday. Otto, Gus, Charles and Parker Haight rode out to Waterloo with Deright in his Locomobile. Noyes and his family also drove and were going to spend the weekend with his parents. Quite a few members of the Auto Association were planning to drive to Waterloo on Monday. H. E. Fredrickson, President of the Association, Burt Murphy, Henry H. Van Brunt, Lee Huff, Deright and Clarke Powell also were planning to make the trip.

No. 27 arrived on time, and Union Pacific Depot Agent Leslie W. Shannon supervised the unloading of the crates. The tent was immediately loaded into one of T. D. Todd's livery wagons and taken out to the Noyes's farm. Scott's men followed it out in his Chalmers 30 and started putting it up. The Baysdorfers helped unload the ship into Todd's other wagon. Then the first aeroplane built in Nebraska went down the main street of Waterloo and over the Elkhorn River bridge. Years later, the original Lincoln Highway would cross on that bridge, but it was torn down in 1990.

They went out the Elk City Road (County Road 84 today), past the Rapp Sheep Farm (whose elevator was torn down in 1991), past the Otte Hill Road, around Billiter Lake, two-and-a-half miles to the hay meadow. The Scott Tent people had the tent up, so the *Hawk* was unloaded; Otto started the assembly with help from Gus, Charles and Haight. A smaller tent was put up inside for the men to live in. It was late November, but so far the weather had been beautiful. Kitty Hawk had been much wilder.

The Noyes had the Baysdorfers over for dinner Friday night and for breakfast Saturday morning. By evening, they had the *Hawk* almost completely put together and planned to test the engine and maybe run her on the ground Sunday.

M. L. Hancock drove out from Omaha and took them all to dinner at the Todd Hotel. The Baysdorfers weren't regular church-goers, but when Mrs. Noyes—whose family was the backbone of the Waterloo Presbyterian Church—invited them to attend Sunday services, they went. Charles wanted all the help he could get. Reverend Aston preached one of his better sermons and said a prayer for success to the aviators. He didn't mention anything about God giving people wings.

For days, most of the folks in Waterloo drove their autos or buggies up to Noyes's meadow to see the aeroplane. Kids rode their ponies or bikes, or even walked. The J. C. Robinson family drove out in their new Pope-Toledo. Half the Waterloo town football team, the "Cornfeds," fresh from a great season, were there, as were town manager A. H. Campbell, Otto Wilson, Homer Payne, Mayor R. P. Nason, H. B. Waldron, and "Doc" Richardson. Town Marshall A. W. Suiter almost had to direct traffic. Folks came from all over—Elkhorn, Valley, Elk City. It was almost like fair time. They weren't disappointed because the Baysdorfer brothers had their plane completely assembled and on view.

The Hawk at Waterloo (left to right): Charles Baysdorfer (at controls), Gus Baysdorfer, M. L. Hancock, Parker Haight and Hyland B. Noyes

Charles said to fire her up. Gus agreed, so he climbed into the seat, and Parker cranked the propeller, and the engine barked into a mighty roar. After checking her all over, Charles opened the throttle and taxied up and down the meadow several hundred feet. He decided not to try to fly her, which was probably a wise decision, seeing that he had never flown an aeroplane. He got the feel of it, though, and later said that he knew she wanted to fly. Sunday was a great day for everyone, and the next day they were going to fly the *Hawk!*

The sun rose at 7:23 a.m. on Monday, November 21, 1910, and the temperature was thirty-seven degrees. It was going to be one of those great Nebraska late fall days. The aeroplane crew was up early. They were almost too excited to eat, but the Noyeses sent over a big pot of coffee and hot biscuits. They spent a couple of hours checking everything, while they were waiting for the Auto Club people to arrive from Omaha. They were bringing a *World-Herald* reporter and photographer with them. Hancock and Haight had gone to Waterloo to get another can of gasoline from Otto Snyder's garage. The *Hawk's* gasoline tank held five gallons.

Anna Otte (now Boettger) lived in the big frame house on her folks' farm on the bluff east of the Noyes's meadow. Her father farmed two hundred acres where Mount Michael Abbey is today. The house had just been built by Bill Meyer, a prominent Elkhorn and western Douglas County contractor. The house is used today for an antique shop, and fine luncheons are served there by the abbey's brothers. Anna and her brothers and sisters went to school at District 41, which was a mile straight north of their house. They would walk through their pasture at the top of the bluff overlooking the great Elkhorn-Platte River valley. They were excited about the aeroplane and had watched Charles run it on Sunday. But they still had two more days of school before Thanksgiving vacation. Ms. Pettycord, their teacher, had one of the boys go outside to listen for the *Hawk's* engine and promised the students she would let them go out to watch if it flew.

Today, Anna Otte Boettger lives in Irvington, Nebraska, about ten miles east of where she was raised. She is ninety-six years old as I write this in 1994. Her memory is fantastic, and we are so lucky that she saw Charles Baysdorfer fly and remembers it in detail.

Finally, at about 11:30, Walt Winterburn, whose turn it was to listen, came running in hollering, "It's started, it's started!" They all ran out and up the hill to the top of the bluff, and, sure enough, the engine was running. Everybody finally had arrived, so Charles said, "Let's fly her." Haight, who was the best at propping, gave it a twist, and the Grey Eagle roared into life—all four cylinders and forty horsepower of her.

She was running great. He gave her the throttle and held her straight. He pressed the right rudder and kept the elevator in neutral. She was picking up speed. He kept her straight. Wanting to fly, the *Hawk* hit a bump, bounced into the air, and came down. He pulled back on the elevator, and then they were in the air and flying! He brought her down, cut the throttle, and kept her going straight. Man! He'd flown!

Haight got to the *Hawk* first. He figured she flew at least 150 feet. He and Charles hugged each other and danced a victory dance. Gus and Otto Baysdorfer, Hancock, and the Auto Club people ran up, and everybody celebrated. Pettycord's school kids were whooping and hollering. They had just seen the first aeroplane fly with a Nebraskan at the controls. Plus, Charles Baysdorfer had become an aviator!

After all the excitement calmed down a little, Charles taxied her back to the north end of the field. The wind was out of the south, but only at about ten miles per hour, so Charles turned her into the wind, hit the throttle, and off she went. In about a hundred feet, she jumped into the air and flew. He held her straight and let her climb to about fifty feet. The wings wavered, but Charles leveled them back. After about six or seven hundred feet, he cut the throttle and brought her down like a bird. He was almost to Billiter Lake. Charles turned her around and started taxiing back to the tent, but he just couldn't resist: He hit the throttle, and the *Hawk* rose into the air. He was learning how to fly. He cut the engine and made a great landing. He stopped her at the tent, to the cheers of all the witnesses. Charles knew he had had help: The air was good; the wind was almost calm; the engine ran; there were no gusts; a wing didn't get down; Gus and Otto's balance calculations were good; the field was long; and there were no obstacles.

It was a little after noon, and the word about the aeroplane flying had spread. People started showing up from every direction. The late Henry Frost, who became mayor of Elkhorn, was five years old at the time. He told me that his mom hitched up their buggy (they lived a half-mile north of the Noyeses) and drove down to see the aeroplane. He remembered it looked like a big bird and it flew through the air.

The Auto Club people had brought sandwiches and coffee, so everyone took a lunch break. The *World-Herald* reporter was getting information from all, while the photographer checked his equipment hoping to get a picture of the *Hawk* flying. Perhaps the reporter who had written the great closing story of the Mid-West Aviation Meet the previous July 27 wrote this one. The only pictures of the *Hawk* are those printed with the following story:

Omaha Daredevils of the Air Find Secret of Air Flying With the New Biplane[.] Flights on the Field Near Waterloo Demonstrate They Will Make It Go. Have Ridden in Everything From the Old Gas Balloon to the Latest Thing. From Waterloo to Omaha is Likely to Be Their First Long Flight.

The people living between Waterloo, Neb., and Omaha may one of these days not very far off, be out in their back yards hunting for the last blade of green grass for a tender tribute to their horses when a gentle purring may cause them to gaze up leisurely in anticipation of seeing one of the horses successful usurpers speeding along the highway.

But their first gaze upward won't be high enough. They will have to raise their eyes to loftier things than a mere puffing, panting old fashioned automobile.

In fact after they have scanned the earth and things horizontally without an explanation, with a louder buzzing in their ears, they will get a higher inspiration, and tilting their noses in a vertical direction may get a final glimpse of the cause of the racket as it swoops off into the blue of the horizon.

The signs of the zodiac, the hypnotic suggestion of the atmospheric conditions, or maybe it's just a little private hunch of one who has seen, point to the fact that Nebraska is about to have its first cross country aeroplane flight. And that, not by some of the professional aviators who are making flights in various parts of the country, but by Omaha men, who have built their own aeroplane. . . .

Charles and Gus Baysdorfer and Parker Haight, all of Omaha, have constructed the aeroplane. Eight days ago they took it to Waterloo and set it up. Two days later they ran it out on the broad level field on the Noyes Farm. . . .

None of them had ever flown in an aeroplane. But Charles Baysdorfer has been up in the air in every kind of machine from a taxicab to a hot air balloon, and he mounted into the seat. He ran the aeroplane across the field twice to see that everything was running smoothly and the third time he elevated the front planes and the aircraft leaped into the atmosphere, and sailed along a couple hundred feet before it alighted and rolled along once more.[3]

The photographer set up his camera at a strategic place where Baysdorfer thought he could get the bird into the air, then the engine was started.

Talk about your wind wagons, the way that bird, like a duck with a broken wing, sped over the ground was something remarkable.

With just the whirling propeller sending it forward it sped from one end of the field to the other in a few seconds. At the end of the field, slowing down a little, Baysdorfer steered it around, just using the "air rudder." . . . One more time the aviator skirted the field and came by the camera, and this time, as he raised the front planes just in front of the picture machine, it once more swept into the air and the camera snapped it. . . .

From the showing already made, with the motor once running smoothly, and a new propeller, it will undoubtedly be an easy matter for the aviators to take a twenty-two mile

---

[3]*Sunday World-Herald,* November 27, 1910.

spin [to Omaha]. The gasoline tank over the top of the machine holds five gallons of gasoline, enough for a 115 mile journey.[4]

Charles made eight flights that day. On the seventh, he felt he could make a turn, and he did, almost ninety degrees to the right. What an accomplishment! Turns were just about the toughest thing to do with a 1910 aeroplane. The *Hawk's* ailerons were controlled by the pilot leaning in the seat. If the pilot wanted to bank to the right, the pilot leaned to the right. The seat swiveled and pulled the wires that moved the ailerons up or down. This was the first time he had landed toward the west, which probably had a crosswind, so he was learning how to fly. He turned the *Hawk* back north and taxied back to the north end of the field. It was almost 4 p.m., but Charles just couldn't quit. He told Gus that he wanted to go one more time and try a 180-degree turn.

School was out, so Anna and some other kids were walking home along the bluff. She probably saw Charles's last flight. She told me,

> He took off south and climbed up until he was right even with me. I could see him plain as day when he went by. He had his cap on backwards. He went almost to Billiter Lake, then turned towards the river. The wings wobbled, but he kept on turning until he was going back towards the Noyes's house. Then he landed up near the tent.

Charles had made his 180-degree turn, a major feat for a guy who had never flown an aeroplane before that day. Charley wisely called it quits. What a great day for all of them! And that ended the day of the first flight of an aeroplane in Nebraska by a Nebraskan. The Auto Club people were so enthusiastic about the events that they were planning to offer a prize for a cross-country flight, like they had done in the East.

Clarke Powell and J. J. Deright were getting the bug and wanting to learn how to fly. W. H. McCord offered $100 to Charles for the first ride. Dave O'Brien, Omaha councilman and candy manufacturer, publicly announced he was going to go up with the Baysdorfers.

---

[4] *Ibid.*

Tuesday was another perfect fall day, so they planned to fly as much as possible. But things didn't start well. First, one of the tires was flat. They had a spare wheel, but it took time to put it on. Next, the Grey Eagle absolutely would not start. Haight wore himself to a frazzle propping her. Otto had gone back to Omaha on the 7 a.m. train—somebody had to run the shop—but Gus finally figured out that something was wrong with the gasoline. They had filled the tank first thing, so they drained the tank—a real project. They had just learned that aeroplane gas must be clean and dry.

Just before noon they got the engine running, but not very well. Gus thought it was the magneto, and he was right. They were super ignition and electrical mechanics, so they worked the magneto over. M. L. Hancock went to town and brought back sandwiches and coffee for lunch. A little after 1 p.m., they were ready to try to fly again. Parker pulled the prop through a few times.

The first time a person hand-props an engine is sort of like soloing in several ways. First, it's dangerous. Second, it can be tough, hard work, and it takes some muscle. It takes good balance and coordination, and it's a real thrill when you pull it through, and the engine says, "Blurp, bang, bang, whoom!" and barks, and tunes into a smooth roar. The first engine I ever propped was a forty-horsepower J-2 Cub. The biggest was a 330-horsepower Jacobs in a D-17 negative-staggerwing Beech. Not very many pilots get the thrill of propping nowadays.

He called, "Contact!" Charles turned the magneto switch on and yelled, "Contact!" Parker pulled the blade down from left to right, and the Grey Eagle fired and ran. When she smoothed out, Charles gave her the gun, and away she went. He let her get into the air, then he cut the throttle and landed. He did this four times before he got to the end of the field. Because he was learning how to fly, Charles was using his head and staying close to the ground. Wilbur Wright had written back in 1904 to stay close to the ground in case something went wrong.

When he flew back, the new birdman climbed a little higher and stayed in the air the whole length of the meadow. Then he landed and turned around. He decided to try turning to the right, then to the left on the next pass. He gave her full throttle, let her get a little higher, held her, and leaned to the right. He had banked too steeply and had to lean back. He leaned to the left. She banked, the engine began to cut out, and the nose went down. He leaned back to straighten the wings. Wham! She hit the ground, bounced, came down, but didn't tip over.

Charles's guardian angel was with him. The only damage was a blown tire and a broken rib (the *Hawk's,* not Charles's). He had survived an engine failure at the worst time in a turn, but he learned about flying. That ended the second day of flying. Then they all went to work fixing the *Hawk.*

The above derives from the recollections and descriptions of Anna Otte Boettger and the late Hans Moeller. Moeller lived in Fremont, Nebraska, and his family were good friends of the Ottes. The Moellers had come down on the train to spend Thanksgiving with the Ottes. Both described how the Baysdorfer engine sounded bad, failed, and then the aeroplane turned, wavered, and hit the ground without turning over. They also described the damage to the *Hawk.*

Wednesday was spent repairing the *Hawk.* Gus had called Otto at the shop on the Noyes's brand-new Bell telephone to tell him they needed help, so Otto came out on the morning train. He brought with him M. A. McLaughlin, one of the mechanics with Curtiss who was a whiz with the Grey Eagle engine. They welcomed McLaughlin because they needed all the help they could get. Gus and Otto's wives came too, hoping to see the bird fly. The women had brought a basket lunch with them, so everyone had a feast.

It took all day to make the wing repairs, though, and there was no flying. The ladies returned on the afternoon train to get ready for Thanksgiving, but the men stayed because they wanted to fly in the morning. Noyes had said they could drive his car to Omaha after the flight. Everything was set to go first thing in the morning.

Thursday dawned on another crisp autumn day. Gould Dietz, Tom Kimball and some other Aero Club members drove out, as did David O'Brien and his wife. The *Bee* sent a reporter. The Baysdorfers waited for all to arrive. A lot of folk from Waterloo, Elkhorn, Valley, Elk City and Bennington came too, since it was a holiday. The Otte and Moeller families had ringside seats up on the bluff. There was a great crowd gathered on the road by the Noyes's meadow. Everyone hoped to see the *Hawk* fly.

And fly she did, pretty spectacularly. At about 11 a.m., Charles climbed into the seat and yelled, "Contact!" Parker gave the prop a swing, and she roared into life. Charles revved her up slowly. McLaughlin, Gus and Otto listened and nodded their approval—she was running great. Charles opened her wide, and the *Hawk* darted forward. He eased the elevator up, and she leaped off the ground. The *Hawk* climbed to about seventy-five feet. She was flying straight as an arrow. When he was almost to Billiter Lake, he made a beautiful turn to the right (west), then another turn right, and was heading back north toward the tent. The engine had evidently given its all, because it blurped a few times and quit. Charles got the nose down, then the plane caught its right wing on one of four haystacks in the meadow. Anna Otte Boettger and Hans Moeller vividly remember the scene: Hay flew; dust flew; and a few pieces of the *Hawk* flew; but Charles rode her out and got only a bruised knee and the heck scared out of him. In spite of the shaky landing, it had been a great Thanksgiving flight, and everyone gave Charles a big ovation.

The aeroplane was in pretty sad shape, though. Its right wing was banged up; the right landing gear was broken; and the prop was splintered. They found out later she had blown a valve. Thus ended the flying for 1910. They gathered up all the pieces, put them in the tent, and planned to crate her up on Friday. Then the whole crew drove to Omaha and had Thanksgiving dinner at Otto Baysdorfer's. What a day—the first Thanksgiving in Nebraska when an aeroplane flew with the turkey, and they both crashed.

Early Friday morning, Otto, Charles, Gus, Parker and McLaughlin drove back to the meadow. They spent all day disassembling and crating the *Hawk*. They loaded her into Todd's wagons for the trip to the Union Pacific station in Waterloo. She would return to Omaha on local *No. 27* at 2:00 p.m. Saturday. After good-byes to all their friends in Waterloo, Noyes drove them back to Omaha.

That week the Baysdorfer aeroplane had flown eleven times; Charles had learned how to fly as well as a lot of guys who called themselves aviators; and he was still alive. The *Hawk* had proven a good aeroplane. Their design was good, and both Gus and Charles had several new ideas to make her better. The biggest problem was the engine—the universal problem with aeroplanes.

Gus was quite disappointed, though. He really wanted to learn how to fly. He did have his picture taken sitting in the pilot's seat (shown on the right), and

Gus Baysdorfer in the Hawk

he kept that picture all his life. He and Charles were the only ones ever to have their photographs taken in the *Hawk*. In the end, however, they were all disappointed that they couldn't fly her back to Omaha. But what if they had? Where would they have landed?

After the Christmas holidays, when they all had some time, the Baysdorfers, Haight and McLaughlin, who had gone to work in Deright's auto shop, sat down and discussed how to make the *Hawk* the best aeroplane around. They knew they had a good design and structure. They knew they had to have a better engine, but so did all aeroplanes. McLaughlin said Curtiss wouldn't sell anybody one of his engines, nor would the Wrights. They wrote to Kemp in Indiana and learned he was working on a six-cylinder, eighty-five horsepower engine, but he, too, wouldn't sell one. If he did, he wanted $3,000 for it.

Otto and Gus decided to rebuild the engine. Charles had two major ideas on the controls, and he was right on target. The first was to have only one wing on the elevator control, in front, and to control the ailerons with the elevator wheel. He also wanted to cut down the size of the horizontal stabilizer in the back. Although Charles Baysdorfer had flown only eleven times, he was suggesting some good ideas. He may not have been the first to suggest controlling the ailerons with the wheel, but it's effective, and it's the way it is done today.

The Baysdorfers repaired the wing and landing gear. Then they made the changes on the elevator and stabilizer. A new prop was ordered from the Aerial Propeller Company at White Plains, New York. The Grey Eagle engine took a lot longer to repair because Otto and Gus machined a new cylinder and other parts and completely overhauled it. Otto revamped the ignition system, and they bought a new magneto. They re-designed the carburetor and built a new gas tank with a sump in the bottom to insure gas flow whether the tank was tipped or level. The aileron control gave them fits, but they finally figured out a system of wires and pulleys fastened to the elevator wheel-control rod. If the wheel was moved (not turned) to the left, the wire pulled the back edge of the right aileron down so she would bank to the left, and vice versa for the wheel to the right. It was a bit crude and required considerable adjusting, but it worked. In 1911, the *Hawk* may have been the first American aeroplane to have the elevator and aileron controls hooked to the wheel. They hoped to have her ready to fly by April or May, so they began looking for a field.

Meanwhile, a group had formed the Omaha Speedway Company and was building a one-mile track for auto racing on West Center Road where the Ak-Sar-Ben race track is today. This was really fortunate because aviators could no longer use the field in the Clairmont addition, where the previous year's meet had been held. That field had twenty brand-new houses in it, and Benson had sold over fifty of the lots. The Speedway people correctly figured aeroplanes would be a great attraction and said that the Baysdorfers could fly from their field at the West Omaha Speedway.

Aviation in 1910 and 1911 was pretty wild and exciting. Aeroplanes and aviators sprang up like dandelions. There were many wrecks, crashes and people killed, but some good came out of it all: Better aeroplanes were being developed and built, and some good pilots were being trained. Still, the aeroplane business was primarily a death-defying show put on by daredevils who made a lot of money, lived fast, dangerously, and often only for a very short time. It was so wild that the public was getting blasé about flying and flyers. The *New York Times* even declared that the sporting element of the science of aviation has practically been eliminated. Charles K. Hamilton, who occupies a prominent place among the leading American aviators, predicts that under present conditions aeroplaning will be in nine or twelve months on a plane with fake ballooning and parachute jumping at county fairs.[5]

Hamilton certainly knew what he was talking about. He flew balloons first, then airships and then aeroplanes. I guess there was just too much money to be made too fast at too high a cost.

The danger of aeroplaning prompted many newspaper editors to call for changes. In July 1910, for example, the *Baltimore American* commented that the number of aviators killed was growing so rapidly that some licensing standards seemed to be in order. Victor Rosewater echoed the thought when he wrote in the *Bee*:

Aviation, like every other science, demands its toll of human life, and in less than two years has caused seventeen deaths. In that time some brilliant feats in the air have been achieved, but little actual progress has been made in the solution of the problem of aerial navigation.[6]

The story of Ralph Johnstone, a twenty-three-year-old from Kansas City, Missouri, tells it all. Johnstone and Arch Hoxsey became standouts on the Wright team—the "Heavenly Twins" or "Stardust Twins" as the newspapers called them. Johnstone was traveling with a vaudeville troupe as a trick cyclist when he saw Hoxsey fly in the first air meet held in the United States in Los Angeles. That did it. All Johnstone wanted to do was fly, so back east he went and signed up at the Wright brothers' flying school in Dayton, Ohio. There Hoxsey taught him to fly, and the pair became the hottest pilots the Wrights ever had.

Johnstone's flying career began in 1910 and ran as follows. In May he took his first ride as a passenger with Hoxsey in a Wright. In June Johnstone made his first solo flight at the Wrights's school. On August 26 he set a new record for a Wright aeroplane, climbing to one thousand feet in ten minutes. The next day, he made a thirty-minute

---

[5] *New York Times,* October 27, 1910.

[6] *Omaha Daily Bee,* July 16, 1910.

flight over Asbury Park and the ocean. On September 12, at the Boston Harvard Meet, the aviator set a new American record for duration flight of three hours five minutes and five seconds. Then he took the American altitude record from J. Armstrong Drexel at Belmont Park, when he rose to seven thousand feet on October 25. On October 27, he climbed to nine thousand feet from Belmont Park, was blown fifty-five miles from the grandstands in a storm, and landed at Middle Village, Long Island. This distance was within 186 feet of the world record set by Wynmalen at Reims. Then on October 31, Johnstone set a new world altitude record of 9,714 feet while flying at Belmont Park. Unfortunately the daring young aviator was killed doing his famous spiral glide at Denver, Colorado, just a few weeks later on November 17, 1910—almost six months to the day that he had taken his first ride with Hoxsey.

> Scarcely had Johnstone hit the ground before morbid men and women swarmed over the wreckage fighting with each other for souvenirs. One of the broken wooden stays had gone almost through Johnstone's body. Before doctors or police could reach the scene one man had torn the splinter from the body and run away carrying his trophy with the aviator's blood still dripping from it at the ends.

> Frantic, the crowd tore away the canvas from over his body and fought for the very gloves that he had protected his hands from the cold with.[7]

Just before going to Denver, Johnstone had told the *Cleveland Plain Dealer* that he lived to fly. He was a fatalist and believed that each person's life was marked out for them. Pilots likely were marked out for short lives. People who attended air shows didn't cry when a plane crashed, they were too busy watching the next pilot to see his performance.

Johnstone was a pretty good philosopher-aviator, twenty-three years old going on ninety, and we are still talking about him in 1994! By the way, Hoxsey, the other "Heavenly Twin," who flew at the State Fair in Lincoln, was killed a week later in Los Angeles.

---

[7]*Omaha Evening World-Herald,* November 18, 1910.

CHAPTER FIVE:
THE MOISANT EPISODE

Alfred J. Moisant and his brother John B. Moisant, both United States citizens of French-Canadian origin, had made a fortune as bankers and sugar planters in El Salvador and other Central American countries. In October 1910 they formed a business known as the "International Aviators." John, one of the world's foremost aviators, was the winner of the Statue of Liberty $10,000 prize at the International Aviation Meet at Belmont in October. He also was the first pilot to fly from Paris to London with a passenger.

The International Aviators planned to conduct exhibitions along educational lines and were willing to enter any meets, open and honestly conducted. The operation was capitalized with $250,000 cash. Its aviators, aeroplane mechanics, ticket sellers and takers, promotional people and other employees traveled around the United States in special trains.

The Moisants had entered the barnstorming business in a big way, and they were adding a little class to the flying business. They put together the best aviation organization in existence at the time: They bought three new Blériots with extra engines and parts, hired the best Blériot mechanics in France, brought them to the United States, and contracted with Rene Barrier (one of Blériot's chief instructor-pilots) and Rene Simon to fly alongside John Moisant. They built a factory and a field at Garden City, New York, and ran the operation from there.

The first exhibition was to be in Havana, Cuba, so they shipped the whole organization to New Orleans, where they put everything together and practiced flying. One of their new Blériots had a much larger gasoline tank and some control changes, and when Moisant flew the plane he was so impressed that he decided to attempt to win the Michelin Cup and to beat the world long distance record of 365 miles set recently by Maurice Tabuteau in France. The International Aviators had been flying from New Orleans's City Park race track, but for some reason John decided to fly from a field in Harahan, Louisiana. He flew the Blériot to the new field, where he was going to land and refuel for the record attempt.

Eyewitnesses (including his brother, Simon and Barrier) said Moisant appeared to be overshooting the field, so he hit the switch.

A Gnome was either on or off. When the engine didn't fire, he had to get the nose down to keep flying speed. He either over-reacted when he shoved the stick forward or the ship malfunctioned, and the nose came down violently beyond the vertical, almost on its back. Moisant was thrown out and killed. The only good that came out of the crash was Alfred Moisant vowed that a belt or strap would be installed in the company's aeroplanes so that the pilot would not have to hang onto something to stay in the plane. Obeying the old adage, "the show must go on," the International Aviators flew their exhibition in New Orleans and then headed for Cuba.

Meanwhile, in Nebraska, 1911 was the biggest year yet in terms of building and flying aeroplanes. The great success of the Curtiss Meet in Omaha and the Baysdorfers's success at Waterloo had all of the natives trying to get into the act. The Baysdorfers were improving the *Hawk,* but all kinds of planes (using the term very loosely) also were under construction.

Deright and Adams, from the Boys Aero Club, wrote in an *Omaha Daily News* story dated June 27, 1910: "Deright and Adams plan biplane works[.] Factory may be started, and at least one machine will be made. Deright to fly First Craft[.]"[9] The following spring, the *World-Herald* declared:

> An aeroplane manufacturing company is in existence in Omaha and has already completed one biplane, and a monoplane will be ready for flight by Tuesday. The name of the firm is LeBron-Adams Aeroplane Company, and the plant is at 313-15 South Twelfth Street. . . . [Their] first flight will be made from the speedway Wednesday. The machine is 200 horse power and is geared to ninety miles an hour.
>
> Bert LeBron is secretary and treasurer of the company and C. F. Adams is vice president and chief aviator.[10]

Bert LeBron had an electric company at the Twelfth Street address, and Clarence F. Adams was the same Adams who was the chief instructor of the Boys Aero Club of Omaha. My research has uncovered no future reports on LeBron and Adams, so I conclude they never built the aeroplanes described, nor did they ever fly an aeroplane in Nebraska. If by some unlikely chance they did, then

---

[9]*Omaha Daily News,* June 27, 1910.

[10]*Omaha World-Herald,* April 8, 1911.

they had the first monoplane in the United States. Moreover, if it went ninety miles per hour, it was the fastest in the world!

According to the *Omaha Daily News*, there were two other inventive aviators in the Omaha area in 1911:

> Mr. [Charles W.] Baker [an Omaha inventor] believes that he and his brother H. [Henry] G. Baker of Harlan, Ia., have invented a form of tiltable wings for biplanes that will revolutionize the conquest of the air.
>
> The first public exhibition of the Baker tiltable wings will be given by C. W. Baker and his brother at the County Fair at Harlan during the week of August 22. They are to get $100 per minute, if they fly for five minutes.
>
> "I expect to make the flight and I expect to succeed," Mr. Baker stated yesterday at his home, 2748 Fort street.
>
> The design of the Baker biplane is the same as the ordinary type, except that where the planes are stationary now the Baker planes may be moved up or down, all together, or four in a series. . . .
>
> . . . The use of the Baker tiltable planes make use of the rudder necessary only to steady the biplane. . . .
>
> . . . The Omaha man taught physics. For eight years he has been a streetcar conductor in Omaha.[11]

When I was a kid, we always thought streetcar conductors were a little dingy. They always dinged the bell twice to tell the motorman to start. Sometimes they sort of played a tune with it. There is no record of Baker ever flying—anywhere. Hopefully he made it to heaven.

M. R. Cole came close to flying at Carter Lake; "Professor" Sorensen survived his glider flight at Broken Bow; and the Savidge brothers flew their glider in Boyd County. As of 1910, however, Curtiss, Mars, Ely and Charles Baysdorfer in Douglas County, and Hoxsey in Lincoln were the only people to have flown an aeroplane in Nebraska.

---

[11] *Omaha Daily News*, July 28, 1911.

The International Aviators Come to Omaha

On April 30, 1911, James Deright, who was still President of the Aero Club of Nebraska, and P. L. Young, Manager of the International Aviators, announced that the aviators would present a flying exhibition, "Rivaling the Birds," at the West Omaha Speedway, May 8-14. Rene Simon (the Swiss "Fool Flyer"), Rene Barrier (a Frenchman who had been knighted for his great flying), "Captain" John J. Frisbie (billed as the only Irishman who flew an aeroplane), and Joe Seymore (the famous English race-car driver turned aviator) would all fly in Omaha. They had three famous French Blériot monoplanes, the same aeroplane that flew across the English Channel first, and Frisbie had his big biplane. The stage was set for the second air meet in Omaha. It also was set for an exciting episode for Baysdorfer number one, the *Hawk*.

Deright talked to Gus and Charles because he knew they were planning to fly the rebuilt *Hawk*. He suggested they talk to Young, the Moisant manager, about flying it at the coming air show. Young told them that he really needed a good biplane and he wanted Frisbie, their biplane expert, to look the *Hawk* over when he arrived in Omaha.

The Moisant Aviators and two Blériots arrived in Omaha and were delivered to the Speedway Field on Saturday, May 6. Frisbie saw the *Hawk* in Baysdorfer's shop at 1620 Capital on Sunday.

John J. Frisbie (or as he was popularly known, "Captain John J. Frisbie") was a balloonist, which gave him the right to any billing he wished. He was born in Oswego, New York, in 1869, so he was forty-one when he showed up in Omaha. His parents were born in Ireland. Frisbie was an aeronaut: He started his flying in a balloon and even flew manned kites before he graduated to aeroplanes. He said he learned to fly in a Curtiss *June Bug* at Hammondsport, New York, in 1909. By 1911 he was supposed to be one of the best, and certainly one of the oldest, pilots in the United States. He was very much in demand by the various flying companies; a real catch, Frisbie reportedly was paid $1,500 a month by the International Aviators. Frisbie knew a lot about biplanes, and he immediately recognized that the *Hawk* was a darn good one. He recommended that Moisant's International Aviators add it to their group. Since their manager, Young, had told Omaha that they would have a new biplane with their Blériots, he started negotiating a deal with the Baysdorfers.

Frisbie quickly realized that the only thing wrong with the *Hawk* was the Grey Eagle engine, even though Otto and Gus had rebuilt it. It just wasn't a very good engine—it was underpowered, too heavy and untrustworthy. So Frisbie suggested they place a Gnome engine in the Baysdorfers's plane. The seventy-horsepower Gnome Rotary, which drove the Blériots, was probably the best aircraft engine in the world. It was manufactured by the French, who were the first to seriously research aeroplanes, engines and flying. The Gnome was a seven-cylinder radial engine, with the cylinders revolving around the crankshaft. It was air-cooled and light. The gas-air mixture was induced into the cylinder through the crankcase, and this eliminated some horrendous fuel problems. The Gnome did have some drawbacks, though. It was either full on or off—there were no in-between throttle positions—so to fly it the aviator repeatedly had to turn the ignition on and off. Even with this problem, though, the Gnome was the best.

The Baysdorfers made a deal. The International Aviators would pay them $1,000 per month to use the *Hawk,* and they would pay Charles $50 a week to maintain it. Frisbie would be the pilot, but he was to teach Charles to fly. On top of that, the Aviators were to loan a Gnome engine to be installed in the *Hawk* as long as it flew for them. It took the entire week to install the new engine. Frisbie was a great help and was anxious to fly her!

Omaha's Second Annual Air Meet wasn't the success that the first was, mainly because it wasn't well promoted. Omaha's Aero Club co-sponsored the meet with the International Aviators, who supposedly were experts at promotion, but their manager was no Clarke Powell, and the Omaha businessmen weren't involved. In addition, the new Speedway Field, out on the west edge of town, was hard to get to: Spectators would have to take a West Leavenworth streetcar, then drive by automobile or walk the remaining half mile to the field. Maybe the biggest reason it wasn't as successful was that it was the second flying meet. Being second in a race, poker game, fight, love or flying is sort of like taking a bath with your socks on.

The first day of the meet was Monday, May 8, 1911. In spite of all of this there was some darn good flying, and the Blériots were

114

as good as any aeroplane around. Omaha town papers gave the meet front-page coverage:

> Flying, sailing up and down, around and across the Omaha Speedway track, Rene Simon and Rene Barrier gave a wonderful exhibition of aviatory skill at the opening of the second annual Omaha Aero meet Monday afternoon. . . .
>
> Barrier made the first rise into the air, soaring straight away into the wind to the south. Continuing on a bee line until a good mile away, he was nearly above the Lane cutoff, [the then brand-new Union Pacific tracks heading west] he turned in a beautiful wide swing to the east and soon was making rings around the big mile race track.
>
> Simon, not content with making fully as pretty a start to the south as Barrier, amused himself and thrilled the crowd by cutting in and around the grandstand, running down with terrific speed to within a few feet of the ground, when, with a sudden start of his motor, he soared away like a huge bird, curving, dipping, and swinging in dangerous angles.[12]

They continued:

> The slight French aviator, had a fine time yesterday . . . . Once he dashed straight for the little group of aviators and newspaper men in the center of the field. Every one of the aviators ducked for the tent and the reporters followed suit.[13]

It was superb flying, but to an audience estimated at less than five hundred people.

On Tuesday, May 9, the wind was wild, blowing out of the south in gusts of around forty miles per hour—we all know the wind can blow like heck in May! There was practically no crowd. Simon and Barrier were disgusted with the small crowd and the wind, so they said no flying—no doubt a wise decision. But the International Aviators's manager, P.L. Young, had invited the newspaper reporters, Deright, and members of the Aero Club to be his guests that day, and he wanted aeroplanes in the air. After much pleading, Barrier consented to fly. The *Bee* reported:

---

[12]*Omaha Bee,* May 9, 1911.

[13]*Ibid.*

Rene Barrier with Rowena Pixley at Ak-Sar-Ben Field

Scarcely had he gotten 100 feet in the air when it was evident from the manner in which the rudder of his machine "jockeyed" that the gusts of wind were hitting the frail craft heavy blows.

The first turn made, the one to the east, was the hardest of the flight. Several times Barrier could plainly and be seen attempting to turn the light machine against the wind, but each time he was evidently warned in time that it would mean grave danger to do so. At last, nearly a mile and a half south, he swerved to the east and went at a fast clip for Omaha.[14]

Barrier made a wide turn to the north and headed back west over the Omaha Field Club at 36th Street. He flew straight into the wind and landed in the middle of the track.

And he likely jumped out of the Blériot and kissed the ground. Barrier quickly recovered his usual charm and assured little Rowena Pixley, daughter of W. A. Pixley and his wife of

---

[14]*Omaha Bee,* May 10, 1911.

116

Dundee, who were at the meet with Deright and his wife, that when
the weather was nicer, he would take her with him to fly.

Wednesday, May 10, was the worst day of all. The wind blew
a gale all day, and a very small crowd showed up. Manager Young
advised them that it would be impossible to fly, but their admission
tickets would be honored the next day. In spite of the bad luck with
the weather, the pilots still expected to fly for the people of Omaha.
In the *World-Herald's* words, Omaha's

> journalistic assemblage, however, voiced its disappointment
> in such pointed and comprehensive language that Rene
> Barrier finally had compassion and took off on a little aerial
> spin for its benefit.[15]

Sounds to me like the press was getting a little bloodthirsty! I know
from experience that, in this country, the wind and weather *cannot
be defied,* so Rene Barrier could have made a fatal mistake that May
afternoon in 1911. He made a huge one when he got in that flimsy
Blériot:

> He buffeted in the whimsical southerly gale for eleven
> minutes, circled the field a couple of times, scared Bill
> Paulson's poultry off of their perches and then accomplished
> a characteristically subtle landing amid immense applause
> from a dozen hardboiled reporters, one lemonade merchant
> and a bored chauffeur. It was really a considerable
> achievement in the wind, and the scribes offered him
> heartfelt thanks for the sorrow he had taken from their
> hearts.[16]

Those reporters could be real pains, but they sure could write! Thus
ended the flying for the day.

The following day, Thursday, May 11, the wind was still
blowing, but the aviators made three great flights and thrilled the
sizable crowd that finally had shown up:

> He [Barrier] bucked into the gale with a right good will
> and achieved a height of probably 200 feet before the real
> strength of the upper air currents reached him. Then the
> spectators were horrified to see his craft dance, pitch and
> stagger under the impact, careening from side to side as a

---

[15]*Omaha World-Herald,* May 10, 1911.

[16]*Ibid.*

feather would perform in the blast of an electric fan. He could be seen struggling desperately with the mechanism which regulates the wings and rudder and there was a sigh of relief when he finally made the perilous turn to the south. . . . He sailed around over Concordia Park, . . . and then wobbled his way to the ground directly in front of the appalled multitude making his characteristically perfect landing. When he stepped from his machine, having been in the air twelve minutes, he said it was the worst scoot he had ever scooted. Manager Young, turned away his face when the first gust hit the aviator and exclaimed, "He's gone!"[17]

An hour later Rene Simon set an altitude record when he

climbed 4,200 feet into the heavens setting a new record for Omaha flights, and one which will not soon be broken. He was up a bare fifteen minutes and the climbing abilities of the new "altitude" wings, fitted to his machine since his accident last week in St. Joseph, brought him up to his great height within eight minutes of his start from the ground. He then headed his machine to fly with the gale. . . . Then shutting off the motor, he came down the "toboggan slide" on a long glide, 4,100 feet in length, not turning his machine nor starting the motor until he was fifty feet from the ground. The glide was new to Omaha and the crowd went wild, hailing Simon with cheers and handclapping.[18]

On Friday, May 12, Rene Simon finally got smart. Anxious not to disappoint the largest crowd that had attended that year's aviation meet, the aviator went up in spite of treacherous gusts of wind whipping across the Speedway Park and against the protests of his manager and fellow aviators. He flew eleven minutes and finally made a beautiful landing in spite of the gale. He climbed out of the Blériot and declared that he would not ride in such a wind again.

The next day, the wind and weather were so bad that no flights even were considered. Manager Young commented,

I think that's the hoodoo. Just say that if there is anything near flying weather we will show Omaha on the last day of

---

[17]*Omaha Evening World-Herald,* May 11, 1911.

[18]*Omaha Bee,* May 12, 1911.

our stay what a pretty sight an aero meet is. There will be races, altitude climbs, aero "joy rides" and other exhibition flights[.][19]

So, on Sunday, May 14, as the *World-Herald* reported:

To please approximately 8,000 people, the largest crowd yet to see the bird-men fly, Rene Simon yesterday again flirted with the treacherous Nebraska winds and made two exhibition flights in his Moisant monoplane.[20]

On the first flight he took off south, turned east, then north, and flew over the grandstands at over a thousand feet in the air. He flew almost a half a mile north, maneuvered a perilous turn into the south wind and then made one of his renowned dips, winging not over ten feet above the heads of the applauding spectators. At the field, he skimmed along the ground a short distance and then went high into the air doing another circle. This time he glided so low that he almost nipped the feathers from the hat of a lady in the stands.

Man! It sounds to me like Simon was a heck of a pilot and had just made a fly-by that would rival an F-16's. He landed as lightly as a feather and got out to a standing ovation from the crowd both outside and inside the grounds.

I know how he probably felt: He had the crowd with him, so he gassed her up so he could show them some real flying. About thirty minutes later he was back in the air. This time,

a fierce upward wind suction caught the south wing and the machine seemed to turn a "flip-flop." Barrier and Captain Frisbie uttered frightened cries and started to run, evidently anticipating a catastrophe. Almost as suddenly the machine righted itself and the crowd burst into wild applause and Simon had once more escaped what seemed to be certain death.[21]

He was fortunate that the wind dropped. If it hadn't, it is likely that the end predicted for him by other aviators and the public would have come true.

---

[19]*Omaha Sunday Bee,* May 14, 1911.

[20]*Omaha Evening World-Herald,* May 15, 1911.

[21]*Ibid.*

Simon sure did something right. The Blériot held together, and the guardian angel (or maybe the old Red-tailed Hawk) must have been close. What a finish to a miserable week for flying! Young attested that the flyers had lost a little money during their stay due to the wind, but W. L. Huffman (Chairman of the Aero Club) said it appeared the Aero Club would break even.

After the International Air Meet ended in Omaha on May 14, 1911, the *World-Herald* noted that Charles Baysdorfer

> left Omaha last evening with the international aviators and his new Curtiss type biplane which he has been working on for the past two years. In company with Captain Frisbie he will make one or two trial flights and from then on he intends to make regular air spins right along with Simon and Barrier. The biplane, which Charles and Gus Baysdorfer have built is credited by the international bird-men to be one of the fastest and most substantially built of any they have ever seen. . . .
>
> For years, Charles Baysdorfer has been interested in aerial touring and his hope for success in the next week at Atlantic City will be the reward of six long years of labor.[22]

With that Charles and the *Hawk* were off barnstorming with the famous Moisant International Aviators.

Barnstorming with Moisant's Aviators

Moisant's Aviators troupe traveled in four railroad cars—three for the aeroplanes, shop parts, supplies and all of the equipment and one for the manager and staff, the pilots, six mechanics, four helpers and sometimes their families. It was quite a sight, almost like the Ringling Brothers Circus. This time the cars were attached to the Northwestern Night Train to Minneapolis and would arrive in Sioux City, Iowa, at 12:45 in the morning. The Baysdorfer brothers and the *Hawk* were on it, looking forward to an exciting barnstorming career.

They quickly found out that barnstorming was no bed of roses. The flying field was the Fair Grounds in northwest Sioux City, and it took all day Monday to load the planes and equipment

---

[22]*Ibid.*

120

into horse-drawn wagons and haul them the six miles to the field. Then, they didn't get the *Hawk* assembled until Tuesday afternoon. The Baysdorfers did almost all the work because the mechanics (who were French) didn't or wouldn't speak English and professed to know nothing about biplanes. Captain Frisbie helped some, and one of the mechanics was assigned to them. Gus had gone along just to help Charles get started, but he planned to return to Omaha as soon as the Sioux City meet was over.

The wind was so wild that nobody flew on Monday. It was still real gusty on Tuesday, but there was a good crowd, so Frisbie decided to fly the *Hawk.* This would be its first flight with the Gnome engine and the new changes made since last fall. Charles and Gus were likely on needles and pins! Frisbie was pretty excited, but he was a seasoned aviator, so this should have been old hat to him.

The field was better than most. It was a good half-mile long, with a big meadow on one side. The *Hawk* was wheeled out and headed into the southwest wind. Frisbie climbed into the seat and signaled to start the Gnome. She started with the first pull through of the prop. The crew held her—the Gnome was either full on or full off—but she was clawing to go.

Finally, Frisbie decided she was ready (or he worked up enough nerve) and signaled to let go. The bird literally leaped forward. Charles thought she got into the air in less than a hundred feet. She was flying, and flying good, but the wind was gusty and probably caused the left wing to drop. Frisbie probably overreacted because she leveled and then rolled to the right, and (bam!) the wing tip hit the ground. He chopped the switch, got her level, landed and stopped right-side up. According to the *Sioux City Journal,* "Great bravery was displayed by Capt. Frisbie yesterday in attempting to fly with his biplane. The machine had just been set up."[23] The right wing was damaged, and a tire had blown, but the aviator told Gus and Charles that she was the best aeroplane he had ever flown. The brothers knew the *Hawk* could fly, so they fixed the wing and looked forward to the next day.

Rene Simon then decided to fly. He took off, with the Blériot pitching like a raft in a storm. Simon found a good field across the

---

[23]*Sioux City Journal,* May 24, 1911.

Big Sioux River in South Dakota (where the old airport used to be before World War II), landed, and decided to fly her out the next day. That ended flying for the day.

Wednesday, May 24, was a pretty decent day. Rene Barrier made a very good flight, and the crowd gave him a great round of applause. But the big story of the day was only half about aviation. Sarah Bernhardt, who was playing at the Opera House in Sioux City, was a friend of Simon's father and wanted to see him fly. She and her entourage arrived at the fairgrounds in a brand-new Firestone-Columbus automobile, but they had to wait almost an hour because of problems getting the Blériot ready. Remember, it was in the field in South Dakota.

Finally, the sound of the motor was waft on the breeze to them. The aviator was still invisible. A few seconds later the Madam stretched out both arms directly towards the rays of the sun. There, surrounded by a sea of gold, hung a tiny bird shaped object. "Bravo Simon," cried the Madam, alternately wringing her hands and throwing kisses toward the daring aviator. The cry was taken up by the crowd of thousands. Simon climbed up to 5,000 feet, then shutting off his motor came down in a glide that was hair raising. "He should not do that. It is suicide," exclaimed the Madam to her physician who was trying to assure her that the birdman had the machine under perfect control. When on his last turn the aviator banked his monoplane until it seemed to fairly stand on end and then headed straight for the stage folks. They scattered. The Divine Sarah gave a gasp of alarm and ducked unceremoniously, remembering no doubt the fate of the Minister of War of France who was killed last year when a plane crashed into the grandstand. But the reckless little aviator, with one hand gaily waving at the terrified spectators, shot past, the tip of his wing not five feet away from where the Divine Sarah stood. When finally he landed in the infield of the race track, Madam Bernhardt grasped both his hands in hers for several moments before speaking. A tear fell on Simon's wrist. He looked at it rather surprisingly and laughed. His mirth seemed to relieve the tension of the moment. Taking a rose from her bouquet and pressing it to her lips, she handed it to Simon.[24]

Ever wondered why those aviators did it? Simon must have thought he had crashed and gone to heaven, and I bet even the old Red-tailed Hawk was impressed!

That ended the flying in Sioux City. The townsfolk wanted them to stay until Sunday, but the Aviators's manager Young couldn't talk the people in Des Moines into delaying the scheduled meet there. So they began dismantling everything and loading it into their railroad cars. This took two days, which was great for the Baysdorfers because they got the *Hawk* completely repaired. Then Gus went back to Omaha, and Charles and the *Hawk* departed for Des Moines with the Aviators.

The Des Moines meet was sponsored by the Hyperion Field and Motor Club and was held on the Club grounds northwest of town where Camp Dodge Road is today. The first day was unbearably hot and very windy. Simon finally flew at 6:45 p.m. His flight was cut short by his engine cutting out, thus the crowd of over a thousand wasn't very happy. Charles had the *Hawk* in perfect shape, but Captain Frisbie decided to wait until the second day.

Rene Simon

The next day was much better. It was hot, but the wind wasn't as bad. Simon put on one of his best shows and had the crowd cheering. Barrier made two good flights, then both pilots made a dandy flight around the field together. The crowd was completely satisfied. Although Frisbie

---

[24] *Ibid.*

fired up the *Hawk* and taxied around the field a short distance, he wasn't completely happy with her, so he didn't fly.

Baysdorfer was fit to be tied. He was sure the *Hawk* was ready, and he probably was dying to fly her himself. He spoke to Manager Young who pointed out that their agreement said Frisbie, not Baysdorfer, was to fly the ship. Perhaps he also had a few words with the Captain because the next day the *Hawk* finally flew. The local paper said:

> For the first time in Iowa, three aeroplanes were in flight at the same time yesterday on the closing day of the Hyperion Club aviation meet.
>
> While Rene Simon, the "fool flyer," was climbing for altitude and Rene Barrier was navigating the calm air at a lower level, Capt. J.J. Frisbie of the United States army corps of aviators, mounted his improved Curtiss biplane and rolled down the runway to take wing for a pretty flight. . . .
>
> Three heavier than air machines, all under perfect control moving like live things, demonstrated to that part of Iowa on the grounds of the Hyperion Club the possibilities of aeronautics.[25]

Charles was so excited he was almost flying, without benefit of plane! The *Hawk* flew perfectly. Frisbie circled the field twice at about seventy-five feet, then flew south along the Des Moines River for about two miles. All total,

> Frisbie was up above six minutes, always within one hundred feet of the ground. He sailed southeast up the valley of the Des Moines river for a distance of two miles, turned in a graceful bank and hurried back toward the starting ground, making as pretty a landing as the more active monoplanes.[26]

The crowd of more than three thousand gave the aviators a mighty ovation. The last day of the Des Moines meet was a great success, but no one was as happy as Charles Baysdorfer. Even though Baysdorfer was beginning to get doubts about Frisbie's flying, the *Hawk* flew beautifully. Frisbie said she was like a bird. The *Hawk* was a flying aeroplane. Now all Baysdorfer wanted to do was fly her.

---

[25] *Des Moines Register and Leader,* June 5, 1911.

[26] *Ibid.*

Best of all Frisbie told Baysdorfer that he could run her a little on the ground when they got her set up at Ottumwa, which was the next town on the circuit. At that point, Charles wrote Gus to come with them. He knew they needed his brother's know-how. Gus replied that he would join them in Davenport, Iowa.

As I see it, the International Aviators were no different from any other barnstormers. They had nerve, too: They made any old claim they thought up, from flying over a battle in the current revolution in Mexico, to Frisbie scaring the heck out of the governor of Texas when his biplane ran away. They advertised that Simon wouldn't fly unless his pet monkey from Mexico or his dog from France was there. They didn't have a biplane in Omaha, even though they advertised they did, and that's why they leased the *Hawk*. They promoted the *Hawk* as being a Curtiss, but when pressed they would say "Curtiss type." They said that the Baysdorfer biplane was built in the Moisant factory in Mineola, Long Island, New York, and on and on. They sometimes acted like two-bit promoters running a sideshow. P. L. Young, the manager, was as good a front man and press agent as there was in the business. Young would tell the news media almost anything that came to his mind. He claimed to have been a very successful newspaper publisher in Waterloo, Iowa, before he was smitten by the marvelous new aeroplane, although I couldn't find any record of him there.

I'm telling you this because Charles and Gus Baysdorfer were beginning to have serious doubts about the Moisant outfit. Although the aeroplanes were Blériots and good ones, it was a foreign concern—all the mechanics and the two chief pilots were French. They weren't sure how Frisbie fit into the picture, but it seemed to them that they, the *Hawk* and Frisbie were treated like second-class citizens. On top of that, as of yet they hadn't been paid a dime. Charles's biggest concern was that he didn't think Captain Frisbie was the great aviator everybody claimed. The birdman had even admitted his inexperience to the *Des Moines* newswriters:

> I am not much of an aviator yet, but I am learning. I was anxious to go up yesterday. I will look for medals and purses after I learn the game more thoroughly. Des Moines

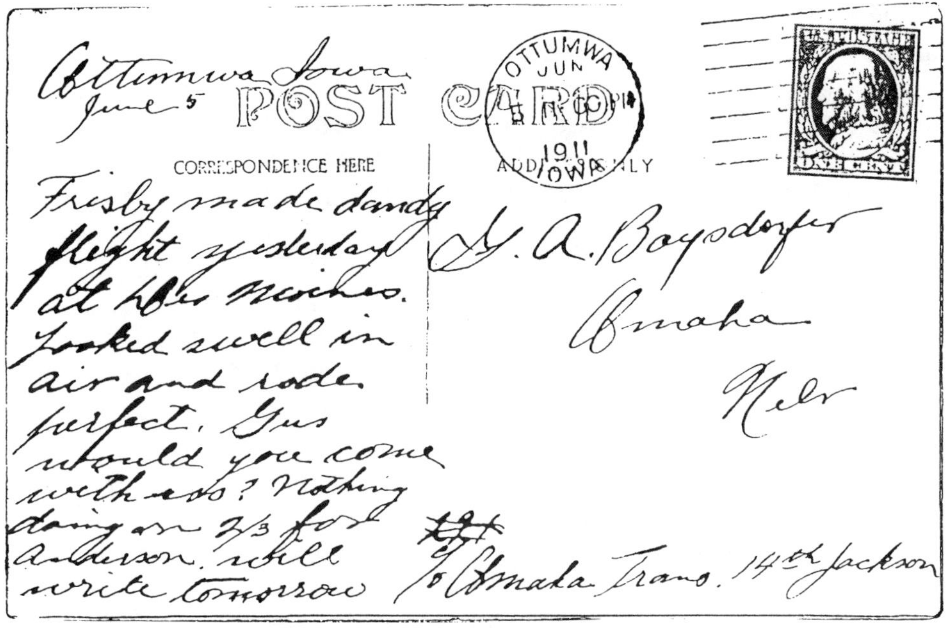

Charles's postcard to Gus

certainly is appreciative. I hope I have good luck so your papers will put me in the headlines later on.[27]

I have been unable to find information about the United States Army Corps of Aviators. The Aviation Branch of the Army Signal Corps just had been established in 1910, but Frisbie wasn't in that service. Another thing the Des Moines papers reported about Frisbie was that he was the "only Irish aviator in the world. . . ."[28] Actually, Frisbie was born in Oswego, New York.

The Aviators arrived in Ottumwa early Monday morning, June 5, and spent the whole day assembling the *Hawk*. The weather was very hot, and the air was very gusty. About 6:30 p.m. the wind calmed down, however, and Frisbie was true to his word. He told Charles to get in and run her a little. That was like telling a fish to swim a little! Frisbie and several mechanics held the plane while another helper propped her. The Gnome fired on the first pull. She wanted to go, and Baysdorfer signaled to let go. They did, then the

---

[27] *Ibid.*

[28] *Ibid.*

Hawk started to roll. There was no way Charles was going to shut her down, and he was in the air in less than a hundred feet. The field was smooth and over a half-mile long, so he let her get up about twenty-five feet before he cut the switch and made a great landing. No turns, just straight ahead, but he had flown his ship.

Frisbie and the crew came running and congratulated Charles as they pushed the *Hawk* back to the tent. Then the roof fell in. Young had seen the whole thing, and he was madder than heck: Frisbie was supposed to fly the machine, not Baysdorfer! The aviator took it good naturedly—in that short flight he had discovered that the *Hawk* was a thousand times better an aeroplane than she had been last fall. He also knew there was no way anyone would keep him from flying her.

The Ottumwa meet lasted two days. Simon was the star, but Barrier flew very well, and Frisbie made two flights in the *Hawk*. He was getting better, and she flew great, so it was a very successful meet.

The troupe arrived in Davenport, Iowa, on June 8. The meet was held at the Tri City Speedway grounds. Friday, June 9, was the first day, and flying was to start at 4:30 p.m. Gus Baysdorfer was there, and he and Charles had the *Hawk* ready to go first thing in the morning.

After breakfast, Charles told Gus that he was going to fly her. Young was in Galesburg, Illinois, and Frisbie was asleep. Gus was agreeable. It was a good morning, clear, with a slight southwest wind. The field was over a half-mile long, clear with no trees. They had their French helper and several mechanics to help Gus hold her, so they wheeled the *Hawk* out and headed her into the little wind. Charles climbed in and hollered, "Contact!" The helper pulled the prop. On the second pull, the Gnome fired. She revved up, the guys held on, Charles signaled to release, and the *Hawk* was moving.

This time Baysdorfer flew straight and level at about twenty feet for a good thousand feet, then cut the switch and made a beautiful landing. It was next to impossible to taxi with a Gnome engine, since the aviator could only turn her on or off, so he waited for Gus and the Frenchman to get to him. They turned her around, and he flew her back to the tent. On both flights the birdman tried very shallow banks, testing the new aileron control, and he was elated. After the flight, the brothers did a victory dance: They knew it a was a good flight and that the *Hawk* was a good aeroplane. Charles told Gus the next time he flew her, he was going to fly around the field each way.

When the Baysdorfers told Frisbie that Charles had flown her, he wasn't upset, so they figured they could handle Young. But they didn't need to worry—Young secretly had gotten married two days before, and the couple was on their honeymoon. His bride supposedly owned two ranches in Wyoming and had a huge spread near Sheridan (where some of my family lived). Again, I found no record of either one.

The flying started on schedule that afternoon. On his first flight, Barrier hit a bad gust just as he was landing, and he nosed over, breaking the prop and damaging the landing gear, but he suffered no injuries. Simon then went up and did his usual "fool flying," which the crowd loved. Frisbie didn't fly, saying the wind was too strong for his biplane. The Baysdorfers wondered about Frisbie's decision not to fly. After all, the *Hawk* had flown beautifully that morning.

Saturday was a spectacular day. Simon flew well, and Frisbie made a great twelve-minute flight. He flew clear to the Mississippi River and back and circled the fairgrounds twice.

Sunday was just as good. Simon thrilled the crowd with his "Dip of Death." Since Barrier's Blériot wasn't repaired yet, he didn't fly. Frisbie made another good flight, though. He never got more than four or five hundred feet in the air, but he made a smooth-as-silk landing. Afterwards, the Captain couldn't say enough good about the *Hawk* and proclaimed it the best biplane in the country.

The next meet was to be in Galesburg, Illinois. It proved to be one of the most successful ever for the International Aviators. On to Galesburg!

The *Evening Mail* sponsored the Aviators's exhibition in Galesburg and announced the fact in the June 3 edition. They did a great job promoting the meet, and it was one of the most successful ever.

They arrived in Galesburg at 8 a.m. Monday, June 12. When Young and his new bride met them at the station, Charles and Gus wasted no time. They confronted Young with a bill for $1,700 for one month's service. He gave them $500 cash from his pocket and said they would have a check for the remainder tomorrow. Five hundred seemed like a million to the Baysdorfers. Their long, hard work was beginning to pay off; their aeroplane had started to make money! Life was great!

The meet was held at the fairgrounds, where the infield of the race track was big, flat, and without obstacles—the best place yet from which to fly. The weather was good, so they were able to make great flights. This was the first time aeroplanes had flown over Galesburg, so the *Mail* issued an "Extra Aviation Special."

The *Mail* had a great new idea. They promoted auto and motorcycle races and other exhibitions in the early afternoon, knowing that flying conditions were better toward dusk. This entertained the people and built up their flying interest.

On the day's first flight, Barrier thrilled the crowd with a forced landing. In the words of the *Galesburg Evening Mail:*

> While the crowd was anxiously watching Rene Barrier, . . . Barrier was fighting hard in his monoplane against death.
>
> He was up over the trees at the corner of Knoxville Road and Farnham Street with his machine dipped at an angle of 45 degrees. . . .
>
> Taking no chances, he gave his machine a turn toward the corn field of Clark, the gardener, and the machine landed in the middle of the soft plowed ground.[29]

Obviously his Blériot hadn't been repaired completely since the crash in Davenport. Now it needed some more repairs, and Barrier needed some help too.

Simon made two flights, which the *Mail* beautifully described:

> Like some gigantic eagle rising from its nest the monoplane with Rene Simon at the helm glided down the track from the east end and rose majestically into the air in front of the grandstand. . . .[30]

Then he flew over the heart of the city at an altitude of two thousand feet. He then returned to the grounds and made his famous "Dip of Death":

> Hanging aloft in midair for a few seconds and apparently motionless like an eagle poising for a swoop at its prey Simon made a half turn. The head of the machine tilted

---

[29]*Galesburg Evening Mail,* June 14, 1911.

[30]*Ibid.*

downwards and in his famous dip of death the daredevil birdman darted downwards at a dizzy speed for nineteen hundred feet in one long nerve-racking glide. Every face in the stands blanched with sudden fear. Had the engine broken down? "Had Simon lost control?" was the unspoken question on every lip.

Strong men set their teeth hard while many women turned their heads away[.] Just when it seemed as if machine and daring rider were about to crash into the earth and certain death had reached out his hand to take the fearless flyer, the wonderful white winged bird raised its head and shot up into the air again and on even planes glided, a thing of marvelous grace and beauty, past the stand at a height of one hundred feet. The pent up excitement in the crowd now burst forth in a spontaneous roar of tumultuous applause.... The reversal of emotions caused dainty ladies to pound the railing with their umbrellas and strong men to clap one another on the shoulder in an abandon of feeling. Whirring about the field for the second time the machine, under perfect control from start to finish, came to an easy landing in the center of the field.[31]

What writing! What flying! Those Frenchmen deserve to take a bow—they flew with emotion.

Frisbie didn't fly the first day, saying that the weather conditions weren't right for the biplane. Charles and Gus talked to Young, expressing their concern about Frisbie's behavior. Young said that Frisbie would fly. He was in a hurry because he, his wife, Barrier, and Simon and his wife had been invited by a representative of the *Evening Mail* for dinner at the country club. This was great for the International Aviators, but not so great for the Baysdorfers and their *Hawk*. And Young didn't have their check yet.

The second day was a repeat. Simon thrilled everybody, and Barrier flew to an altitude of six thousand feet, just under the world record. Both aviators flew Simon's Blériot.

Meanwhile, a crowd had gathered around the *Hawk,* and there was much talk that the biplane wouldn't fly. Charles was fit

---

[31]*Ibid.*

130

to be tied and threatened to fly her himself, when finally Frisbie said he would fly her. It was about 6 p.m. He flew around the race track four times at about a hundred feet, while Barrier was flying above at a thousand. It was a great sight, the two ships in the air at the same time. The *Hawk* flew great, and Frisbie made a beautiful landing.

On Thursday, the last day, Young asked Captain Frisbie to fly first. I guess he meant it when he told Charles and Gus that Frisbie would fly. The Captain acceded to Young's wishes. Barrier, Simon, and the Baysdorfers watched with special interest.

> The huge machine [the Hawk was almost twice as big as a Blériot] was a beautiful sight gliding over the country and the height was not so great, but the Captain could see the upturned faces of many busy farmers in the corn fields and women and children as they rushed from the houses to admire the great flight. . . .

> Captain Frisbie rose about the center of the enclosure and flew south for almost a mile and a half, when he started on a wide swing and then circled around in a large circle, over five miles in diameter, traveling all the time at a rate of about 65 miles per hour. The huge machine came clear around to the south side of the track and then one more smaller circuit was made and the landing was made from the west, the machine coming to a most beautiful and graceful stop about in the center of the field. . . .

> . . . Capt. Frisbie's exhibition was little short of marvelous when the conditions are taken into consideration.[32]

When the captain landed, Young and the aviators congratulated him. The Baysdorfers knew it was a great flight, but they were proudest of the *Hawk.* Darn, they probably felt good—two Nebraska boys whose hand-built aeroplane was flying over Galesburg, up there with the famous French Blériots, where the Wrights and Curtisses had never flown!

The meet ended on a positive note, with the only complaint being the familiar one that more people watched the flying without paying than paid. The *Mail* declared that

---

[32]*Galesburg Evening Mail,* June 16, 1911.

artistically and financially the meet was a success. Probably thirty thousand people saw these bird men in the air and enough of that thirty thousand were sufficiently public spirited to help pay a share of the amount of money it took to advertise and produce such an exhibition.

To those who did not pay, *The Evening Mail* has no ill feeling. This newspaper is glad that folks who did not leave their homes were given the privilege of seeing the birdmen soaring miles through the air. The small and large boys and girls who took advantage of the [lack of]. . . fences were welcome to the exhibition, for well do we realize that in these days of high cost of living there are many to whom fifty cents is an almost impossible tax. As for the many who sat outside the grounds in their carriages and automobiles to see the flights, they too were welcome because their next payment on their vehicle will use up the last half dollar they have. To the businessmen who endorsed this enterprise of the *Evening Mail* we give our thanks.[33]

I think they summed it up pretty well.

At these airshows, Simon's "Dip of Death" was the star, but it wasn't all "show." Remember the Gnome engine was either full on or off, sort of like your two-cycle lawnmower or weed-trimmer or chain saw. The fuel-air mixture and lubricant were induced into the crankcase of the Gnome and injected into the cylinder head through slip valves. Then the mixture was fired by a spark plug charged by a magneto. This unique system was made necessary because the cylinders rotated around the crankshaft, which was fixed, and the propeller was fastened to the cylinders. It was therefore impossible to have an intake and exhaust manifold or cams like our modern reciprocating engines. When power was needed, the ignition switch was on, and when it wasn't needed, the switch was off. The Gnome was a good engine—light, air-cooled, very reliable, developed fifty- to seventy-five horsepower, and burned only five to eight gallons of fuel an hour.

For take-off, climbing and straight and level flying, the Gnome was great. In those days planes didn't have much power, so aviators flew full-bore until they wanted to come down. The Blériots

---

[33]*Ibid.*

were pretty high wing-loaded tractors (propeller in front). From all the pictures I have seen of them gliding or landing, I'm reminded of those awful Waco Gliders in World War II. They "glided" like a concrete block, and I'm sure the center of gravity must have moved catastrophically. I'm sure all Blériot pilots that survived used about the same technique on final and landing. They came in high, cut the switch, and got the nose down in a really steep dive, for two reasons: first, to keep the prop windmilling so they could start the engine, and second, to stay over the field, then flare out and land. If they were under- or overshooting they could hit the switch and pray that she fired, and pull in or go around. Simon was likely a heck of a pilot because he had the technique down the best of all Blériot drivers: He would start his dive from as high as he could get, pull her out at the last second, and very seldom have to use power or go around, thus the "Fool Flyer" and the "Dip of Death!"

Once when I was slow-timing a Fairchild PT-19, I accidentally almost performed the "Dip of Death." It was one of those great summer days when there were many cumulus clouds cruising across the sky at about four thousand feet, and the air was like silk and satin. I climbed up as high as she would go, cut the switch, and glided like a hawk. It was *real flying*. I decided to rejoin the world at about four thousand feet when I hit the clouds. I laid the PT-19 over into a dive, and the prop never budged until I was in a heck of a steep one. Finally, almost at redline, it started windmilling. I hit the switch and got on the wobble pump, and she finally fired. I was under a thousand feet when I had her running and flying like an aeroplane again. Hats off to you Gnome drivers!

On the train to another show at Terre Haute, Indiana, the Baysdorfers finally got to talk to Young. First, they talked about money: The Nebraskans had $1,200 coming and almost another full week's pay. Young made a few excuses, peeled off $700 from his money belt, and promised a check in Terre Haute. Second, he agreed that Charles could fly the *Hawk* as soon as Frisbie gave the OK. They were satisfied, but Gus wanted to return to Omaha as soon as they got the *Hawk* set up in Terre Haute.

The Moisant special train carrying the flyers from Galesburg arrived in Terre Haute at 8:20 Friday morning, June 16, 1911. There, the ground operation was well organized. Wagons and sometimes motor trucks waiting at the station were loaded up, and all of the equipment was taken to the fairgrounds direct from the

Union Station. Crews put up the tents; tools were set up in a shop; and the aeroplanes were uncrated and assembled. The planes were kept in separate tents called hangars, another aviation term coined by the French. Spare parts were stocked, and gasoline and oil always seemed to appear as if by magic. It was a lot like the way a circus is moved and set up. Incidentally, the movement of a large group of people, machines and animals by train wasn't the invention of P. T. Barnum, but of Col. William F. "Buffalo Bill" Cody, who moved his first "Wild West Show" from North Platte, Nebraska, to Omaha in May 1883. In Terre Haute, this efficient ground crew had all the tents up and the machines assembled by noon. Gus Baysdorfer caught the 1:30 p.m. Wabash for Chicago and would arrive in Omaha the next day on the Northwestern.

The *Terre Haute Tribune* reported:

Captain Frisbie, whose Irish countenance is recognizable on sight, talked briefly with a *Tribune* man.

"We are somewhat tired from our work at Galesburg," said Captain Frisbie, "but a little rest will put us in condition for the flights this evening. Simon, Barrier and possibly myself will go up this afternoon, weather permitting."[34]

Well, the weather was bad, so nobody flew that afternoon at the Terre Haute meet, which was being sponsored by the local Elks' Lodge. Ninety miles away, however, the Elks' Lodge in Lafayette, Indiana, was sponsoring a meet of Curtiss Flyers, and the Terre Haute lodge had the idea of challenging the Lafayette lodge to a race between the Moisant Aviators and the Curtiss Flyers. One aeroplane from each group would take off from each town at the same time, and the first one to get to the other town, a distance of ninety miles, would be the winner. Manager Young was immediately enthusiastic and put up a purse of $3,000 to be awarded to one of the Curtiss Flyers should a successful flight be made from Lafayette to Terre Haute, provided the Curtiss management also put up a purse of $3,000 for a Moisant Flyer's successful flight from Terre Haute to Lafayette. Young wasn't sticking his neck out very far because a ninety-mile flight cross-country in 1911 was a heck of an accomplishment, and he knew the Curtiss Flyers were leaving Lafayette on Thursday to start an exhibition on Saturday in

---

[34]*Terre Haute Tribune*, June 16, 1911.

Evansville, Indiana, about eighty miles south of Terre Haute. Nonetheless, the *Terre Haute Tribune* excitedly plastered the story on the front page.

Meanwhile, Lincoln Beachey, the star aviator of the Curtiss team, stopped in Terre Haute on Friday to visit with Barrier, Simon and Frisbie, the latter of whom he had helped teach to fly at the Curtiss school the previous February. Beachey himself had been flying aeroplanes only since October 1910, when his old friend "Captain" Tom Baldwin (then a Curtiss Flyer) suggested that they get Beachey in their school. Beachey took to flying like a Red-tailed Hawk and immediately became one of the best. At Terre Haute, Beachey asked Frisbie if Charles and Gus Baysdorfer were there. He had heard about their new biplane in Omaha. Curtiss had told Beachey that he had seen it when he was in Omaha the previous year and that it looked like it could be a good one. Beachey said that he had beaten Charles in an airship race in St. Louis a couple of years ago, and they had a great ship and knew how to fly it. Well, this was almost like getting an endorsement from the good Lord himself. Someone found Charles Baysdorfer, and Beachey greeted him like an old friend. He looked the *Hawk* over, especially the aileron control, and said he'd sure like to fly her, but he didn't have time right then. Afterwards Beachey left for Evansville.

At that point Charles had a brand new status with the Moisant International Aviators. Even better, Young gave him a draft for $500. Charles ran, or probably flew, to the fair office and called Otto on the Bell. When Otto answered, he about fainted, and yelling, he asked if Charles had crashed. Charles told him the great news, and that Gus would be home the next day. That was the good news on Friday, June 16, 1911.

Saturday morning at 9:30, Rene Simon took off from the Shelton Addition, northwest of the fairgrounds, driving the first aeroplane ever to fly over Terre Haute.

> The "fool flyer" then flew over the city west, circled, ascended and dipped, and after twenty minutes in the air returned to the Shelton field lighting within fifty feet of the starting point. . . .
>
> . . . The biplane [the *Hawk*] made a cleaner getaway than the single machine and Frisbie gave a fine exhibition. He circled over Edgewood and came down town as far as Thirteenth street. . . . Frisbie ascended 1,500 feet and

speeded. Rapidly the big double-planed machine went from north to south, east to west, up and down, the noise of the exhaust being heard for half a mile.

After Frisbie had been in the air for fifteen minutes he was signaled to come down by the waving of a big American flag. Instead the nervy captain sailed toward the fairgrounds, two blocks distant. When directly over the big trees on the ground the machine was seen to dip in a crippled way and disappear. The spectators rushed to the fairground to find Frisbie smilingly lighting a cigarette directly in front of the grand stand. He had made a fine flight and an absolutely correct landing. . . .

"Just thought I would give you a thrill," was Frisbie's greeting to the newsmen when they approached.

Though impromptu, the flights Saturday morning were seen by hundreds of people. The noise of the engines attracted merchants from their stores, women from their kitchens and kids from everywhere. The streets were lined in the eastern part of the city with neck craning people watching the maneuvers of the air crafts.[35]

Charles felt Frisbie finally had lived up to the *Hawk* and wished Gus was there to see the great flight.

Saturday afternoon, after the motorcycle races and a concert by the Ringold Band at about 5:30, Simon flew for about fifteen minutes, doing his famous "dip," and landed in Shelton Field west of the ballpark. He thought it was smoother than the center of the racetrack. Then Charles and his crew wheeled the *Hawk* out into the infield, and Captain Frisbie, who was very popular with the newsmen and the locals, announced to the newspapermen, officials and fans that he couldn't fly because there was a valve problem with the engine. Disappointed, they started walking back to the stands.

Baysdorfer was flabbergasted—there was no problem with the engine! Then he got mad. He was getting tired of Frisbie's ups and downs. After the crowd had moved away from the *Hawk,* Charles told Frisbie that he was going to fly her. Frisbie told him to go ahead as he walked away.

---

[35]*Terre Haute Tribune,* June 17, 1911.

Imitating Captain Frisbie, Charles Baysdorfer climbed into the *Hawk* and yelled, "Contact!" to the crew. When the Gnome fired and wound up, he signaled "release" and in eighty or ninety feet was in the air. This was Charles's *Hawk*, and she liked him and behaved herself. He climbed her up to about three hundred feet, flew straight east, made a ninety-degree turn to the right and then another. He flew back over the fairgrounds, cut the Gnome, and made a great landing on Shelton Field where Simon had landed. He wisely quit while he was ahead.

The crowd gave Frisbie a big hand and started heading home. But the French mechanics and Rene Simon were floored when they got to the bird. The mechanics thought it was great because they liked Charles. Simon said that Beachey told them that Charles could fly.

The crews pushed both the *Hawk* and Blériot back to the hangar tent. Charles helped push, and (wonder of wonders), Simon grabbed a wing, too. This was the first time one of the aviators did anything but fly.

At the hangar, they were met by a fuming Young. Charles was ready and planned to have it out with the manager, but, before the explosion ignited, Simon slyly said he thought that they had a good aviator and good ones were difficult to find. Young shied off. Simon was a favorite of A. J. Moisant, and Young knew that Frisbie had his moments. He also had been impressed with Beachey's comments, and he owed Charles money. Being a press-agent promoter, Young quickly adjusted and told Charles that it was a great flight, but to keep quiet about it for a while.

That was great with Charles. He was still flying a couple of feet off the ground. He burned a path to the fair office to call Otto on the "Bell." Again, Otto was shaken by the call. Charles was so worked up that Otto handed the phone to Gus, who had just gotten in from Chicago. After hearing what had happened, both Baysdorfers did a victory dance on the "Bell." The call cost $4.78, and both the Terre Haute and Omaha long-distance operators were cheering and clapping, even if they didn't understand what the brothers were talking about. And all the hawks were happy—Red-tailed and Baysdorfer. An aeroplane built in Nebraska, flown by a Nebraskan, had made a beautiful flight in Indiana on June 17, 1911. Darn, that was some accomplishment!

Sunday was the last day of the meet. It rained off and on. Finally, at about 5 p.m. they flew, and the *Tribune* reported:

Captain John J. Frisbie, the biplane operator, and Rene Simon, the monoplane driver, fulfilled their Sunday contract with the Elks by making flights at the fairgrounds. . . . The two flights were cheered by the tired crowd who had waited three hours for the sight of the machines in the air for fifteen minutes. . . .

Immediately after the flights Sunday the machines were returned to the hangar and dismantled, ready for the trip to Logansport where the aviators show next.

Barrier was the disappointment of the meet. The Frenchman refused to make a flight, first on the pretense that the grounds was a death trap. . . . A spectator suggested that this was not Barrier "the bird-man," but Barrier the groundhog.[36]

Frisbie had barely gotten out of his biplane after a flight when he was surrounded by a crowd. His inconsistent and temperamental personality was further revealed when a small boy from the crowd asked him for his autograph. Frisbie took the boy's pencil and gave him his autograph. You see why the man's behavior would have tried the patience of a saint?

They weren't scheduled to fly until Friday, June 23, in Logansport, Indiana, so the cars carrying the planes and supplies didn't leave Terre Haute until Wednesday. Charles had some time on his hands. The first thing he did was go to the bank to cash the draft that was drawn on the International Aviators's account at the bank. He was referred to a person named Root, who was very pleasant and took the draft to one of the teller's windows. After some time, Root returned and informed Charles that the account didn't have enough in it to cover the $500. Fortunately, Charles was pretty solvent, but he was getting a little tired of the money situation. He had lunch with a couple of the French mechanics and asked them if they had trouble getting their pay. They said they had always received their pay on Friday each week.

Young had gone to Indianapolis, and he didn't know where Frisbie or Simon or Barrier were. So he called Omaha and talked

---

[36]*Terre Haute Tribune,* June 19, 1911.

138

to both Gus and Otto. Gus said he would meet him in Logansport, and they would get the money and other matters worked out or have somebody's hide.

The meet was sponsored by the *Logansport Tribune* and the Commercial Club, and it was promoted well. There was a lot of interest when their train arrived Thursday afternoon.

Charles and his French helper had the *Hawk* all ready to go when Gus arrived Friday morning. The program wasn't scheduled to start until 2:30 p.m., but a huge crowd was at the driving park by 10 a.m., milling around, looking at the planes and talking to the flyers. These were the first aeroplanes ever in Logansport. The *Logansport Daily Tribune* exclaimed:

> See the birdmen fly.
>
> Everything is in readiness for the first flight of three big flying machines at the driving park this afternoon. The American built biplane which Captain Frisbie rides is ready for the word. . . . This huge machine which bears aloft the intrepid navigator was built by the Baysdorfer Biplane company and two members of the firm, G. A. Baysdorfer and Charles Baysdorfer, are here superintending its preparation for the flight. The car [fuselage] is American built, but is equipped with a Gnome motor built in France. . . . The Baysdorfer Bros. stated yesterday that they confidently expect Captain Frisbie to eclipse [*sic*] his former records. . . . Consequently something spectacular may be expected from the biplane this afternoon. . . .[37]

The *Tribune* also noted that "Simon and Barrier were objects of great interest about town. Both were men of striking appearance and wear the air of reckless bravado incident to their hazardous calling."[38]

Promptly at 2:30, the Citizens Band, all forty-four of them, struck up a stirring rendition of the "Washington Post March." The motorcycle races started at 3:30 and were thrilling, but the throng, estimated at over four thousand, was waiting for the birdmen. It was a good field, over a half-mile long, and smooth, thanks to the efforts of a corps of workmen.

---

[37] *Logansport Daily Tribune,* June 23, 1911.

[38] *Ibid.*

Simon flew first. The crowd went wild with his "Fool Flying" and his "Dip of Death," and his landing was frosting on the cake. Barrier then climbed up and out of sight. Obviously his "no-flying" problem at the Terre Haute meet was forgotten. He came back in sight at about 3,500 feet, came down in a spiral, and landed. The crowd loved it.

Gus and Charles had the *Hawk* all ready to go, but Frisbie wasn't there. Neither one of them remembered seeing him at any time during the day. Young finally arrived, and that was the first time they had seen him. He was fit to be tied. Fortunately, there weren't too many people in the infield, and none of the local committee people were there, so Charles casually told Young that he had himself a German Irishman. He already had a cap, and he didn't need any fancy flying tunic.

Although flying was dangerous, the Baysdorfer brothers said that it was amusing when the aeroplanes were about to fly. They always warned the crowd to keep back because of the force of the wind from the propeller, which could reach a speed of about ninety miles per hour. The crowd usually ignored the warning. As a result the wind often blew away their umbrellas, hats and other items, and people had to pursue the flying objects.

Charles climbed in the *Hawk* and said, "Contact!" The Gnome fired right off and revved up. He dropped his hand, and the *Hawk* was in the air in about seventy-five feet. There wasn't much wind, so he climbed straight up to what he thought was about a thousand feet. He hadn't been that high since the *Comet*. He made a ninety-degree turn to the right, although most aviators turned to the left all of the time because of torque. The *Hawk* flew beautifully. He made another ninety-degree turn and headed back to the field. He felt out the aileron control and was able to keep the wings level with no trouble. He had never been so far away from the field before, so he had to work his tail off getting her back and down. He flew back across the field, a little to the south, and practiced letting down with the switch-on-off technique.

When he got down to five hundred feet, he made a 360-degree turn to head back to the field. His approach was good, but he was too high. The Logansport folks saw one of the first "fly-bys" ever flown. He was about fifty feet in the air as he went by the stands, and Gus (who was dying a thousand deaths) said he must have been going eighty miles per hour. He climbed her back to about two hundred feet, did a 360, and came back over the stands heading east. The Gnome was really cackling, and the crowd loved it. Charles was sweating her out when he came back on final. He stuck with his switch-on-off technique and came over the east turn of the track perfectly, cut the switch, and made a beautiful landing. All the aviation people surrounded the *Hawk* and got Charles to the hangar before any of the crowd or newspaper people could get there. That ended flying for the day, and everybody was singing praises about the flyers. They had witnessed a pretty spectacular flight.

Young agreed that they should get the situation straightened out, so a meeting was held in the hangar. Gus said they wanted the money straightened out first. The brothers had been with Young a month now and should have been paid $1,000. Actually, Charley had been with them five weeks; that alone was $1,000. To date, Young had paid them $1,200, so he owed the Baysdorfers $800. They wanted it today. Young said that he would give them a check on the International Aviators account at Mineola, New York. They reluctantly agreed.

Now to the flying. After Beachey's revelation that Frisbie had been flying aeroplanes only since February, they felt that Charles was as qualified, and from then on they would make the decision on who would fly the *Hawk*.

This went over with Young about like a lead balloon. He said that the Aviators had a contract with them to lease the airplane, but Frisbie was the pilot.

When the Baysdorfers asked him where his pilot was today, Young couldn't answer, but he pointed out that they had fifteen bookings and all the promotion said that Captain Frisbie would fly his big biplane. Gus and Charles put their heads together and came up with letting them use the Frisbie billing and promotion but deciding themselves who was going to fly the *Hawk*. Charles, for one, didn't mind being billed as Captain Frisbie.

Young figured he'd better fall back and regroup, so he agreed, but he wanted to talk to A. J. Moisant first. The brothers were happy. The meeting ended, and Young hurried to town looking for a telephone. Nobody yet knew where Frisbie was.

Saturday, the second day, was a complete failure. It rained off and on all day, so there was no flying. Young finally heard from Frisbie. He had gone to Rochester, New York, his hometown, and had missed connections to Logansport. He would be there the next day.

Sunday, the last day, started with a wicked wind from the south. The race track was muddy, but they still ran some motorcycle races. All of the flyers agreed that it would have been fatal to fly, so they delayed as much as possible, hoping it would clear up. The crowd of about 3,500 was very restless and clamored for a flight. Finally, at about 5:30, Simon fired up his ship, but he shut down just after she got into the air and landed. This probably saved his life because a horrendous thunderstorm hit the fairgrounds, soaked and blew around everything, and turned over his Blériot.

That ended the Logansport meet, and the people, at least

LOGANSPORT, IND.   TUESDAY, JUNE 27, 1911.   All the News First   PRICE TWO CENTS

CROWD OF 3,000 HOWLS STUNG!

ATTORNEY-GENERAL DECLARES POST CARD CRIMINAL OFFENSE

The Journal is in receipt of the following official notice from the Treasury Department Secret Service Field Force at Indianapolis:

Editor Journal, Logansport, Ind., June 26th, 1911.

DEAR SIR:-

Patrons of Aviaton Meet at Spencer Park Declare Management Purposely Awaited the Coming of Rain to Avoid Making Airship Flights as Advertised

"AVIATORS" DID NOT FLY

Marion, June 26—Aviators Barrier, Frisbie and Simon did not fly today.

They state that the machines were broken in attempting to fly at Logansport Sunday.

The daring aviators mounted their machines at Logansport in

some of them, were very unhappy. The *Logansport Journal* recounted in a story on June 27 that people were very upset by lack of flying during weather that seemed favorable for flying and felt that they had been cheated out of their money.

Remembering that the *Journal* didn't sponsor the air meet and newspapers at that time were in tremendous competition with each other leads me to think some of the howls were exaggerated. As a matter of fact, the weekly *Logansport Pharos* took a few shots at the flyers, too. They gave the people some great flying the only day they could fly, and summer cumulo-nimbus developments are just as wild and dangerous today as they were in 1911.

They were booked to fly in Marion, Indiana, about seventy-five miles southeast of Logansport, on Monday. The storm had raised ned with them, damaged a Blériot, blew the hangar tent down and turned the grounds into a quagmire. Young talked to the people at the Mineola factory, and it was decided to bring both Blériots there to get them in good shape. They were booked in Detroit on July 1 and Rochester, New York, on July 8, so they would try to fly the *Hawk* at Marion and Detroit.

Gus and Charles took the *Hawk* to Marion on Monday. Frisbie met them there. Young went with them. They really hit a bee's nest because they were supposed to fly that afternoon, and they didn't get there until 6 p.m. Young told the people that the day's tickets would be good the next day, and Frisbie said he would fly, weather permitting.

Tuesday the weather was wild—thunderstorms all over and all day. Young canceled the meet and was almost immediately picked up by the sheriff and hauled before a local magistrate for swindling the locals. An amicable settlement was finally reached whereby the sum of $528.90 was placed on deposit at the Marion National Bank by Young. The amount was to be refunded to all holders of coupons which were detached from their admittance tickets for Monday and Tuesday.

The Baysdorfers hadn't taken the *Hawk* out of the crates, so they put it on the train for Detroit and arrived there late Wednesday night. They had a meeting with Young and Frisbie while on the train. Frisbie had no problem with Gus and Charles being in charge of the *Hawk* and agreed that Charles should fly her whenever he couldn't. He also was very willing to give Charles flying help and pointers. Young was quite relieved because he had a million problems, one of which was the meet in Detroit, without having to worry about who was flying. Since both of their Blériots were in the factory, they were going to have to fly the *Hawk* and hoped they could borrow one of St. Croix Johnstone's Blériots for Simon and Barrier to fly. Johnstone was an American who had lived in France most of his life. He had worked for the Blériot Company in Paris and owned the two ships that were in Detroit. They made a deal, and the meet started Friday afternoon, June 30, at the Michigan State Fair Grounds. That day

> an inclination on the part of flyers to outdo one another in
> bizarre feats . . . gave the 7,000 who filled the grandstand and

lined the fences at the State Fair grounds . . . a vivid notion of what may be necessary some day in the way of aerial traffic regulations. Gov. Osborn added his vigorous share to the enthusiasm.

St. Croix Johnstone was first up. A $25 prize is offered each day for the aviator who first leaves the earth, and Mr. Johnstone evidently intends to add this to his other honers [*sic*] and prerequisites each day. It had been hinted that the American might do something hazardous and spectacular, but everyone was bewildered when, swooping close to the grandstand, he let flutter earthward an envelope which contained two verses constructed by himself, aided, it is reported, by Mrs. Johnstone.

No copyright notice was attached. They were dated, "From the clouds," and were too ephemeral and romantic in tone to be included in a mere newspaper story. Oh, well, here is one:

"There is no place on earth for me,
And I don't care about the sea.
So I soar high up in the air,
Because it seems to suit me there."[39]

The brothers thought it was the best flight ever made by Frisbie, and the crowd loved it. He upstaged Simon and Barrier that afternoon.

On Saturday, Johnstone again won the $25, getting his Blériot in the air first. Barrier took off right behind him, and the two ships in the air at the same time thrilled the crowd. Simon had them going wild with his "Dip of Death" and beautiful spot landing. Governor Osborn strode on to the field eager to see and admire everything, even asking Frisbie if he could have a ride in the *Hawk*. Captain Frisbie had to turn him down and then made a great flight circling the field both ways four times. Charles and Gus were elated. The talk had sure got Frisbie off dead-center. So far the meet was a tremendous success.

Sunday, the last day, was the best. They "snookered" Johnstone. Johnstone knew that only Simon and Barrier flew Blériots and that Frisbie never flew the *Hawk* until toward the end

---

[39]*Detroit Morning News,* July 1, 1911.

of the day. So, when the two Renes were standing talking to the crowd a long ways from the other Blériot and Frisbie was nowhere to be seen, he had no competition for being the first to get into the air. He relaxed and was in no big hurry to get started. Man, was that a mistake! The crew had pushed the *Hawk* out; they had her all ready to go.

Charles didn't see Johnstone anywhere, so he decided to fly the *Hawk*. Charles got in; the Gnome was propped; and (bingo!) Charles won the $25. He climbed up to about five hundred feet, flew several miles east, then did a 180-degree turn and came back over the infield.

Johnstone was ticked. He jumped in his ship and looked like he was going to chase Charles out of the sky. Simon got into the act and had the other Blériot in the air right behind Johnstone. The crowd of about nine thousand went nuts with three aeroplanes in the air at the same time. Fortunately, they stayed away from Charles, and he brought her in to a dandy landing using his switch-on-off technique. And only the Baysdorfers knew who really had flown the *Hawk*.

The reporter from the *Detroit News* wrote:

This twenty-ninth day of June in the year 1911 brought here in Detroit such a marvel as the earth has not witnessed since the time when the Israelites conquered the land of Canaan. . . .

The rays of sunlight that fell athwart the state fair grounds looked down on four monoplanes cavorting and kicking their heels, where in reality the highest possible number of machines aloft that day was three—and one of these a biplane. This wonderful twentieth century is the real age of miracles.[40]

The next meet was in Rochester, New York, under the auspices of the Aero Club of Rochester. The two Blériots had been overhauled at the Mineola, Long Island, factory and arrived in Rochester on the New York night train early Friday morning July 7, 1911. Captain Frisbie was fit to be tied because this was his homecoming. He was so anxious to show the hometown folks that he was a flyer that he made a flight Friday afternoon, even though

---

[40]*Ibid.*

the meet wasn't to start until Saturday. He hounded Gus and Charles to get the *Hawk* ready, and they did. Frisbie flew the Baysdorfers's *Hawk* in spite of the headline, which said it was his own plane that he flew. The *Rochester Democrat and Chronicle* story read:

> Captain John Frisbie of this city, now with the Moisant International Aviators, flew over his hometown late yesterday afternoon and proved to his old friends and neighbors that he is worthy to be classed among the foremost airmen of the world. . . .
>
> Not only was his flight a surprise to the peope [*sic*] of Rochester, but it was the first time an aviator ever made a flight over Rochester, and one of the prettiest over city flights ever made in this country.[41]

How about that! The *Hawk* from Omaha was the first aeroplane to fly over Rochester. The story continued:

> It is not so very long ago that "Jack" Frisbie was not rated especially high as an aviator by his townspeople. There was never any question about his nerve and his determination, but somehow he did not get very high in the air. . . . On one or two memorable occasions he landed on the and in tree tops and again he dropped into the chilling waters of the Erie canal.
>
> In those days there were persons in Rochester uncharitable enough to tell Frisbie that nature had never ordained that he should be a birdman and advised him to take up some other occupation that would hold him closer to mother earth.
>
> But Frisbie though[t] he knew best. . . . So he left Rochester determined someday to return and fly over the old town.
>
> And that is precisely what he did last night. There were scarcely a score of people on the flying grounds yesterday afternoon when Frisbie made his flight. Those who were there were not a little skeptical about the matter when

---

[41]*Rochester Democrat and Chronicle,* July 8, 1911.

the big biplane was rolled out of the tent, and Frisbie said,
very modestly, that he was going to fly over the city.[42]
Thousands saw him as he soared over the parks, streets, and
downtown buildings, at about seventy miles per hour. A baseball
team stopped and watched his big plane; automobiles stopped and
pedestrians looked skyward. Aviator John J. Frisbie flew over his
hometown.

Man! I'm getting as big a thrill out of this as Frisbie
probably did. There never has been an aviator who didn't want to
fly some fancy dadoes over his hometown.

The first day of the meet was a great success. They all flew.
Simon climbed to four thousand feet and thrilled them with his "Dip
of Death." Frisbie had arranged with the Commander of Company
B of the Rochester State Militia to have a sham battle with them the
next day, according to a *Democrat and Chronicle* story. The Captain
believed the aeroplane would be a major weapon in future warfare.
You know Frisbie was right, but I don't think he really knew what
he was talking about. I think it was a lot of blarney for the
hometown folks.

Sunday was the best day. The weather was beautiful, and,
although the paid crowd inside the field at Highland Avenue wasn't
too large, an estimated eight-thousand men, women and children
thronged Cobb's Hill. Simon told the Aero Club officials he would
show those who hadn't paid a thing or two. He took off right at
Cobb's Hill and had just gotten into the air when his engine quit.
He set right down and got stopped immediately. The mechanics
pushed the ship back to the tent and in fifteen minutes said she was
ready.

Again Simon started towards the hill and, lo and behold, the
Gnome quit again. They returned to the tent and finally decided to
use Barrier's ship. Simon was satisfied with the warm-up, so he
signalled release. The Blériot rose into the air, and Simon flew her
directly at the crowd on the hill. They were a little nervous from
the two bad starts, so, when he whizzed over their heads, you'd
better believe they hit the deck. He flew over the top, did a 180, and
came back down the side again. The freeloaders got a thrill, and the

---

[42] *Ibid.*

paid stands cheered lustily. Simon was wound up and, according to Charles, did the best flying he had ever seen.

Frisbie then took off in the *Hawk*, flying out of sight behind the hills and trees of the Pinnacle and then reappearing off in the direction of Elmwood Avenue and Goodman Street. He came back over the field at a terrific speed—he said at least seventy miles per hour—circled two times and made a beautiful landing. He was sure knocking himself out for the folks.

The feature of the day was the battle. Frisbie took off first with Simon right behind. They disappeared behind the hills of the Pinnacle. The soldiers were deployed, and they hid in the wheat field near the southwest corner of the grounds. The aviators charged over their territory, dropping flour bombs which exploded when near the earth and covered the soldiers with the white substance. Nobody got hurt, and the Rochester folks saw the first air raid. It was a fine air show, and the *Hawk* from Omaha flew in it.

The biggest thing that happened Monday, the last day, was that Charles flew the *Hawk*. It's sort of a long story. First, Barrier had been sick for several days and had not flown. Simon's Blériot was still ailing, so he was flying Barrier's. He was the only Blériot pilot, and Simon had never flown a biplane. Now, the final straw: Frisbie had evidently met a bunch of his old friends Sunday evening, and they seem to have had a great reunion and celebration in true Irish style. It was a fact that there was a bit of leprechaun in Frisbie. He had a great penchant for Chinese food and was known to appreciate a touch of the rye on occasion. It was a great occasion, his homecoming.

Anyway, Frisbie couldn't be found at flying time, and was Charles tickled. He made his best and longest flight. Gus said he was in the air for twelve minutes. He stayed close to the field and shot two landings, the last a beauty. Charles had watched and talked to Simon enough that he was ready to try a high-approach, switch-off landing. On his second landing he came in at about eight hundred feet, cut the switch, and dropped the nose (not as far as Simon but the steepest dive he had ever been in). The prop windmilled; he levelled her out at two hundred feet and glided in to a beauty of a landing. Simon gave him a big hug and told him he "had the touch!" That was like getting approval from the old Red-tailed Hawk. Charles and Gus Baysdorfer—and the *Hawk*—never forgot that day in Rochester, New York.

The next meet was scheduled for July 21 and 22 at the Chemung County Fairgrounds in Elmira, New York. Young arranged for Charles and Gus to visit the Moisant factory and school at Garden City, Long Island. Did the two cornfeds from Nebraska ever get a kick out of New York City! Gus said they really were rubber-necks taking in the sights. Moisant himself took them around the factory. He told them that he wanted to see the *Hawk* because he had heard so many good things about her and that they should bring her there when she needed an overhaul.

The Moisants also had a flying school, and Matilde Moisant, their sister, and Harriet Quimby were taking lessons. Quimby became the first woman flyer in the United States and was issued a license by the Aero Club of America in 1912. Also attending the flying school were Edward and Katherine Stinson.

The factory was building a radically new biplane, a tractor with a fuselage, named the Scout for Katherine Stinson. Eddie Stinson helped design the new ship. Eddie Stinson became one of the best aeroplane designers and builders in the United States. He and his aeroplanes will appear in this story many more times. Charles worked on the Scout during the winter of 1911 and 1912.

The brothers spent two more days at the plant, then they were off to Elmira. The *Elmira Advertiser* sponsored the air meet, which started Friday, July 21. The advance promotion was excellent, with many stories, photos, advertisements, and the like.

Simon was the star, as usual. He was the only one to fly on Friday, the opening day. The weather had been threatening all day, but it cleared enough for flying at about 3 p.m. He climbed to about 2,500 feet, flew over East Hill and the state reformatory, came back, flew his famous "Dip of Death," and landed just ahead of a real summer rainstorm. The *Elmira Advertiser* ran this story:

> Skimming lightly over the aviation field within the circular track at the Chemung County Fair grounds, then gracefully lifting his Moisant monoplane into the air, Rene Simon, the French "Fool Flyer," began the first successful aeroplane flight ever made in Chemung County.[43]

Elmira, New York, was the turning point in the aviation career of the Baysdorfer brothers and the *Hawk*. Charles had

---

[43]*Elmira Advertiser,* July 22, 1911.

decided that an aviator's life was for him. He knew he was a pretty good pilot and that he would get better. He was fascinated with the Moisant factory and knew he could work there. The *Hawk* was one of the best biplane pushers in the country, and, with their latest front elevator modification, it was far ahead in its controllability.

Gus had had about enough of the flying business. He was ready to go back to Omaha, to the shop and to his family. On the negative side, they were still having trouble getting paid—it seemed that the International Aviators always owed them money. Still, it

AVIATION MEET
AT ELMIRA
TO-DAY   TO-DAY
FAIR GROUNDS

Company L in Battle With Airships
Auto and Airship in Speed Contest

Gates Open at 2 O'clock   -   -   -   First Flight at 4:30 O'clock

SIMON-FRISBIE-BARRIER
MOISANT INTERNATIONAL AVIATORS

ROUND TRIP ON TROLLEY LINES 15 CENTS

Admission 50c   -   -   -   Grand Stand 25c Extra
CHILDREN UNDER TWELVE YEARS OF AGE HALF PRICE

seemed only a matter of time until Charles took over from Captain Frisbie and flew the *Hawk* all of the time. Young had already talked to him about it. Young figured that they could eliminate Frisbie and his pay, keep the *Hawk* in the show, and save some money.

Then Charles K. Hamilton, who was now one of the most noted birdmen in this country, showed up in Elmira. He was fresh from his great New York to Philadelphia flight and had come to

Elmira to greet his old friend Rene Simon. He was the guest of the Elmira Elks. Hamilton had been flying for Curtiss, but he had managed to make enough money that he now owned two Curtiss Flyers and was flying exhibitions and barnstorming on his own. He told Charles that they ought to take the *Hawk* to Chicago for the Grant Park meet the first week in August because he thought she was fast enough to win some big prize money. He also told Frisbie that Curtiss needed a pilot to fly exhibitions that he had booked in Kansas and Oklahoma in the last of August and first of September. He knew this because he originally was going to do the flying, but he was too busy now and told Curtiss that he couldn't fly the bookings. The Moisants had only three more bookings left, in Binghamton and Troy, New York, then Passaic, New Jersey. After these they were taking the aeroplanes and equipment to Garden City, New York, for complete overhaul.

Everything came together for the Baysdorfers. They flew the meets in Binghamton and Troy. Then Charles, Gus and Frisbie crated up the *Hawk* and headed for Chicago. Young even paid the fares and asked Charles to bring the *Hawk* to Garden City after the meet. Frisbie had contacted Curtiss and made a deal to fly the exhibitions in the west, beginning at the Norton County Fair at Norton, Kansas, on September 1.

Gus wrote Otto and brought him up to date. He asked him to come to Chicago for the meet. Otto's reply caught up with Gus in Chicago.

The Grant Park meet was one of the best ever held to date in the United States. All of the famous aviators were there: Paulhan, Johnstone, Beachey, Drexel, Hoxsey, Hamilton, Frisbie and Charles Baysdorfer. There were over thirty aeroplanes—Wrights, Curtisses, Farmans, Blériots and the *Hawk,* among others.

They entered the fastest take-off event which had a first place prize of $1,500. It was decided to have Captain Frisbie fly because of his greater experience. Three Wrights, four Curtisses and the *Hawk* were entered. The weather was great—a south wind of about ten miles per hour. Arch Hoxsey flew his sixty-horsepower Wright first and got off of the ground in exactly 112 feet. Hamilton, flying his Curtiss June Bug, left the ground in a much shorter distance. The other three Curtiss planes then took off and didn't come close to Hamilton.

Now it was the *Hawk's* turn. Gus, Charles, Frisbie and a couple mechanics wheeled her out to the starting line. They had just run the Gnome and had her adjusted like a watch. Frisbie climbed into the seat and shouted, "Contact!" Charles propped her, and she fired like a bee. The ground crew held on for dear life. Frisbie waved his hand; they let go; the *Hawk* leaped forward like a shot and was in the air before you could say "Jack Robinson."

The judge stretched the steel tape to the exact spot of take-off and announced the that *Hawk* was in first place. When Frisbie landed and taxied back to them, all three—Gus, Charles, and Frisbie—did a victory dance. Nobody else came close, so they won the fastest takeoff and $1,500.

They crated the *Hawk,* and Charles headed to Garden City with her. Gus went back to Omaha on the Northwestern night train, and Frisbie went to Rochester to visit his family before going out to Norton, Kansas, for the 1911 Norton County Fair.

The Norton County Fair

In June the Agricultural Association contracted with the Curtiss Amusement Company to fly exhibitions each day of the fair, August 30 through September 2, for a sum of $5,000. Much publicity was given this great event, and the *Norton Telegram* printed a story with a picture of Charles Hamilton, the famous Curtiss aviator, who would do the flying.

The Board was quite surprised when they greeted Captain Frisbie, his wife and youngest son when they arrived on the Kansas City train Wednesday, August 30. The Curtiss people had not advised them that Hamilton wouldn't be there.

Unbeknownst to Frisbie, the Curtiss aeroplane he was to fly was on the same train. The plane was supposedly shipped from Philadelphia in time to arrive on Tuesday, the day two Curtiss mechanics arrived to assemble it. It took the mechanics, Frisbie and a local garageman until Thursday afternoon to get the ship ready to fly. They finally got the engine running at about 4 p.m., and Frisbie took off from the west end of the infield of the race track. He got about fifty feet in the air when she nosed up and fell on the right wing. Frisbie got the nose down, but she hit the ground with a bang and broke and bent a few parts. He wasn't hurt. So it was back to the shop again.

They worked all night and the next morning to get her back together. The Fair Board was a little unhappy, because there were supposed to be three flights per day, and the fairgoers were loud in their comments. The mechanics told Frisbie that the plane was about two-years old and had been used by Curtiss, who put pontoons on it and flew it off of the water in a demonstration for the Navy. Evidently Curtiss had already committed all of his good aeroplanes to other exhibitions and hurriedly put wheels on the old ship and sent her to Norton.

Frisbie, in his short flight, knew that she wanted to turn to the right, nose high. So when they got it back together he wanted to trim and balance her, if possible. They fired her up, and Frisbie taxied back and forth on the infield, stopping and making wing and control adjustments. Finally, at about 3 p.m. he thought he could fly her, but the wind had gotten pretty wild (like it can in Kansas in August), and it was obvious that a cumulo-nimbus development was building in the southwest. Frisbie wisely said there would be no flying then. They had the ship at the northeast corner of the infield and were sitting on the ground by it. The fair officials respected Frisbie's judgment. He assured them he would fly as soon as the wind settled down.

The big fair crowd was busy with the exhibits, judging and horse racing. When the racing ended, a large crowd drifted down to the aeroplane. They were told there would be a delay because the winds were unfavorable. The *Norton Telegram* reported the story on Wednesday, September 6, 1911.

Almost two hundred people gathered around the ship, and some began to taunt Frisbie for his unwillingness to fly. Finally, against his better judgment and the better judgment of his mechanics, the aviator decided to fly. He drove the plane onto the field, and in about two hundred feet it was in the air. Around sixty or seventy feet up, the plane dipped, but it headed up again and flew over the track at about a hundred feet. Frisbie turned north. The turn gave the south winds a chance to get under the plane. The plane seemed uncontrollable; it turned on edge; and it fell upside down. It hit the edge of a row of barns. Half the plane was demolished, and Frisbie was thrown forward, beneath the engine. The plane finally landed on its back. Physicians arrived instantly, and emergency treatment was given, but to no avail. His head was bruised; four ribs were broken; and his left arm was broken near the wrist. The engine crashing down on him caused fatal internal injuries. Removed to the hospital, the birdman never regained consciousness and died at about 7:30 p.m.

A local woman started a collection for the surviving members of Frisbie's family. The collection drive was very successful. Five hundred dollars came from the Agricultural Association after the Curtiss Airplane Corporation received the association's bill. The nobility of the association's act was praised in the *Norton Courier* of September 7.

The *Omaha Daily News* printed the story of Frisbie's death on September 2, 1911:

> Captain John Frisbie had been using the biplane designed and built by Gus and Charles Baysdorfer since the Moisant aviators left here last summer. . . .
>
> That Frisbie's fatal fall was taken in the Baysdorfer machine was remote in the belief of Gus Baysdorfer, who returned to Omaha just before the Chicago meet. . . .
>
> "I expect to hear from my brother Charles, who was the mechanician for Frisbie, in a few hours. . . ." [44]

Well, the fatal crash wasn't in the *Hawk*. As a matter of fact, it's too bad Frisbie wasn't flying the *Hawk*—because he might well be still flying if he had been. John J. Frisbie played one of the most important parts of the Baysdorfer aeroplane and aviation story, since he figured largely in the development of the *Hawk* and Charles's pilot training. Without him, the story would have been different.

My research strongly indicates that Frisbie was the first aviator killed in Kansas, so perhaps someone should place a plaque in Elmwood Park in Norton telling the world so. Captain John J. Frisbie, aeronaut and aviator, may you always have tail winds as you fly into the Wild Blue Yonder!

Gus quickly got back into the shop and was swamped with work, as the automobile was booming. Gus got out of the aeroplane business until World War II, when he worked at the Martin Bomber Plant in Omaha. He confessed that he never really lost his desire to learn how to fly, but he never did learn.

Charles and the *Hawk* traveled to Garden City, and he went to work in the Moisant factory. Because of his engine and shop

---

[44]*Omaha Daily News*, September 2, 1911.

know-how, he worked with Henry W. Ashmusen, who was building revolutionary horizontal-opposed cylinders—air-cooled engines. The two hit it off right away and became good friends. He also worked on the *Scout* being built for Edward and Katherine Stinson.

The Moisants took back the Gnome engine that was in the *Hawk*, which was a blessing because they installed an Ashmusen eight-cylinder sixty-horsepower engine in her, covered both sides of the wings, put shocks on the landing gear, and overhauled and put her into A-1 shape. This made the *Hawk* just about the fastest and best pusher in the U.S.A.

It was 1912, and aeroplanes were changing. The engines were in front. Fuselages were being built to hold the pilot, engine, wings, and tail, and wheels were being fastened to the fuselage. Monoplanes appeared. All of the European governments were in the aeroplane business because they all saw the tactical value of the aeroplane. The development in the United States, though, was done by the citizenry and really didn't get too far out of the barnstorming stage.

The Moisants were involved with a motion picture company, so in the spring Charles and the *Hawk* flew for the cameras. Charles's flying got better and better, and in July 1912 the Aero Club of America issued him aviator's license No. 183. I guess you could now call him a full-fledged pilot. He was traveling in pretty exclusive company, a long way from the Noyes's meadow.

Charles returned to Omaha only twice, once in 1919 and again in 1925. His aerial activities from 1911 to 1916, when he quit flying, are very sketchy and can be pieced together only from bits of information. He flew for the Moisants, the movies and exhibitions. He barnstormed and taught flying—all of this back east in New York, Pennsylvania, New England and South Carolina. He was supposed to have flown in California and may have when the movies moved there. After all, Charles was a real birdman, as well as an aeronaut, which put him with a special breed that flew in those days.

His flying ended in Vermont in the summer of 1916. Charles still was flying the *Hawk*. By 1916 a biplane pusher was sort of a rare bird, so it was in demand for shows, meets and movies. A New York movie company was making an epic titled *Battle in the Clouds*. All of the aerial scenes were filmed in Vermont in the southern end of the Green Mountains near Brattleboro. The *Hawk* still weighed only six-hundred pounds, would go about seventy-five miles per hour, get off fast, and climb very fast. In other words, *Baysdorfer No. 1* was still a darn good aeroplane.

Charles and the *Hawk* flew as the villain of the show. The script called for the villain to steal the rare empire jewels and drive madly over the back roads in a Pope-Toledo Runabout to the field where he had his escape plane hidden. Of course, he was being pursued by the Prince, probably, in a Locomobile. The villain drove up to the *Hawk*, leaped out of the auto with the jewels, and climbed into the seat of the *Hawk*. Somebody (Charles) propped the engine, and he was off. It was neat flying. Charles got to buzz the pursuing hero. He ran the big Locomobile off the road, and the *Hawk* flew away toward the mountains.

Lo and behold, the hero's Scout was parked right where he ran off the road. He jumped into the cockpit; someone propped it; and he was off chasing the *Hawk*. As they approached the mountains, the villain flew into the clouds, and the hero lost him. Remember, this was 1916, and no one had developed guns or aerial combat. They had a camera in the chase plane, so they had aerial pictures. Charles was supposed to fly out of the clouds smack into a mountain, which he darn near did. They had the camera on the ground as Charles flew out of the clouds and headed straight for the mountain. He was to pull up and fly through a valley, which would look to the camera like he had hit the mountain. Charles said that when he pulled up, the Ashmusen gave a big blurp and quit. All he could see were trees. He came down in them and didn't remember anything until the crew lifted him into the backseat of an auto.

They took him to the hospital in Brattleboro. He was lucky: He had a headache from a crack on the noggin, some burns on his arms and legs from the exhaust stack, and numerous cuts and bruises. Charles thought his back was broken, but it wasn't. He lay in the hospital for a few days, and that was it. The *Hawk* was practically demolished, though. The movie company paid Charles for his flying and his hospital bill. With the plane gone, however, Charles had lost everything. The sad part is that no one saved any part of the *Hawk,* so Nebraska's first aeroplane was scattered over a Vermont mountain. Darn! Maybe we ought to go look for pieces.

With Charles's back killing him, he couldn't fly or work, so he took a friend's invitation and rode with him to St. Petersburg, Florida. The year 1916 was the right time to move to Florida. His

No Other Spark Plugs Can Do This!

No other spark plugs in the world can *do* the things AFFINITY SPARK PLUGS can do!

No other plugs can fire a spark gap of more than .020" (twenty-thousandths of an inch) *at all car speeds.*
AFFINITY SPARK PLUGS fire an .035" gap (nearly twice as long) at a car speed too slow to move the speedometer!

No other spark plugs in the world will make easy starting on weak ignition.
Yet AFFINITY SPARK PLUGS will start almost any car on a *low-tension magneto*—or on one *single dry cell.*

No other spark plugs in the world will fire perfectly at any car speed on a *single dry cell.*
Yet AFFINITY SPARK PLUGS ran a 4-cylinder Regal car for *three months on one dry cell!*

No other spark plugs in the world will give you as much power—on hills and on the level.
No others will allow you to throttle your car so slowly.
No others will give you such quick, snappy "pick-up" and fast "get-away."

No others in the world will fire such a lean mixture—and thus save 10 per cent or more on gas.

No others in the world will stay clean so long in oily cylinders —particularly the front cylinder of a Ford.
Yet AFFINITY SPARK PLUGS do all this—and more. They are able to do it because they embody for the first time the well-known fact that a current will jump a gap with almost *no resistance* if allowed to jump from a *surface* to a point. All other plugs fire from one point to another, but—

AFFINITY SPARK PLUGS are so constructed that, no matter how they are connected up, they invariably fire from *surface* to point.

You can't get all these benefits with one AFFINITY SPARK PLUG—therefore:

Buy Them by the Set (4 for $4) of

OTTO BAYSDORFER,
Ignition Specialist,
210 No. 18th St., Omaha, Nebr.

What a Few Pleased Users Say:

Affinity Spark Plug Company,
Omaha, Nebraska.

After making a thorough test and inspection of the Affinity Spark Plug I can say that it is my belief that this is the best spark plug on the market to my knowledge. Its longevity, sure sparking and nonfouling characteristics are what the automobile public have been looking for, for a long time. I take great pleasure in recommending the Affinity Spark Plug to all users of gas engines.

Yours very truly,
H. E. Schmidt (signed)   Willys-Overland, Inc.,
Service Manager: OMAHA BRANCH

Wheatland, Iowa,
Affinity Spark Plug Company,
Omaha, Nebraska.
Gentlemen:

Witte Brothers sold me a set of your Affinity Spark Plugs on the following assurance:

"That they would start car with starter any time — day or night, high or low atmospheric temperature, dry or moist atmosphere, any time between January 15th and February 15th, 1917."

Result — absolute fulfillment. Started car at twenty-two degrees below zero. With old style plugs I always primed car in the morning. I do not now. During the day, with yours, I had to thin my mixture. I had two other makes of standard plugs and in extreme cold could not start car at all with starter.

I consider your plugs have saved me the price in one month in gasoline saved and saving of wear on batteries — say nothing of the wear and tear on my temper, which is not of the best.

I thank you for bringing them to me. I drive the Maxwell car. You should call their attention to your plug. It will sell cars for them.

Yours truly,
J. S. Dean, M. D.

back got better, and it wasn't long before he was working in a garage and chauffeuring. Tampa Bay fascinated Charles. Fishing, I believe, was his real love, even before aeroplanes and flying. He soon acquired a piece of land on the west side of old Tampa Bay, built a boat, and spent most of his time fishing. He never flew again, but became one of the best guides and fishermen on Tampa Bay. His knowledge of the bay was so good that the Coast Guard and others consulted with him. Otto's son, the late Gerald Baysdorfer, told me of many vacations spent at Charles's Florida place and the wonderful fishing they did. Gus's daughters (Betty, Bonnie and Judy) said their families spent many great vacations there.

Charles was married several times, but never had any children. He died in 1962, the last of the "Wright brothers of the Noyes's meadow." Several old Red-tailed Hawks flew the "Missing Man" formation over the Noyes meadow that day, and Nebraska's first successful aviator headed into the Wild Blue Yonder!

Otto and Gus ran their shop for three more decades in Omaha and continued to invent things and do super machine- and service-work. Their most successful invention was the Affinity Spark Plug, which they manufactured and sold mainly during the 1920s. It was a dandy plug, and they sold thousands. They also invented an electric wall outlet and a clinker tong for removing clinkers from heating furnaces. Otto died in 1944. Gus followed him in 1952. They are both buried in Omaha.

In the Baysdorfers's honor, on August 5, 1990, a historical marker commemorating Charles's first flight was dedicated. The marker is located on Douglas County Road 84, two miles north of Douglas County Road 64, at the field where the flight was made. It was quite an affair, attended by several hundred people, with fly-bys and parachute jumps. Best of all, the plaque was unveiled by Anne Otte Boettger, who saw the first flight, and Betty Baysdorfer Moucka and Judy Baysdorfer Foisey, daughters of Gus. The flying events were arranged by Jim Pollack, the airport operator at Wahoo, Nebraska. Michael Tabler flew Charles Casey's *Breezy*, and Roger Haney and Jim Dager flew two Stearmans that day. The parachute jumps were made by Clark Bedner and Tom Smith of the Lincoln Skydivers. Tom Doyle, Douglas County Engineer, arranged for the construction of the marker and the site. Con Agra provided one of the Stearmans and financial aid, as did Bill Hamsa, pilot and airplane builder. I am the president of the Aero Club of Nebraska that sponsored the event. The flight, location and marker have the approval of the Historical Society of Douglas County and the Nebraska State Historical Society.

BOOK THREE:

What's to Be Done with This Flying Machine (1911 to 1919)?

From 1911 until after the Great War, there was really very little flying or interest in flying in Omaha, in Douglas County or in any of Nebraska. Charles Baysdorfer was the only bona fide aviator Omaha had, and he never even flew in Omaha—he flew only at Waterloo in Douglas County in 1910. There were scattered reports of aeroplanes being built, but most of them never flew. They either never got into the air or crashed immediately if they did.

Sgt. Clarence Adams, now in partnership with Bert LeBron, was still trying to fly. Adams had just finished his revolutionary monoplane and took it out to the West Omaha Speedway for trials. When he and several friends got it assembled, he fired up the engine and taxied—right into a fence, which broke the prop and gave him a big bump on his head. That ended things for the day. They pushed the ship back beside the tent, which they used as a shop and a place to live. Adams lit the brand-new gasoline stove to cook their dinner. It blew up. The fire spread quickly and completely destroyed the aeroplane in spite of all their efforts to save it. The September 13, 1911, *World-Herald* reported: "Aviator scorched and aeroplane is destroyed."[1] Alas, Sergeant Adams just wasn't meant to fly. I believe he finally realized that, for I never did uncover any more information on his flying efforts. Good try, Sergeant!

The Army flew balloons at Fort Omaha every May until 1913, when all of the Signal Corps equipment was transferred to Fort Riley, Kansas. On May 30, 1912, *U.S. Army Balloon No. 11* made a successful eighty-four mile flight to land in Missouri.

Of greater note, this was the same day that Wilbur Wright died. President Taft said,

> I am very sorry that the father of the great new science of aeronautics is dead . . . and that he has not been permitted to live to see the wonderful development that is sure to follow along the primary lines of the new science which he laid down.[2]

---

[1] *Omaha World-Herald,* September 13, 1911.

[2] *Omaha Evening World-Herald,* May 30, 1912.

Across the rest of the great state of Nebraska, I have found pictures and a little information about a couple of aviators. R. D. Herzog, a garage owner in Harvard, Nebraska, built a biplane he named the *Meteor* in 1912 or 1913. Whether or not it ever flew, or what happened to it, I do not know.

Everett D. Morrow arrived in Burwell, Nebraska, sometime in 1911. He was supposed to be a successful aviator from the West. What "West" I have never found out, but he did build an aeroplane.

If it flew, it must have given some ride. In the photo it looks like a clipped-wing Curtiss *June Bug,* with a great wide landing

gear. Morrow is supposed to be at the wheel. Note the neat auto with the buggy wheels and Nebraska license plate No. 18251. In 1912 Morrow built his second aeroplane—still looking like a Curtiss but with some wing and landing gear modification—at Carlton, Nebraska. Though I couldn't find a flying record of it, it looks like it could have flown. In late 1912 Morrow flew number three in Ord, Nebraska.

Morrow finally built what looked like a good aeroplane, with wings and a cleaned-up front control. In 1913 Morrow was back in Burwell to build what must have been the first tractor (propeller in front) in the state of Nebraska. Burt Young, who had been serving under Morrow for a year, flew it in April. After a ten-minute flight, the ship crashed on landing, and it was almost destroyed. I never did find out what happened to Young, and Morrow seemed to vanish into the Wild Blue Yonder, too.

CHAPTER TWO:
THE SAVIDGE BROTHERS

There were some very successful flyers up in Boyd County. The seven Savidge brothers, who lived on their folks' ranch eleven miles southwest of Ewing, Nebraska, were for real. They were a talented bunch of singers and performers, about as popular and well known as anybody in the country. They all got the flying bug watching Red-tailed Hawks in the Sand Hills. George, Matt and John built a glider that did fly. In the spring of 1909, they started building their first aeroplane, which, like most of the home-built ones, sort of resembled the Curtiss *June Bug*. The Detroit Maximotor engine they used couldn't even get the heavy ship to roll on the ground. They started to build another, much lighter ship and hoped to get a Hall-Scott engine for it. They really wanted to be the first to fly in Nebraska, but Glenn Curtiss flew at the meet in Omaha in July 1910. The Savidges drove down to Omaha to see Curtiss fly, got a real good look at his aeroplane, then went to the Baysdorfer shop and saw the *Hawk*. They returned home vowing to beat the Baysdorfers or anyone else in Nebraska who was building an aeroplane. But it wasn't until May 1911 that they got the new ship built, and Matt successfully flew it. It was a very good aeroplane. I call her the *Sandhill Crane*. On May 7, Matt flew the *Crane* several times in their pasture, and it seemed everyone in the countryside came to watch. In the next few days, John also flew her, and his sister Mary, sitting on a board next to him, became the first woman in Nebraska to fly in an aeroplane.

Matt took to flying like a hawk, barnstorming throughout the Midwest from the Dakotas to Texas. As one of the first stunt flyers, he could do and would try anything. At the 1915 Orange Days Celebration, he flew over Alvin, Texas, writing "M A T T" across the sky with smoke. He was probably one of the first—if not the first—skywriters ever. The letters seemed to stretch over the whole town and got people's attention.

The Sandhill Crane

That reminds me of once when I was about six and just learning to read. Coming home from school one noon, I looked up into the sky. There, in huge letters, was "L U C K Y." It scared the heck out of me. I ran into the house and yelled at my mom, "Come quick. God is writing something in the sky!" She ran outside and by then "S T R I K E" was visible. No message from God.

Well, Matt was one of the best acrobatic pilots around, particularly at looping. George, John and Matt were probably as knowledgeable about aeroplanes and flying as anybody, including Curtiss, and maybe even the Wrights. They were constantly modifying and improving the *Crane*. In 1916 while in Texas, they bought or had built for them a new biplane—a pusher that had a lot of their ideas built into it. One of the best ideas was ailerons in the trailing edge of the top wing. Much of its structural-strength members were metal, and it had a wing design that probably was way ahead of the times. It was powered by a brand-new, $2,700 gyro-radial engine, probably either a Gnome or one built in the United States (France was busy fighting Germany). It developed close to one hundred horsepower, and the cylinders rotated around the crankshaft. Both Matt and John flew it and felt it was too unstable, but they crated it up and shipped it to Ewing, so brother George (the aeronautical-engineering genius behind all of the Savidge aeroplanes) could work on it. George made some modifications and explained to his brothers about torque (which the Gnome certainly created much more of than any other engine they had flown). Finally, in early June 1916, George had her ready to fly. John flew it first, then Matt. Like any new machine, it needed a lot of adjustments, which were made, and then it was tested by both John and Matt flying it.

On June 17, John flew it and said that she didn't fly right. There was still something wrong with her, and he just couldn't figure out what it was. After supper Matt said he was going to fly her and figure out the problem. Almost everybody in town turned out to watch. John told Matt not to try anything fancy as he got ready to prop her. She fired off on the first pull. Matt grinned and signaled them to release her, and she was in the air in 150 feet. He took her up to eight or nine hundred feet in a spiral climb. She flew like a hawk. He leveled off and did a perfect 360 both right and left, obviously trying her out. He then dove her for speed and pulled her up into one of his beautiful loops. As he came around and started his pullout, the nose stayed down. The engine was running—there was no apparent structural failure; Matt fought her with everything he had, but he couldn't get her out of that dive. After his death, the surviving Savidge brothers gave up flying.

I think Matt saw firsthand a high-speed stall caused by the center of gravity moving disastrously when he tried to pull out of the loop. Once it happened to me in a North American P-51 Mustang in a tight turn, but I had a lot of air below me.

Matt and John Savidge were the second and third Nebraskans to successfully fly an aeroplane in Nebraska, and they built it themselves. Elizabeth Savidge Horn, whose first husband was a cousin of Matt, John, George and Mary, gave me some of the Savidge history. As you enter Ewing from the east, there is a marker on the south side of U.S. Highway 275 that tells the world that the Savidge brothers flew here a long time ago.

Though I have investigated numerous leads, I haven't been able to confirm any other Nebraska aviators at that time, nor any native-Omaha aviator other than Charles Baysdorfer. Three photos have driven me nuts for about five years. Mrs. Halsey Noyes, who still lives where the Noyes's farms were north of Waterloo, has the original glass plates of two aeroplanes. She says that her father-in-law took the pictures in the field that Charles Baysdorfer flew from in 1910, but she doesn't know what year they were taken—maybe

Loch-Coleman tractor

between 1910 and 1914. I thought that one might be the *Hawk*, but after much studying and comparing newspaper photos, I am positive it isn't Baysdorfer's aeroplane. The other plane really boggles my mind: It is a tractor (prop in front), and there weren't many tractor aeroplanes in the United States at the time. I checked it with the Morrow Burwell photo, and it wasn't the same.

In the Fremont, Nebraska, public library in January 1992, I was researching this photo of an aeroplane supposed to be at the Nebraska National Guard maneuvers held near Waterloo, Nebraska, during August 1913. Mabel Campbell Wilson of Waterloo (the mother of Ralph Wilson, dean of all the historians of western

Nebraska State Guard with plane, 1913

Douglas County) took the photo of a plane sitting in a pasture on the east side of the Elkhorn River just north of Waterloo on the Sumnick farm.

In 1913 the Nebraska National Guard began the development of an aviation section. Two Fremont men, Adolph Jensen and Vernon Knox, members of the Fremont Signal Corps Guard company, built an aeroplane. It very closely resembled a Curtiss Model B. By April they had it done but lacked an engine. The April 29 *Fremont Tri-Weekly Tribune* reported that a Burwell birdman named Burt Young would fly an exhibition in his plane on May 10 and 11 at the Fremont Driving Park to raise money for Jensen and Knox to buy a $650 engine for their bird. (This never happened because, as we know, Young wrecked his aeroplane on its first flight.) But *Tribune* stories reported that the Guard became interested.

Jensen and Knox couldn't raise the funds to buy an engine for their bird, but Capt. Henry Jess, commander of the Fremont Signal Corps, really wanted the plane for his outfit and brought it to the attention of Gen. Phil Hall, commander of the Guard. The War Department would buy the engine as long as the machine would be at the maneuvers to be held near Waterloo. Jess, Parks and the others bought a fifty-horsepower engine in Council Bluffs and had it shipped to Fremont. The August 7 *Fremont Tribune* reported:

A big crowd of Fremonters saw the first trials of the Fremont-made aeroplane Sunday afternoon in a field near Inglewood where the biplane fitted for use of the Fremont signal corps was operated for the first time. No member of the signal corps ventured to shoot into the air with the bird wonder, but the machine was given several lively runs by means of the engine and fly-wheel, which performed in business-like fashion. That the winged frame would have risen with its load had its beak been turned upward seemed certain. However, Aviators Knox and Jensen and Cap. Jess of the signal corps thought best not to experiment too extensively at first.[3]

Without a doubt, that's the smartest thing the two did because they didn't know how to fly. The two fired her up the next day, and Knox taxied around the field. Wham! The crankshaft broke, and there was no way the engine could be fixed. The Fremont aeroplane was loaded on the Union Pacific morning freight for Waterloo and arrived for the big maneuvers (minus the engine). And that's the story about Mrs. Wilson's aeroplane photo.

The August 7 *Tribune* carried another story, one reporting the death of Samuel F. Cody, the Texan who built the British Army's first aeroplane. The newspaper also reported that an Omaha aviator named De la Roche was killed, and this opened up a whole new avenue of research. I call it "the Loch-Coleman-Turner and De la Roche Episode."

The Loch-Coleman-Turner and De la Roche Episode

Around the turn of the century, wrestler Pete Loch was big time, was even—briefly—the World Champ, and wrestled them all—Frank Gotch, Farmer Burns, Strangler Lewis, and tough middleweight Frank Coleman. Loch retired to open a restaurant-bar on the Crounse block in Omaha, which became famous and profitable. Ex-middleweight Coleman decided that his future was

---

[3]*Fremont Tribune,* August 7, 1913.

Loch-Coleman No. 1

brighter as an automobile mechanic; he worked in Oklahoma, Kansas, and Nebraska and ended up in Havana, Cuba, with Glenn Curtiss. He took to working on aeroplanes like a duck to water. In 1910 he was in Omaha with ideas about building a real good aeroplane and talked Loch into putting up money to build an aeroplane. Loch introduced Coleman to Otto and Gus Baysdorfer. Their shop was on Capitol Avenue, just around the corner from his restaurant. In 1911 Coleman started building his plane in a shop in a building at 117 1/2 North 16th Street. The Baysdorfers had no interest in the venture, I think, because I never found any mention of Coleman in my research. But Coleman probably asked them a million questions.

Coleman first bought a brand-new Curtiss eight-cylinder, eighty-horsepower engine. He knew Curtiss, who was now selling engines, and his plane resembled the Curtiss Model B except for the ailerons. And here was where Coleman's design was real good: Coleman's ailerons were mounted between the wings behind the trailing edge, sort of like the Travel Air 2000, and they were controlled with the steering wheel. This was revolutionary! Remember the Wrights's and Curtiss's fights over bank control and that the Baysdorfers controlled the *Hawk* with the wheel by moving it side to side? Coleman controlled it by turning the wheel, just as it is done today. He also had the rudder controlled by the wheel. Though he claimed to have it patented, I think he was just "shooting the breeze." His control system was good.

By summer 1911, his first ship, which I have named *Loch-Coleman No. 1,* was almost done, and he was building *No. 2,* which was a tractor. He exhibited *No. 1* at the Ak-Sar-Ben Fall Festival, but he didn't attempt to run it. By late spring 1912, *No. 1* was taken to the pasture on the Winter's farm west of Papillion, Nebraska, about ten miles south of Omaha, where it was assembled and tested.

Enter now Glenn Turner. Turner was born somewhere in the South and came to Omaha when he was sixteen. He took up motorcycle racing and raced at the West Omaha Speedway, East Omaha Speedway, Lincoln, Fremont, and Auburn, Nebraska, and Sioux City and Red Oak, Iowa, and many other tracks. He turned to auto racing as a driver and finally as a mechanic and went to work in Coleman's shop in 1911. He had made up his mind that he was going to be an aviator, and Coleman (who never flew) told him he was his pilot.

Turner spent several days taxiing *No. 1* around the pasture. He let her get into the air a very short time, and Coleman made a lot of adjustments. Finally, on June 27, Turner was ready. There was a big crowd, led by Loch and an Omaha delegation, many local citizens and over a hundred farmers. The weather was perfect. At 4:30 p.m., Turner hit the throttle and was in the air in less than three hundred feet. He took her up to two hundred feet and flew about two miles straight north. He did all of the right things, flew her straight and level, and felt her out. Then he made a turn to the west. The *Omaha Daily News* reporter wrote, "Turner held the new machine easily under control and in making his first turn performed a miniature dip of the variety that made Hoxie [Arch Hoxsey] famous."[4]

Then Turner flew back to the field, as reported by the *Papillion Times.*

> He had circled the field twice in a splendid manner and when attempting to descend he found that he had slightly misjudged his position and that he would not be able to make a landing in the pasture and he therefore again took to the air and cleared the mass of telephone wires just south of the road attempting to make a landing in Schaab's alfalfa field. A gust of air sweeping down over the hill drove his machine downward and he landed violently, smashing the rudder of

---

[4] *Omaha Daily News,* June 28, 1912.

the machine and injuring the steering gear, propeller and several other parts.[5]

I say a darn fine landing for the first one he ever made!

Turner joined the ranks of Charles Baysdorfer, John and Matt Savidge, Everett Morrow and Burt Young as the first Nebraskans to make successful aeroplane flights in Nebraska.

Repairs were quickly made because they had a booking to fly at the big Fourth of July celebration at Morehead, Iowa (about fifty miles northeast of Omaha), for which they were to receive $900.

Loch-Coleman No. 1 was put on the Illinois Central freight and arrived in Morehead the morning of the Fourth. Turner flew her out of a field just south of town. His first flight was a beauty—over town and around the field three times. There was a huge crowd at the field when he hit the throttle to take off for his last flight. By now he probably thought he was one of the Wright brothers. The wind had come up and the field was rough. He picked up speed and bounced every time he hit a swail in the ground. Finally he hit a big one at the same time as a big wind gust. The nose went almost straight up, she stalled, came down like a rock, hit the ground in a cloud of dust, and Turner was thrown out. That saved his neck because the front end was demolished. He was knocked out, bumped and bruised, but nothing worse. This ended his barnstorming and convinced him that there was a lot to be learned about flying.

This also ended flying exhibitions for the summer. *No. 1* was repaired in a little over a month, but the engine was badly damaged, and the prop was gone. Coleman carved the prop himself out of one block of native Nebraska ash using what he called "secret know-how," which took time. A new engine wasn't delivered until October.

Loch's restaurant was a popular watering and dining spot for newspaper journalists. Many a post-putting-to-bed session was held there. Sandy S. Griswold (the legendary sports editor of the *World-Herald*), Miles Greenleaf (who wrote the very popular column "Hot Off The Bat"), and staff photographer Pat McAndrews were regulars. So were the *Bee* and *News* reporters and all the famous and not-so-famous sports, stage and political people and other characters around

---

[5] *Papillion Times*, July 4, 1912.

town. Otto and Gus Baysdorfer stopped for a sandwich and a beer now and then. Loch's was the clearing house for what was going on everywhere. Everybody knew about *Loch-Coleman No. 1* and followed its rebuilding with great interest. McAndrews wanted a ride in it to take a picture from the air.

Coleman and Turner hoped to find a bigger and better field to fly her from when she was repaired. Along came Hyland B. Noyes and suggested they use his folks' pasture out at Waterloo, the same that Charles Baysdorfer flew the *Hawk* from. On October 21, they took *No. 1* and the new tractor *No. 2* out to Waterloo on the Union Pacific.

McAndrews's editor approved his story-photo idea, so he, Griswold, and Greenleaf drove to Waterloo in John Dugan's new Carter Car. On that beautiful fall afternoon, Tuesday, October 22, 1912, almost two years to the day after Charles Baysdorfer first flew the *Hawk,* Glenn Turner flew the *Loch-Coleman No. 1* over the same Noyes pasture north of Waterloo, Nebraska, and *World-Herald* photographer Pat McAndrews became the first newspaper man in Nebraska to ride in an aeroplane and take a picture from it. It was his pictures that solved the aeroplane-identification mystery in the Noyes photo I mentioned earlier.

McAndrews wrote:

According to my friends, family and about everybody else, I am the biggest simp in the world today. To go up in an aeroplane, especially in a new and comparatively untried machine, and with an aviator who had never before carried a passenger—all this seemed to them to be attempting suicide. As a matter of fact, those few minutes up in the air were the pleasantest I have spent for a long time and I would give a good deal to do it all over again. . . .

Then Coleman said he was ready, Turner said that he was, too, and I just gulped. There was a terrific deafening racket as the motor started and I waited. . . .

All of the sudden we commenced to bounce along the ground, which was pretty rough. . . . I saw Turner grinning at me for a second and then—it seemed as if we were just floating.

I really couldn't tell when we left the ground—but we left it. I know because I saw it going away from us. They say we were traveling about sixty miles an hour—that we

Turner and McAndrews before their flight near Waterloo, Nebraska

had to in order to keep up—but it didn't feel like that. It seemed as if we were just droning along like a couple of winged fairies—although we didn't look much like fairies—and it was the most delightful sensation I had ever experienced. . . . I didn't care if we ever came down or not. I saw our *World-Herald* machine and the boys down there between two haystacks and took a shot at them with my machine. They were waving their hats, so I waved mine. . . . So I decided to ask Glenn to try one [a turn]. I opened my mouth and started to talk, but I couldn't. The air rushed in and I pretty nearly strangled before I could shut my mouth again. Pretty soon I tried it once more and managed to get out a wild yell, but Turner was too busy to talk—and couldn't if he wanted to— so he just grinned and nodded his head.

He made a half turn to the left, and then decided that he might have to land in a cornfield if he kept on, so he came down at the edge of the field.[6]

That is a firsthand description of a 1912 aeroplane ride. The great day ended with Turner taking Loch for a ride around the field. Loch, the backer and promoter of the local aviation company, was quoted as saying:

I certainly admire your man McAndrews. . . . I'd never gone up myself if I hadn't seen how coolly he took it and how tickled he was when he came down. And Pat McAndrews was the first newspaper man in the state to go up in one of those things. That's going some![7]

The next day, the repaired engine was finally installed in *No. 2*, the tractor, and it finally ran well enough for Turner to try taxiing, which for some reason was very difficult. *No. 2* needed some changes, so it was decided not to fly it. And that ended the flying season of 1912.

I became well acquainted with Miles Greenleaf, the editor of the *Dundee News* (a weekly paper published by David Blacker in west Omaha), in the late 1930s. The *News* evolved into the *Suburban Sun Papers* and won a Pulitzer. Greenleaf asked me to write a column about sports every now and then, which I did. Greenleaf named the column "Sporting with Bob Adwers," saying athletes and aviators were sports. It was 1938, and he knew I was doing a lot of flying. One day he told me about a story he wrote in early teens about an aeroplane flight. Turned out it was the McAndrews flight. I didn't remember that until all of this developed.

Tony Jannus, a pilot for the Tom Benoist Company of St. Louis, Missouri, flew a Benoist Hydro-Aeroplane off of Carter Lake (formerly Cut-off Lake) to New Orleans. Jannus made several flights before a large crowd at Courtland Beach. Even though he was a staunch supporter of aviation, Mayor Dahlman declined a ride, saying he'd better stick to horses. However, Gould Dietz took a ride. On Thursday, November 7, Jannus and his brother Roger took off for New Orleans, made St. Louis on November 16, and finally arrived

---

[6]*Sunday World-Herald,* October 27, 1912.

[7]*Ibid.*

in New Orleans on December 1. Both the Jannus brothers later were killed flying: Tony demonstrating a Curtiss in Russia in 1916 and Roger in a SE-5 training flight crash in France in 1917.

In the spring of 1913, Turner flew *No. 1* almost every weekend out at the West Omaha Speedway. *No. 2* was still in the shop—Coleman was having a tough time revamping the wing-tip design. Turner flew many exhibitions in eastern Nebraska and western Iowa and was gone with the plane most of the time. Loch was trying to sell *No. 2* to Walter Moise, President of the Willow Springs Distillery, Omaha's famous bourbon whiskey distillery, who was very interested in the aeroplane for advertising purposes. Coleman finally thought he had *No. 2* ready to fly, so about the first of August they took it out to the West Omaha Speedway.

Meanwhile, a good-looking, personable, twenty-eight-year-old Frenchman and ace mechanic Henri de la Roche arrived in Omaha in the fall of 1912. He immediately was hired by Fire Chief Charles A. Walter to keep the fire department's new automotive equipment in working order. He was one of the most popular regulars at Loch's food and sustenance emporium (now you know how he got hooked up with the Loch-Coleman aeroplanes). De la Roche was born and raised in France and somewhere learned to be a great mechanic, first with automobiles and then with aeroplanes. He claimed to have been the King of Greece's chauffeur and to have worked for the best—Louis Blériot, Henri Farman, Charles Voisin. He said he had been the mechanic for the famous auto race driver Barney Oldfield. His *claimed* sister, Raymonde de la Roche, was taught to fly by Voisin himself. In 1910 Raymonde, a self-styled baroness, passed the qualifying tests and obtained from the Aero Club of France the first pilot's license issued to a woman anywhere in the world. She was flying in a July 1910 endurance race at Reims when she hit the new phenomenon called "prop-wash" and crashed. Though de la Roche was severely injured and even reported dead, she recovered and flew again. Ironically, she was killed while riding as a passenger in the crash of an experimental plane in 1919.

Henri de la Roche helped Frank Coleman assemble *No. 2* at the West Omaha Speedway. Turner was flying *No. 1* in Iowa, so Coleman asked de la Roche if he thought he could fly her. De la Roche said sure, so on August 6 Pete Loch, Walter Moise and Charles E. Fanning drove out to the Speedway in Wayne Burbank's big Page touring car.

According to the *World-Herald,* de la Roche
had often told of his experience as an aviator, and claimed to
have flown the English channel. . . . His evident knowledge
of gasoline motors and machinery in general while employed
by the city fire department led to the belief that he was really
an experienced airman. . . .

De la Roche made several runs with the biplane, which
was successfully flown all last year by Glen Turner. The
Frenchman claimed he could not take the air and that the
machine would not steer. Experts on the ground said that he
had apparently no knowledge of aircraft and that he did not
even try to rase [*sic*] the machine. Some of the witnesses
suggested that he was afraid to go up, and all agreed that the
Frenchman did not understand the game.[8]

It should be noted that Turner had only taxied the plane the
previous year and had had difficulty controlling it.

At last Frank Coleman asked De la Roche if he wished
to "try it once more." The Frenchman said he would, and
kicked off his shoes to get a better grip on the controls. He
then made a run, jerking his elevator suddenly and the craft
swept violently upward, swooped to the right and fell with a
crash upon its right wing. De la Roche was picked up
bleeding and unconscious and was hurried to Clarkson
hospital.[9]

Trapped by the jeers of the crowd into flying (like "Captain"
Frisbie and many others), de la Roche suffered serious injuries. De
la Roche fooled them and was recovering (maybe Raymonde *was* his
sister—she survived injuries almost as bad). In a few days he was
able to talk to Coleman, and they agreed that it was the split wing-
tips that caused the instability. The crash probably was due to a
catastrophically unstable combination: a tractor and a biplane with
a radical new wing design, Coleman not having enough experience
in aeroplane design, and the aviator not having enough experience
in flying. Unfortunately, after a sudden relapse, de la Roche died on
August 14. He gave it his best shot—fly with the Hawk!

---

[8]*Omaha Evening World-Herald,* August 14, 1913.

[9]*Ibid.*

Loch had spent several thousand dollars and a couple of years' work and had had enough. He gave *No. 1* to Coleman, who took it to Texas, where both he and the aeroplane sort of faded into the sunset. Loch moved his place to the east side of the courthouse on 17th Street and remained there until his death in 1934. Glenn Turner (much to his new bride's joy) gave up flying and worked in the Curtiss plant in Buffalo, New York, making Jennies during World War I. He came back to Omaha and worked for the Billings Dental Supply Company for twenty years, then moved to California. He died when he fell from a ladder in 1959.

The next big flying event in Omaha wasn't until the Ak-Sar-Ben Fall Festival of 1914. The newspapers wrote stories, printed many pictures, and featured aeroplane and aviation articles. Many companies used aeroplanes in their ads. The Jetter Brewing Company of South Omaha used them, even though the plane in their advertisement was inverted.

World War I started in Europe in August 1914, and aeroplanes and zeppelins became military tactical weapons. In the United States, the military still had them as a branch of the Signal Corps. Aeroplanes and flying might have been too darn tough for most people, but people adjusted to the automobile quickly enough. Omaha loved auto racing. There were two tracks, the West Omaha Speedway out on West Center Road and the East Omaha Speedway by Carter Lake. Cars were on all the city streets, country roads and any place one could drive. The automobile had arrived, and Henry Ford was making sure that America really was spelled "Model T."

Aeroplane pioneers Otto and Gus Baysdorfer went with the automobile and always made a good living from their service business. Charles Baysdorfer never came back to Omaha. None of the other early aviators built or flew planes in Omaha again, either. Omaha business leaders directed their talents to other things, too. Gould C. Dietz, although always an aviation enthusiast, had many other interests. Clarke G. Powell became completely involved with the Omaha Chamber of Commerce and was its long-time commissioner. James J. Deright, who was the guiding force for the Aero Club and aviation and many other activities, tragically died in 1912.

Most Omahans had seen the Curtiss Flyers or Moisant's International Aviators, and they knew flying was very difficult. I think the old Red-tailed Hawk was sort of sad because he knew he could fly better than those human birds, but he got a kick out of having them around.

A famous aviator-to-be arrived in Omaha in January 1910. Edward V. Rickenbacker was twenty years old when he came from Columbus, Ohio, to work for the Racine-Slattery Company, the Omaha distributors for the Firestone-Columbus automobile, also known as the CBC (Columbus Buggy Company). John M. Larsen,[10] who had sold Columbus Buggies for years in Illinois and Iowa, was the sales manager. It wasn't long before Rickenbacker was branch manager for CBC, selling carload after carload of the 1910 models. As a publicity stunt, he took up auto racing and won four races at the new West Omaha Speedway (where Ak-Sar-Ben is today) when it opened on October 1, 1910. He was a pretty slick operator, sold cars like crazy, and was a heck of a race car driver. He got into big-time racing when the new East Omaha Speedway was built next to Carter Lake, just north of where 13th and Locust Streets are today.

Now to Lincoln Beachey. The Nebraska State Fair had booked Beachey to fly on September 7-11, 1914, for $1,000 per day. Beachey was without rival as an airman. He was America's—and probably the world's—greatest acrobatic pilot. The Fair Board advertised that their contract with Beachey prohibited him from flying within two hundred miles of Lincoln.

Lincolnites underestimated the Omaha and South Omaha contingency of 250 who arrived in Lincoln on the morning Burlington the first day of the fair. Chaperoned by Everett E. Buckingham, Bill Shelberg and A. F. Stryker, stockyards and Ak-Sar-Ben drivers, the group was so impressed by Beachey's aerial performance that they had to meet him. The *Bee* story on September 11, 1914, reported:

> He [Beachey] was found in the Administration building, waiting for his day's wages, said to be $1,000, a few minutes afterwards. It was explained that the uncertainty surrounding an aviator's life was the reason for the haste.

---

[10]Although John Larsen's name is occasionally found in primary sources as Larson, Larsen is the preferred spelling.

182

A Lincoln paper carried an item that Beacher [*sic*] would retire after the Lincoln flights, but the aviator denied this to many Omahans who asked him about it.[11]

Before Buckingham and Shelberg left, they had a handshake agreement with Beachey to fly at the Ak-Sar-Ben Fall Festival.

Buckingham commented that he didn't care about the two-hundred-mile limitation in the contract with the Fair Board. On September 17, the *Bee* ran this story:

Beachey to loop over Omaha for Ak-Sar-Ben[.]

He is to loop for the Ak-Sar-Ben multitudes at the carnival grounds, October 5, 6, and 7.

This is considered a ten-strike by the board of governors. . . . Yesterday his manager was in Omaha and signed the contract.[12]

This was the same Lincoln Beachey who beat Charles Baysdorfer and the airship *Comet* in St. Louis in 1907. He learned to fly aeroplanes at the Curtiss School in 1910 and quickly became the best acrobatic pilot in the country. His one ambition was to be the first to fly a loop, but, alas, a Frenchman named Adolphe Pégoud was first in 1913. Beachey about had a fit and talked Curtiss into building him a beefed-up Flyer so he could loop. Curtiss did, and Beachey became the best looper in the world. He not only looped, but flew inverted and dove from several thousand feet doing gyrations never before done.

The *Bee* story dated September 20, 1914, quoted him:

I made up my mind if I do tumble out from the sky I do not want my final drop to stamp me as a piker. When it does come my time to bow to the scythe-wielder, I want the drop to be thousands of feet. I want the grandstands and grounds to be packed with a huge cheering mob and the band to be crashing out the latest rag, and when the ambulance or worse hauls me away, I want them all to say as they file out of the gates, *"Well he was certainly flying some"* [author's italics]. . . . But I know as sure as fate that some day there will be a tiny flaw in a piece of steel, just a little speck, but it will not take longer than a tick of a watch, and then it will

---

[11]*Bee*, September 11, 1914.

[12]*Bee*, September 17, 1914.

A GOOD TIME FOR EVERYBODY!

AK-SAR-BEN

FALL FESTIVAL

OMAHA

SEPT. 30 to OCT. 10, 1914

LINCOLN BEACHEY THE WORLD FAMED AVIATOR

The man who outflies the birds, will give flights daily, October 5, 6 and 7, both morning and afternoon—rain or shine

Electrical Parade, Evening October 7th.   Fraternal Parade, Afternoon October 6th

ON THE CARNIVAL GROUNDS—Afternoon and Evening

THE WORLD AT HOME

Twentieth Century Shows With Advanced Ideas.

SOME OF THE EXTRAORDINARY FEATURES

THE GARDEN OF ALLAH—A visualization of life and customs in Arabia.

CALIFORNIA FRANK'S WILD WEST AND INDIAN CONGRESS—The Passing of the West in stirring scenes and incidents.

THE PANAMA CANAL — The only working model officially endorsed by the U. S. government.

THE HUMAN BUTTERFLY — Omar Sami's East Indian Phantasy.

THE MARVELS OF THE UNIVERSE—A coterie of East Indian Magicians, Conjurers, Illusionists, Hindoo Fakirs and Necromancers.

MAZEPPA—The horse with a human brain.

ARMSTRONG'S 10 in 1.

ARMSTRONG'S FAT AND LEAN CONVENTION—Big and little folks in friendly rivalry.

MOTORDROME—Sensational contest between motorcycles on the largest portable saucer track ever constructed.

THE TANGO WAVE.

CARRY-US-ALL—Ten Thousand Dollar Riding Machine.

BIG ELI FERRIS WHEEL—Largest ever transported.

THE WORLD AT HOME PREMIER CONCERT BAND.

HOME COMING WEEK, Oct. 5th to 10th

Territorial Pioneers' Reunion, Sept. 30th to Oct. 3d

184

> be all over, that is if I stick to it forever. When I make
> $1,000,000, I quit.[13]

Remember he was an aeronaut, aviator and barnstormer. He flew a biplane pusher. Beachey was the toast of Omaha. The *Omaha Daily News* story from October 6, 1914, described him thus:

> Beachey himself is small physically and quite modest and retiring in spite of his big stunts and his large signature on the Henshaw [hotel] register. . . .
>
> He wears his regular street clothes while doing his hair-raising stunts, and even keeps his monster diamond pin in his necktie.[14]

Beachey was entertained royally by the Ak-Sar-Ben Governors during his stay and completely wowed the assemblage at a dinner at the Country Club Tuesday night. His brother and manager of the operation, Hillery Beachey, arrived in Omaha on October 5. The first thing he did was call Otto and Gus Baysdorfer because he knew they would be able to help him, and help they did.

The festival grounds were located on Harney, Howard and Jackson Streets from 16th to 20th Streets, which meant there was no place to fly from. Gus Baysdorfer suggested the Omaha Field Club golf course located on 36th and Woolworth. There was a long clear stretch that a plane could easily fly in and out of. The Omaha Field Club's directors were agreeable, so the special Curtiss Flyer was assembled in a tent just north of the clubhouse. Beachey arrived on Sunday and the Beacheys, Baysdorfers and George E. Yager had a swell get together. The Baysdorfers—Otto and Gus—assisted the mechanics and marveled at the Curtiss, which was really beefed up with many metal parts and a brand-new seven-cylinder, eighty-horsepower Anzani radial engine with a new carburetor system. Being around it for three days almost got them back into building aeroplanes, but better judgment prevailed.

The festival opened Monday, and Beachey captivated the thousands in attendance. The October 6, 1914, *World-Herald* exclaimed:

> Birdman Beachey gives thrilling exhibitions here[.] Great American Aviator Gives Spectacular Flights for Ak-Sar-Ben

---

[13] *Omaha Sunday Bee,* September 20, 1914.

[14] *Omaha Daily News,* October 6, 1914.

Visitors. Wonderful stunts of looping the loop[.] Flies twice
again today and tomorrow. . . .[15]

My mom and dad saw him fly on Tuesday. I was there, but
I don't remember it. Mom took pictures with her Kodak, but they
didn't come out. Dad said she was moving the camera too much,
trying to follow Beachey's flying. I have never seen a photograph of
him flying, but his looping, diving and flour-bombing the battleship
at 18th and Howard Streets certainly impressed people. The late
Bill Meyer of Elkhorn (the last veteran of World War I in western
Douglas County) and a couple of his buddies saw him fly. "He must
have looped a dozen times," Meyer told me. Charles Martin, Omaha
and Nebraska's well-known historian, saw Beachey takeoff and land
because his kindergarten teacher at Columbian Elementary School,
Ms. Hutchinson, took the class over to the field so they could see the
aeroplane fly. Martin told me the plane made a lot of noise, took off
towards the railroad tracks, flew back over their heads, and after a
while came back and landed by the tent. Beachey made a big
impression on Omaha in October 1914.

The only other flying in 1914 was when Brandeis Stores
sponsored young Michigan aviator Art Smith. His plane was
displayed in the famous corner window at 16th and Douglas, where
Smith gave lectures and autographed post cards for several days
during Thanksgiving week. Then on Sunday, November 29, he did
some great flying at the East Omaha Speedway.

On March 14, 1915, in San Francisco, the spectacular career
of Lincoln Beachey came to its end. He plunged to his death at the
Panama-Pacific Exposition in sight of over fifty thousand spectators.
Beachey entrusted his life for the first time to a new monoplane.
The machine was at an altitude of three thousand feet when he shut
off the power and dropped head-on for the earth. When the aviator
grasped the control levers to adjust the plane for the graceful
descent which had characterized his previous flight, the wing
crumbled like a collapsed umbrella, and the aeroplane twisted over
in its fall. Beachey was seen frantically trying to get it under
control, even holding one of the flying wires, but it plunged into San
Francisco Bay. Divers from the battleship USS *Oregon* recovered
the body from the wreckage. Dr. David E. Stafford, autopsy surgeon,

---

[15]*Omaha Morning World-Herald,* October 6, 1914.

expressed the opinion that Beachey was still alive when he struck the water and had sustained no major injuries except a broken leg as a result of the fall. His conclusion was that death was caused by drowning. *Well, he was certainly flying some!*

In the fall of 1914, a group of businessmen took over the board track at Carter Lake. The new Omaha Auto Speedway Company planned to completely rebuild the track and stands and make it one of the best in the United States. Bert LeBron was president and Adolph Storz was treasurer. The company expected to spend $100,000 on the facility. All of the work was completed in late June of 1915, and the first race was scheduled for July 5. The

View from grandstand of first turn at East Omaha Speedway

mile-and-a-quarter track was one of the best in the country. The turns were completely banked, and the surface was made of fir two-by-fours laid on edge. The builder, C. R. Vaughan, claimed it could take a hundred miles an hour. The covered grandstand would seat 15,000 people.

On July 1, 1915, the *Bee* ran the headline "Omaha Speedway invites the World on July 5th,"[16] and all the great drivers came, including Rickenbacker, who had become one of the best drivers in

---

[16]*Omaha Daily Bee,* July 1, 1915.

the country and always said he was from Omaha. A throng of thirty thousand (including my big brothers Ardon and Jack, who had sneaked in under the fence) saw Rickenbacker win the big race. The Speedway continued to be a success until the war, but it never came back again. I remember there were still footings and some boards left when I was a kid.

There just wasn't much flying going on in Omaha or Nebraska, but the military was seeing how the aeroplane was developing in the European conflict, and Glenn Curtiss built his famous JN-4D Jenny, which helped chase Pancho Villa around on the Texas-Mexican border.

About this time, the people of the United States were becoming aware of the horrible war going on in Europe. The papers were full of stories. They also carried marvelous photographs of French, German and British aeroplanes, which were impressive and made ours look sick.

The United States entered World War I in April 1917. Many Omaha and Nebraska boys signed up for the Army and Navy Air Service. A few of them became pilots. Only a handful made it to combat. The United States, where the aeroplane was invented, didn't make one tactical aircraft during the War. The closest we came was to assemble the British de Havilland DH-4, which was not a front-line tactical aircraft. We did make five thousand of them by the war's end. We made about ten thousand Curtiss J-5 trainers. Although that didn't do the country much good in the war, it started aviation in the United States after the war. We assembled a few British Handley Page 0/400 bombers and a few Italian Caproni three-engine bombers. We did make a pretty good engine, the Liberty twelve-cylinder, four hundred-horsepower monster that became the backbone of our military aviation until the mid-1920s.

We had about 150 pilots and fewer than 250 aircraft when we entered the war. All of the tactical aircraft flown by Americans were either British or French, and all of the tactical-pilot training was done in England or France. About 2,500 Americans received advanced training, most of them in France, and by April 1918 American fliers were in combat flying British and French planes. Under the command of Col. (later Gen.) William "Billy" Mitchell, Americans carried an ever-increasing share of the aerial offensive in the last six months of the war. By the early summer of 1918, American names were being added to the list of aces.

Eddie Rickenbacker enlisted in the Army in June 1917 and was immediately assigned as Mitchell's driver. He went with him to France in September. Eddie took one look at those French aeroplanes and thought, "Flying's for me." He took to flying like an old Red-tailed Hawk and graduated from tactical training in March of 1918. He was assigned to the U. S. 94th Aero Squadron under the command of Maj. Raoul Lufbery, the leading ace of the Lafayette Escadrille (a group of American college men who fought for the French before America got into the war). Rickenbacker, whom other pilots described as a big, tough-as-nails, roughneck, older race-car driver, stuck out like a sore thumb. He was humbled when he flew his first patrol over a real battlefield with Lufbery, the most celebrated aviator in American uniform.

When I instructed Single Engine Advanced (AT-6s) at Randolph Field, San Antonio, Texas, in World War II, we entered the pattern flying a Lufbery Circle, which was a tactical formation created by Lufbery when he was in the Lafayette Escadrille. I was pretty thrilled and a little awed to be flying a Lufbery because I knew who Raoul Lufbery was from reading about him in Driggs's book *Heros of Aviation* at the Northside Library (it used to be a church) on the southwest corner of 25th and Ames Avenue, when I was a kid.

Rickenbacker learned well. Two weeks after joining the squadron at Villeneuve les Vertus, Lieutenant Rickenbacker scored his first victory on April 29 against an Albatros. His sixth victory came on May 30. He did not fly in June, July or August because of a bad ear infection. In September and October, however, he shot down twenty more planes and ended the war with twenty-six to his credit. Nobody flew better, shot better or knew how to run an engine better. His hat-in-the-ring SPAD became almost as famous as Manfred von Richthofen's red-nosed Fokker. He shot down Fokker D-7s, Pfalz 12s and Albatros D-5s. Not bad for the car salesman and race-car driver who claimed he was from Omaha! He was now Capt. Eddie Rickenbacker, America's ace of aces and Medal of Honor recipient.

Nebraska's native son, Lt. Orville A. Ralston, became an ace with five victories. Ralston was born and raised in Weeping Water, graduated from Peru State and went into the Army Flying Service in May 1917. He trained in Canada and Texas and then was sent to England for tactical training.

Ralston was attached to the 8,559th Royal Air Force Squadron near Amiens in France in July 1918. They flew the famous Farnborough SE-5. His first victory came on July 24 and his fifth on October 3. Ralston took many photos and kept a great diary, all of which are now in the archives of the Nebraska State Historical Society in Lincoln. This diary entry from August 22, 1918, recounts a patrol flown during World War I:

> There is a rather thick ground mist today and a strong west wind. Our troops have been pushing along the lines here in several places. There are hundreds of Yankees going past by the trainload to the front.
>
> About 3:00 p.m. we start on patrol in squadron formation with an impossible strong west wind and a heavy

ground mist. We climb to 13,000 feet and go over the lines southeast of Amiens, then towards Peronne where we go north. We meet nine Hun Fokker biplanes with dark green camouflage and wide yellow bands around the fuselage. "Mack" is leading our patrol and we are the first to meet them being a little above them. They dive on our lower formation and start for home but we catch them and the usual half-roll and dive takes place by the Hun. I put a good burst into one that spins away then tackle another and follow him down to 3,000 ft. when he flattens out and I fire 50 rounds at fairly close range. Bits of fabric and wood fly away from him and his right top extension seems to fly in bits when he goes into a spin and crashes just northwest of Peronne. I have a narrow escape from the usual ground fire that the Hun "archie" and ground fire

Lt. Orville A. Ralston

put up, all kinds of flaming onions and tracer bullets go past me, but I get back O.K. Our bunch got four crashed and ten out of control. There is a terrible boom of guns all night.[17]

Ralston came home, graduated from the University of Nebraska Dental School, and practiced in Valentine and Neligh,

---

[17] Orville A. Ralston, "Diary," Nebraska State Historical Society, Lincoln, Nebraska.

192

Nebraska. He went into the Army Air Corps in World War II, as a supply officer, and was killed in the crash of a B-17 while he was "dead-heading" home on leave in December 1942. The plane was on a tactical training flight from Great Falls, Montana, to Ainsworth, Nebraska. Fly with the Hawk, Orville! Lt. Gerald K. Beem, from Omaha, was flying co-pilot. The cause of the crash was never determined. My friend Judson L. Hansen was given leave from Peterson Field, Colorado Springs, Colorado, to be a pallbearer at Beem's funeral.

After ferrying new planes to France and bringing beat-up ones back for repair, Omahan Jarvis Offutt (who was in Ralston's outfit all of the way and was his good buddy) transferred to a tactical front-line squadron in July. The first military aviator from Omaha to be killed in France, Offut died while flying an SE-5 during gunnery training near Valheureuv, France, on August 13, 1918. He was twenty-three years old. Offutt Air Force Base in Omaha, the headquarters of STRATCOM, was dedicated as Offutt Field in 1924.

Another Omaha native, Lt. Earl W. Porter, had a different, but distinguished, flying career in France. Porter closed his architect's office in the Peters Trust Building at 17th & Farnam in July 1917 and went to basic ground school at Austin, Texas. He was shipped to France in November and underwent five months training as a bomber and aerial gunner. Porter flew his first combat mission on May 19, 1918, in a SPAD M-X (a two-seat, scout-bomber version of the famous fighter). SPADs were built by the French government. The mission was a dandy: Porter's pilot got lost; they never found the target; the door the bombs dropped through wouldn't open; Porter finally reached down and got the four bombs and threw them over the side. Where the bombs landed, no one knows. His last mission was flown August 9. Porter's words were later reported by the *Omaha Daily News:*

I went up with Blake in "M-X" (the plane) on a bombardment on August 9, in the afternoon.

We flew northeast of Beauvais, then southeast so as to cross the lines with the wind. The clouds were very low and we flew over at 1,600 meters. We were the lowest in the formation and lost the rest of the planes before we crossed the lines. . . .

We were very careless in watching for Hun planes and all at once I felt something hit me in the neck and work around my body and lungs and chest.

The shock numbed me. . . .

When I looked around one Hun was diving on me from above and four climbing on the tail. I started to shoot and one of them went down in flames and another in a tail spin.[18]

Lt. Jarvis Offutt

Both kills were confirmed by ground troops, so Porter had two German kills to his credit. His flying career was over after thirty-two days in combat. The bullet that hit him entered his neck, broke his lower jaw, slid around his chest and lodged in his left armpit. He walked into the hospital. After the operation the doctor gave the bullet to him, and I am sure Porter kept all of his life.

Lieutenant Porter returned to Omaha and reopened his office at 596 Brandeis Theater Building in 1919. He became active in the Chamber of Commerce and the Aero Club of Nebraska.

Stuart D. Culley, son of Mr. and Mrs. Walter J. Culley, who lived at 4906 Underwood Avenue in Dundee, enlisted in the air service of the British Navy in 1916. He was trained in Canada and England, where he won his wings and was commissioned a lieutenant in the Royal Air Force in 1917. Early Sunday morning,

---

[18]*Omaha Daily News Magazine*, March 16, 1919.

August 11, 1918, word was brought that a German Zeppelin, the *L-53*, had been sighted near the Frisian Islands. Culley, the most experienced flyer, was dispatched in his Sopwith Camel to intercept and destroy the Zeppelin. The way he was launched is almost as exciting as his mission. The Royal Navy's first aircraft carrier was a lighter stripped of gear and engines and topped by a deck thirty feet long and twenty-four feet wide. The lighter was towed behind one of the navy's fastest destroyers, HMS *Redoubt*, at full speed into the wind. The Camel was fastened to the deck with special chocks that locked the landing gear to the deck, where it stayed when the Camel became airborne. Culley fired up his 150-horsepower Bentley Rotary, gave her a quick warm-up, opened her wide, signaled to release the gear lock, and prayed. The Camel lifted shakily on the ragged edge of a stall, wobbled out of the turbulence of the destroyer's superstructure, and finally started gaining altitude. (I'll bet a Tomcat carrier takeoff isn't any more exciting.)

When he reached ten thousand feet, the Nebraskan saw the Zeppelin a considerable distance away, headed for the German coast. He gave chase hoping to get above her, but the Bentley at full bore just couldn't get the Camel above the Zeppelin. Then for some strange reason, when the Zeppelin sighted him, it turned, heading directly at him. He was not able to climb above, so he attacked from below. He emptied his machine gun—all two hundred rounds—into the bottom of the airship. Spurts of flames leaped from various parts of the dirigible, and as Culley darted below she blew up, rolling the Camel on her back. The flame could be seen for fifty miles. The sky was full of wreckage that fell into the North Sea seven miles north of the island of Ameland. Culley returned to a navy land-based field and made his first-ever belly-landing successfully. His victory was confirmed by officers of the Royal Navy, and Culley later received the Distinguished Service Order for shooting down the last Zeppelin of the war. I think he should have been given two more medals—one for taking off from the lighter and one for the landing. Young Culley retired from the Royal Navy and lived in England until his death in the late 1960s.

I personally knew two other Omaha men who flew in World War I: Reed Davis (long-time Omaha realtor and the author of a book on his flying experiences) and Lawrence I. "Shorty" Shaw (general attorney for Northern Natural Gas Company). Shaw was piloting a flying boat at the Navy Station, San Diego, California, when the war ended. There must have been others that flew, but their flying did not get into the news. I'd like to know about them.

During the war, the Balloon Corps School at Fort Omaha was a going Jenny. The school was re-opened in November of 1916 under the command of Capt. Charles Chandler, the same Chandler who commanded the balloon at the Mid-West Aviation Meet in 1910. He was promoted to major and ordered to Washington on May 1, 1917, as the Chief Balloon Officer of the Army. On May 24, 1917, Maj. Frank P. Lahm, the veteran balloon pilot who won the First International Gordon Bennett Balloon Race starting from Paris in 1906, took command. This is the same Lahm who, as a lieutenant, flew the American balloon (alongside the Baysdorfer's *Comet*) in the Bennett Cup Race in St. Louis in 1907. He became the Army's first aeroplane pilot when Orville Wright soloed him in October 1909. I flew in a review for him in 1941 at Randolph Field when he retired as the Commanding General of the Army Air Corps Training Command. Lahm was soon ordered to France and his aide, Maj. H. B. Hersey, took the command.

Two Goodyear kite balloons were put into operation at both Omaha's North Field (Florence Field, the area west of 30th between Vane Street on the south and Forest Lawn Drive on the north) and the South Field at the Fort. The Second Balloon School Squadron was made up of four companies and was formed on September 20, 1917, under the command of Capt. John A. Pagelow. Its first squadron was shipped overseas on November 27, 1917.

Many more squadrons were trained, even after the war ended. My father-in-law, Lester L. Stepanek, served in the Balloon Corps at Fort Omaha. He told me he took one look at one of those balloons flying and immediately put in for chauffeuring duty. The only other man I knew that served at Fort Omaha was my high school physics teacher, Frank H. Gulgard. Of interest is the fact that one balloon squadron, the 61st, was trained at Fort Crook and shipped overseas on November 9, 1918. This fact, I think, is very significant because the headquarters of one of the biggest and best aeroplane operations in the history of flying, STRATCOM (formerly the Strategic Air Command), is located at Fort Crook today. The aeronauts were there first, just like I guess they always have been, along with the old Red-tailed Hawk.

BOOK FOUR:

Wild, Wooly but Starting to Fly
(1919 to 1925)

CHAPTER ONE:
FLYING NEBRASKA AFTER THE WAR

The "War to End All Wars" ended November 11, 1918, and the boys came home. Almost seven thousand of them now had flying in their blood, and about seven hundred of those aviators had gained combat experience. About ten thousand Jennys and a bunch of de Havilland DH-4s soon were declared surplus, and these aeroplanes began percolating down to the public in the post-World War I years.

Two Jennys had found their way to Omaha by 1919—Andy Nielsen's and Henry W. Ashmusen's. They were both flown from Ashmusen Field at 63rd and Center Streets. Then, in April 1919, Nebraska's Dr. F. A. Brewster of Beaver City became the state's first doctor to own an aeroplane to call on his patients. He bought a Curtiss civilian-version JN-4D-2 for $8,000 and hired Wade Stevens, former U.S. Army Air Service pilot and son of Furnas County Attorney John Stevens, to fly for him. Stevens may have been Nebraska's first commercial pilot. Brewster planned to use the plane on his

The Martin Bomber my dad
flew in in Indianapolis

long trips and his new Ford on short calls. Shortly before the plane arrived, Stevens announced, "We are going to give everybody an opportunity to have a ride in the clouds. Our charge will be $25 for a short exhibition flight; and you get your money back, if you don't come down."[1]

Even my Dad flew. He was the Captain of the Tangier Shrine Patrol that went to the national convention in Indianapolis in 1921. Because he was a Spanish-American War veteran, he got to ride in

---

[1]*Omaha Sunday Bee,* April 6, 1919.

a Martin bomber that had been built too late for the war. He took the picture above.

Some of Nebraska's flyers became quite famous, some were well-known, but some were just people who flew without much fanfare. One of the latter was Errold Grover Bahl. Errold Bahl, the aviator, was born and raised in Humboldt, Nebraska. He was a real "free spirit." After graduating from Humboldt High School in 1914, he and two friends took a canoe trip up the Red River to the north end of Lake Winnipeg in Canada. He attended the University of Nebraska for two years, then he joined the U.S. Army Signal Corps and took up flying. After completing flight school in Texas, Bahl was commissioned a second lieutenant in 1917. He served as an instructor until the end of World War I. This Nebraskan was a heck of a pilot, and he and two high school friends (Brookes Harding and Joe Zook) formed a flying company. They bought a surplus Standard and started barnstorming out of Lincoln. While they were busily flying and repairing airplanes in their shop at 107 North 9th Street in Lincoln, Bahl began working on an idea for a new type of plane.

Omaha's interest in aviation steadily grew, and flying often made the news. As a recruiting promotion, on November 8, 1919, Lt. H. P. Applegate, a Navy recruiting officer based in Kansas City, landed a Boeing float plane on the Missouri River between the Union Pacific Bridge and the Omaha Street Railway Company's Douglas Street Bridge. It took him two hours and forty-five minutes to make the trip from Kansas City, with stops at Atchison, Kansas, and Nebraska City, Nebraska.

At that time, Omaha had a very good chance at becoming the aeroplane capital of the United States: It lay right in the center of the country, and the Plains is great country to fly over. No other city in the United States had staked a claim to the title. So, over dinner in the Fontenelle Hotel on October 25, 1919, eighty-five of Omaha's business and civic leaders, World War I Air Service veterans and others laid plans to make Omaha the air center of the United States. Harley G. Conant, Chairman of the newly-formed Aerial Navigation Committee, declared the Chamber of Commerce would work towards that goal; Mayor Dahlman promised the full support of the city; and pioneer aviation enthusiasts W. A. Pixley and Gould Dietz spoke on the subject as well.

The group resolved to help establish airways, flying fields, mail- and passenger-service and aeroplane manufacturing plants in the city. It was decided to revive the Aero Club of Nebraska, last active in 1913 when Gould Dietz was its president. A group which included the lawyer Ralph G. Coad, aviation service veteran Earl Porter, and World War I veterans W. R. Coates, Sam W. Reynolds and Al Scott were charged with the latter task. The Aero Club of Omaha was incorporated in January 1920.

In Lincoln, Ray Page was putting together an aeroplane, and Warren P. Kite of Grand Island ran an advertisement in the December 5, 1919, *World-Herald* telling the world that he had Curtiss Orioles for sale. Aviation was picking up in Nebraska.

In 1908 New Yorker Henry W. Ashmusen had set up a plant on Long Island to study aeronautics, but he devoted most of his time to designing and building an aeronautical motor. He came up with a horizontally-opposed, eight-cylinder, air-cooled engine (about thirty years ahead of its time). He installed the engine in a Blériot owned by Stanley Y. Beach, aeronautical editor of the *Scientific American,* and it was flown very successfully by Horace Kemmerle in 1909. Ashmusen also installed one of his engines in George W. Beatty's Wright twin-propeller pusher. Considering how notoriously unreliable and underpowered aeroplane engines were in those days, Ashmusen probably had the best engine around. The Wrights and Curtiss, however, built their own engines and weren't about to give him the time of day, so he was pretty much on his own.

In 1911 Ashmusen built a tractor biplane with his engine in it. It had shock absorbers on its wheels, and the wings were covered on both sides. He flew this ship himself over the sand flats of Smithtown Bay, New York.

A pretty darn good aeronautical engineer, engine- and aeroplane-builder and pilot, Ashmusen was long on ideas and ability but short on money. So, the Moisant brothers, who were looking for a place to build their shop and a field for the International Aviators, made Ashmusen a deal. Charles Baysdorfer met Ashmusen in the fall of 1911 when he went to work in the Moisant factory, and they became good friends. Ashmusen built engines for the Moisants and installed them in the four Scouts built for the Stinsons. The Moisant aeroplane operation was very busy in 1912 and 1913 building, repairing and servicing aeroplanes, but the International Aviators operation came to a halt early in 1914.

One of Moisant's friends in Washington, D.C., was John M. Larsen (the same John M. Larsen who sold autos in Omaha and had become a real Washington lobbyist). At the time, Larsen was spending a lot of time in the Moisant factory. It was said that he had connections who were going to order fifty Scouts, an order which would have kept the factory busy for more than a year. This never materialized, so Alfred J. Moisant closed down his aeroplane operation. It was rumored that Moisant was in the munitions business (and 1914 was a real good time to be in munitions).

Ashmusen then went to work at the Curtiss plant in Hammondsport and helped build Jennys during the war. He and his wife Clara K. Ashmusen, who was from Omaha, spent Christmas 1918 with Clara's brother and family, who lived at 2114 Fontenelle Boulevard (exactly where Glenn Curtiss flew in 1910). While in Omaha, Ashmusen visited the Baysdorfer's shop and told them that he had the best engine around and wanted to start building them. Gus Baysdorfer (who was still a member of the Aero Club and had read about the recent decision to make Omaha the nation's air capital) suggested that he contact Dietz.

To make a long story short, a group of Omaha businessmen led by Dietz invested in Ashmusen's new enterprise. Ashmusen bought ground at 64th and Center Streets and built a small factory

Ashmusen brochure, 1919

for the manufacturing of aeroplanes and engines. There Ashmusen designed and started building the third aeroplane that really flew in Omaha. His *Bluebird,* when it was finally completed, was a darn good aeroplane.

Ashmusen's real claim to fame was his revolutionary engine—a high-grade, self-cooled aviation power plant years ahead of its time, which was manufactured in his plant at 63rd and Center Streets.

Unfortunately, Henry Ashmusen died of pneumonia in February 1920. The aviator was buried in Forest Lawn Cemetery and had the distinction of having the first fly-by in Nebraska at his funeral. R. W. Wagner, who was his chief pilot, flew overhead in a

Bluebird. Andy Nielsen, one of Omaha's best-known pilots during the twenties and thirties (even though he lived in Council Bluffs), rode with him and dropped several floral wreaths over the gravesite.

Ashmusen's wife, Clara Ashmusen, operated the company for several years, ran a flying school, and serviced the air mail planes. She closed the business when the Air Mail Service moved to Fort Crook in 1924. She was quite a lady—one of the women who watched the old Red-tailed Hawk.

Thor Bronderslev, another Nebraska birdman, was raised on a farm near Funk, Nebraska. He was a good mechanic and worked in the Ford Garage in Kearney, Nebraska. In 1920 Bronderslev bought a Lincoln Standard from Ray Page for $1,500. Actually, it was a surplus Standard that was made in New Jersey during the war and assembled in Page's factory in Lincoln. Flying lessons came with the plane, and his instructors were Edward V. Gardner and I. O. Biffel. Somehow Bronderslev survived and soloed after three

hours of instruction. He flew his Standard back to Kearney and started a flying service. He barnstormed around the countryside and became the first pilot to land an aeroplane in Ainsworth, Nebraska. He and his wife, the late Irene Bronderslev, once flew the Standard from Kearney to Lincoln in eighty-eight minutes. Bronderslev once owned a Longren built in Wichita, Kansas. I knew him in the 1930s when he was a pilot and mechanic at Omaha Muni Field (Honest, that's what we called it!).

CHAPTER THREE:
THE LARSEN
AND FETTERS EPISODES

John M. Larsen (who had met Eddie Rickenbacker while working for the Firestone-Columbus Automobile Company) was Alfred J. Moisant's friend and also knew Ashmusen. Larsen was building a thirteen-passenger all-metal monoplane in his plant near Philadelphia, and he called Ashmusen when he was having trouble with his design and engine. In November 1919, Ashmusen went back east to help.

Ashmusen encountered a very complicated situation in Philadelphia. Larsen was a real promoter who had somehow acquired twenty German Junkers, but he couldn't figure out how to get them together and flying. The Treaty of Versailles specified that the Germans could not build any more aeroplanes or fly them, so Larsen shipped one of the planes to Omaha for Ashmusen to assemble. Ashmusen had just started working on it when he became fatally ill. After his death, a couple of Larsen's men (who may have been Germans who came with the planes) and Ashmusen's mechanics continued the work on the plane.

Enter now Ashmusen's good friend Arthur H. Fetters, the Union Pacific mechanical engineer who designed the famous Union Pacific 9000 Type steam locomotive, the fastest and most powerful locomotive used by the railroad at that time. Fetters had an interest in the Ashmusen Company, and he had designed and was building a small biplane in Ashmusen's shop.

My dad fired with Carl R. Gray on the Union Pacific St. Joseph short-line from Grand Island when they were young men. When I was twelve, Gray was president of the Union Pacific Railroad, and he arranged for Dad and me to ride in the cab of a 9000 from Omaha to Grand Island. I was wavering between being a locomotive engineer or an aeroplane pilot at that time.

Unfortunately, I never knew Fetters. Fetters became acquainted with the Larsen men and helped them with the Junker-Larson JL-6, although he was in no way connected with Larsen. They probably helped him with the biplane too. Fetters's biplane, the *Bluebird Junior*, had a wing-span of

> only twenty-four feet as against the forty-three-foot wing-span of the air mail De Havilands [*sic*]. . . .

208

<blockquote>

The diminutive all-white Fetters plane, said to be the lightest biplane ever built, is planned by the builder to revolutionize the air industry by making possible an individual plane for the business man, practicable for limited used [*sic*] in the limited compass of his backyard. Groovings, and the use lightest wood and metals, make for maximum strength with minimum weight.... Preliminary engine tests have proven satisfactory.[2]

</blockquote>

The plane was first flown on September 5, 1920.

When the JL-6 was completed in May, Capt. Harold H. Hartney, Larsen's chief pilot, came out to test it. Hartney, an ace with seven victories, had been Rickenbacker's commanding officer in France and was a colonel then. He was a captain in the U.S. Army Flying Reserve in 1920. Hartney flew the JL-6 first on May 5. By June 1 they had worked out the bugs and flew it to Grand Island and back. On June 26 Captain Hartney, with Lt. Charles R. Colt as his mechanic and Larsen as a passenger, flew the JL-6 nonstop (that's what they claimed) from Omaha to Philadelphia, which would have ranked close to a world record.

By late July the Larsen Company had three JL-6s ready to try a New York to San Francisco flight. The Army was interested, and Larsen was hoping to blaze a transcontinental route for airmail and passenger service. Omahan Gould Dietz was a passenger in one of the planes; Larsen and Eddie Rickenbacker rode in the others. The pilots were Bert Acosta, S. D. Eaton and Hartney. The three ships arrived in Omaha on August 1, five days after leaving New York, with Rickenbacker flying one of them. It was rumored that he was financially interested.

The cross-country group was entertained royally by the Chamber of Commerce and Ak-Sar-Ben. From Omaha, Mr. and Mrs. T. J. O'Brien were going to fly as Larsen's guests to San Francisco. A large group of people assembled at the field Tuesday morning, August 3, to watch the departure, among them Dietz, Charles Gardner, Oscar G. Lieben, Henry W. Dunn and Albert Cahn.

The JL-6's big 185-horsepower Mercedes engine was fired up with Hartney at the controls and Lt. Charles Colt as mechanic. The O'Briens, Rickenbacker and Larsen were passengers. They taxied

---

[2]*Omaha Daily News,* June 30, 1920.

to the north side of the Chamber of Commerce Field. Hartney turned her into the wind, hit the throttle and the ship gathered speed. She barely cleared the fence. Spectators ducked as she skimmed over them, glided above Center Street, and plowed into Charles Jensen's house at 6507 Center, wrecking the house and the furniture inside. Fortunately, the Jensens were not at home.

Larsen, Rickenbacker, and the O'Briens smiled as they stepped from the cabin, as if nothing had happened. The damage to the plane was slight: a broken prop, smashed radiator and a dent in the left wing. According to Hartney and Rickenbacker, the Mercedes engine simply didn't wind up to the required speed, but there was some thought that too many people and too much baggage had fouled the flight.

Ready to board for California—Rickenbacker (second from left) and Hartney (on right)

Larsen wasn't disturbed by the crash, but Mr. and Mrs. O'Brien cancelled their trip to the Pacific Coast.

The plane was taken to the Ashmusen shop for repairs, where it remained for several months. The other two took off (without the

O'Briens) at 12:30 p.m. for North Platte. They arrived in San Francisco almost two weeks later, and both planes returned to New York in the first part of September.

Actually the Junker was a pretty darn good aeroplane, far better than anything we had in the United States at the time. It flew faster, carried more payload inside the cabin, and used much less fuel. It was made out of that newfangled aluminum, corrugated for strength. It's really too bad Ashmusen didn't live to get a chance to work on it.

Meanwhile, Errold Bahl still had his Standard and kept it at the Aviation Field located at 20th and High Streets, which was just west of where the Lincoln Country Club is today. I never found out whether Page was associated with Bahl in his enterprise, but I don't believe he was. Page was busy building Standards and running a flying school, and he was always capital-poor. Bahl did make a little money with the Standard, flew some parts for a Valparaiso, Nebraska, implement dealer, and may have run the first aerial delivery service in Nebraska. Unfortunately, Bahl banged up his Standard on a bad forced-landing while hauling passengers at a Fourth of July celebration at Crete, Nebraska, in 1920. They hauled it back on a truck to Page's shop in Lincoln, where Page fixed it in return for half-interest in the plane.

As mentioned earlier, Bahl had some good ideas for an aeroplane and talked to O. W. Timm about them. Like Otto Baysdorfer, Timm was a whiz at building aeroplanes. He built his first one in 1911 at his home in New Jersey before he came to Lincoln to show Page how to convert the U.S. Army Standard to the Lincoln Standard. It wasn't long before Timm and Bahl had designed a revolutionary parasol-wing, two-place, semi-monocoque-fuselage plane (which they called the *Lark)* and construction started. A monocoque plane has no internal or external bracing: All of the fuselage's strength comes from itself, like a pipe. Their *Lark* was one of the first, if not *the* first, monocoque fuselages made of thin plywood. Timm became the chief engineer for the Lincoln Standard Aircraft Company in 1921. Later he worked at Lockheed on the famous Vegas—all monocoque fuselages and full cantilever wings—and also designed the all-plywood Timm Army Primary Trainer (which I flew at Randolph Field during World War II).

Bahl and partners completed their *Lark,* and Bahl flew her to Shubert, Nebraska, late in the spring of 1920. Then in July, he flew her to Humboldt to show her off to the hometown folks. The *Humboldt Standard* described his visit in detail:

Plane Obliterates Distance[.] Makes Flight from Lincoln to Humboldt in Forty Minutes. In what is declared to be the lightest two-passenger monoplane ever built Errold Bahl "took the air" at Lincoln Tuesday evening and just forty minutes later was gliding across the landing field just west of town. This machine, jocularly dubbed the "jitney of the air," is built in its entirety with the exception of the engine—in the Harding, Zook & Bahl factory at Lincoln. It weighs but 600 pounds, is propelled by a three cylinder 60 horsepower engine and will travel three hours on a gallon of gasoline, or

*Errold Bahl and the **Lark** in Humboldt*

about twenty miles per hour. Its advantage in [t]his respect is known [since] . . . that from two to seven miles is average mileage attained by other machines on a gallon of gas. Last Saturday night Errold attempted an altitude ascension and rose to a height of 12,000 feet but found it too cold to go higher. Wednesday afternoon, after the machine had been exhibited on our streets, the big propeller was given a start and after gliding along the paving for about 150 feet the plane took the air and in ten minutes time had attained an altitude of 6000 feet, a feat never accomplished in the same distance and time by any other type of plane. Messrs. Harding, Zook & Bahl are pioneers in the aeroplane building

212

industry in Nebraska and since they are all Humboldt boys a lively interest is taken by the people of this city in the success of their venture.[3] Charles Marburger, who saw the take-off, said Bahl had her in the air at the west side of the square, wobbled a little, then headed for the blue. Today a pilot would be grounded, fined, have his or her card lifted, or maybe even thrown out of the country for pulling such a trick.

Harriet Long Stotts

Interestingly enough, Harriet Long Stotts from Barada, Nebraska, near Falls City, could possibly have been the first woman who soloed a plane in Nebraska. During the time she worked for Harding, Zook and Bahl in 1920, in Lincoln, Bahl gave her flying lessons. She had been hired to demonstrate the *Lark,* and her letters indicate she flew the plane during her lessons. Whether she actually soloed is still unknown.

---

[3]*Humboldt Standard,* July 2, 1920.

In the fall of 1920, the Crawford Chamber of Commerce staged the first air race in Nebraska. The main feature of the Dawes County Fair, the race began at Ak-Sar-Ben Field in Omaha and ended at the Dawes County Fairgrounds in Crawford, Nebraska. Its course ran via Grand Island and spanned a distance of 470 miles. The prizes were $1,000, $500 and $300.

The official starters in Omaha, appointed by Crawford Mayor Arah Hungerford, were Nelson B. Updike, Walter W. Head, Paul Skinner, Henry Doorly and Joseph Polcar. The three best aeroplanes in the state were entered. Early on the morning of September 16, 1920, before a crowd of almost a thousand, Updike fired his starter pistol. Clarence C. Lang, the government air-mail pilot who had established the record run between Chicago and Omaha, hit the throttle and took his plane into the air at 8:31 a.m. He was flying the beautiful little biplane ship designed and owned by Arthur H. Fetters of Omaha. It was powered with a revolutionary eighty-horsepower Ashmusen horizontal-opposed engine that had been installed just two days before. Warren P. Kite of the Grand Island Aero Company left the ground at 9:09 a.m. in a brand-spanking-new Curtiss Oriole, which his company distributed. It was powered by a new factory version of Curtiss's famous ninety-horsepower, eight-cylinder OX-5 engine. Edward Gardner (Bronderslev's instructor, mentioned above), from the Nebraska Aircraft Corporation of Lincoln, started at 9:12 a.m. in his red Lincoln Standard. Its mighty eighty-five-horsepower French Hisso engine rattled the air as he took off. A number of other planes had entered the race, but for various reasons it was impossible to get them ready (especially after their pilots had seen the three ships that were entered).

All three planes reached Grand Island safely. Gardner was in the lead, arriving in one hour and ten minutes. Kite followed thirteen minutes behind the leader, and Lange was third at one hour and thirty-three minutes. The pilots duly paid their respects to Grand Island and were off for Crawford. A crowd of eight thousand people were gathered at the fair grounds to see the finish of the race, as well as a ball game and horse races. People were there from as far north as Sheridan, Wyoming; Deadwood and Rapid City, South

Dakota; and from as far west as Casper and Lusk, Wyoming. They came from Valentine, Lincoln and Omaha, Nebraska; Denver, Colorado; and Sioux City, Iowa. It was a great day for the Pine Ridge country—cowboys, Native Americans, horses, cattle and the new birds (aeroplanes) were everywhere.

The crowd was buzzing; the air was electric. Then, a speck was spotted over Twelve Mile Butte—it must be a hawk; no, it's growing larger; it's a plane! Lange's beautiful Bluebird biplane roared over the fair grounds, circled the racetrack, then landed first at exactly 12:41 p.m. Edward Gardner arrived at 1:36 p.m., and Warren Kite landed at 2:04 p.m. They had flown from Omaha to Crawford in four hours and twelve minutes (pretty fast stepping for 1920)! Moreover, the first two aeroplanes to finish had been built in Nebraska, and only ten years had elapsed since Charles Baysdorfer flew the *Hawk*.

A short time later, the Pulitzer Race (the first major air-racing event held in the United States) was held at Mitchel Field, Long Island, New York, on November 1, 1920. Ralph Pulitzer and his brothers Herbert and Joseph, Jr., publishers of the *New York World* and the *St. Louis Post-Dispatch,* had received the official approval of the United States government and the aviation industry to award a Pulitzer Trophy each year for the fastest flight of an aeroplane over a controlled course. A $5,000 prize and a beautiful trophy said to be worth $10,000 would go to the winner.

Almost all the engine and aeroplane manufacturers in the world entered the event. U.S. Army Lt. Corliss Moseley won with his American-made Verville-Packard 600 racer and a world record speed of 178 miles per hour for the 132 mile course. Omahans Gould Dietz and his wife were guests of John Larsen at the first Pulitzer Race. They sat in the box of Alan R. Hawley, long-time aeronaut, aviator, charter member of the Aero Club of America, organizer of the Lafayette Escadrille in World War I, and vice president of the Aerial League of America. I can see the idea for the location of the Pulitzer Races zipping through Dietz's mind as he watched at Mitchel Field.

The aeroplane and aviation, in my opinion, really started growing up in the United States in 1919 when the Post Office Department got serious about establishing air-mail service. Very few hundred-mile cross-country flights had ever been made, but the Post Office was planning to fly mail from the Atlantic to the Pacific and from Canada to the Gulf. It was a huge and hazardous undertaking, but they had a large number of surplus de Havilland DH-4s and a whole bunch of guys who had the flying bug.

Of the first forty pilots hired, thirty-one were killed before the operation was turned over to private airlines. Remember the

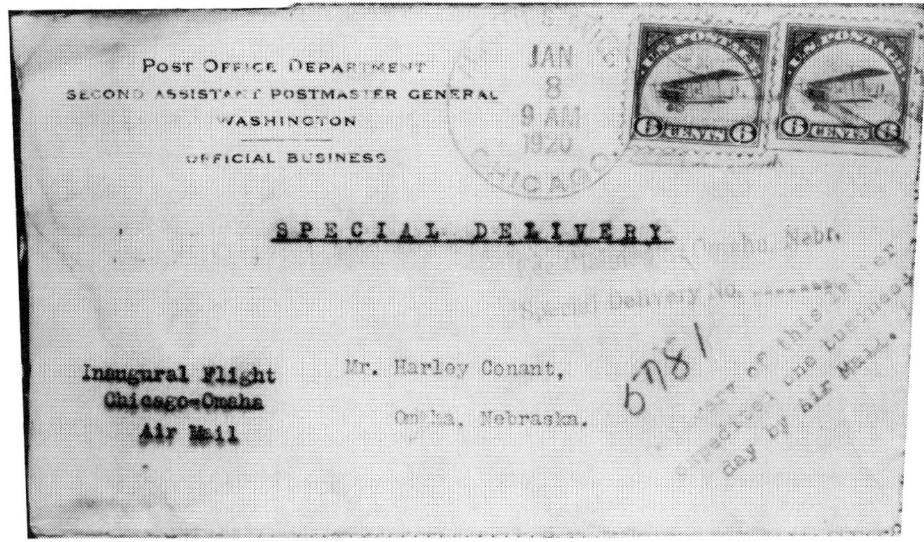

Envelope on first Chicago-Omaha air-mail flight

aeroplanes weren't the greatest; the pilots weren't really cross-country flyers; the fields were poor; and there were no marked airways, no radios and no blind-flying instruments.

By the fall of 1919, service was inaugurated between New York and Chicago. On January 8, 1920, pilot Walter J. Smith flew the first mail from Chicago to Omaha and landed at the field where Ak-Sar-Ben is today. That was the first air-mail flight west of Chicago. His time was three hours twenty-seven minutes. Harley

Conant, chairman of the new Aerial Navigation Committee of the Chamber of Commerce, received the priceless inaugural-flight envelope pictured above. Later his family presented it to the archives of the Historical Society of Douglas County.

On the same day, a consignment of mail was flown from Omaha to Chicago by pilot Farr Nutter. Both of these flights were preliminary ones, as a regular daily schedule wasn't inaugurated until May 15, 1920, when the Chicago-Omaha Division became the third link in the transcontinental system. Omaha became the only operating center outside of Washington, D.C. The air-mail field was located at 63rd and Center Streets, where the Ak-Sar-Ben racetrack is today. The field was situated just south of the newly-built grandstands (still there today) and ran from 63rd Street on the east to Papio Creek on the west and Center Street on the south. The center field of the race track was sometimes used as a landing strip, too. The Chamber of Commerce raised $50,000 to rent the field from Ak-Sar-Ben and to build a hangar.

A staff of pilots, mechanics and office workers moved to Omaha. Air mail was here in a big way, and Omaha loved it. The fourth and final link, Omaha to San Francisco, was completed on September 8, 1920.

The Great Pacific to Atlantic Flight

Omaha was right in the middle of what may have been the most important flying event in the history of the air-mail service. In the House of Representatives in Washington, D.C., congressmen were debating the Post Office appropriation bill, which contained an item of $1,250,000 for the air-mail service. The word was that it was hopeless. But the nay-sayers were thwarted by Otto M. Prager (third assistant postmaster general and highest official of the airmail system), ten of his pilots, a dozen ground crews, lots of luck and great help from above. They vowed to fly a load of mail from San Francisco to New York in one day. This meant flying almost three thousand miles cross-country in the daytime and at night with no lights.

On Washington's Birthday, February 22, 1921, at 4:30 a.m. Pacific Time, pilots Farr Nutter and Ray Little hit the throttles of the big Liberties in their DH-4s and took off from San Francisco carrying three hundred pounds of mail each. They landed at Reno,

Nevada, within four minutes of each other around 6:50 a.m. After the planes were refueled and serviced, William F. Lewis and Jack Eaton took off at 7:25 a.m. They landed at Elko, Nevada, at 10:15 a.m.

Lewis's engine had started running roughly about a half-hour before getting to Elko. The mechanics worked on the carburetor and seemed to have it running well, so he took off at 10:45 a.m. When he was about three hundred feet high and still over the field climbing, the Liberty burped, shot out a cloud of black smoke, and the DH-4 stalled. She hit the ground nose first in a cloud of dust and debris, and a small fire started. Lewis was dead, and some of the mail had been charred. If there isn't a marker there today, there sure as heck should be, dedicated to William F. Lewis, airmail pilot and World War I aviator. The mail was transferred to another plane, and William F. Blanchfield took off at 11:30 a.m.

Eaton had taken off about a half-hour before and didn't learn about the crash until he got to Salt Lake City, Utah. Blanchfield landed at Salt Lake at 1 p.m. There were two DH-4s at Salt Lake, so they divided up the mail. James P. Murray and Topecliffe Paine then took off at 1:20 p.m. on the toughest and longest run—over the Continental Divide and the rugged Medicine Bow and Laramie Mountains to Cheyenne, Wyoming. The weather was good and the visibility CAVU (ceiling and visibility unlimited). Murray landed in Cheyenne at 4:50 p.m., and Paine followed at 5:20.

Then things started getting hairy. After the three hundred pounds of mail was transferred to his plane, Frank Yager took off for North Platte, Nebraska, at 4:55 p.m. At 6:48 p.m., it was completely dark—except for a burning bucket of gasoline at the North Platte hangar's edge. Yager landed really hard and broke the tail skid off. The plane required over three hours to repair.

H. G. Smith took off for North Platte at 5:35 p.m. and landed at 7:15 p.m. He had a cup of coffee, refueled, and took off for Omaha at 7:55 p.m. Smith followed the railroad tracks to Central City, Nebraska, made a bee-line for Omaha, and passed over Wahoo, Nebraska, on his way. The night was ideal for flying. Smith said the lights on the Omaha field became visible shortly after he passed Wahoo. He also could see the Platte River, the lights of Lincoln, bonfires burning on the fields at North Platte and Grand Island, and many towns along the course. The pilot arrived at the Omaha field at 11:35 p.m. and made a perfect landing within the big electrically-lighted "U" laid out to guide the aviators. He planned to get some rest and take off for Chicago at daylight.

Back in North Platte, Jack Knight (who normally flew the round-trip Omaha-Cheyenne flight) resumed the flight after the tail skid on the first plane finally was repaired. His report said that he left North Platte at 10:44 p.m. It was cloudy, and the moon was shining intermittently. He had no plane trouble. He followed the course using a compass and occasionally looking at the ground. He arrived Omaha 1:10 a.m.

What a modest description for the first night cross-country flight Knight ever had made! Being an aviator, Knight added a little bravado to his landing by playing his usual "jazzy tune" with the Liberty for the benefit of the ground crew as he taxied to the hangar. About a hundred people stayed at the field to witness the arrivals. Mrs. Andrew Bahm, known as the mother of Omaha's Air Mail Service and who lived across Center Street from Ak-Sar-Ben Field, served lunch and hot coffee to the pilots.

Chicago had just advised Omaha that the weather was socking in there, and they were going to close down. When Knight heard this, he said that he'd continue, because one man had already died flying the mail, and he couldn't stop and let that man down. Superintendent Votaw agreed and gave his O.K. After refueling, three cups of coffee, a sandwich, warming up, studying the map (Knight had never flown the Omaha to Chicago run in the daytime, let alone at night in winter), and sending a telegram to his wife in Cheyenne that he was going on to Chicago, Knight took off at 1:55 a.m. Over Des Moines it started to cloud up, and he ran into snow. He flew for another hour and figured he was close to Iowa City. By then the visibility was rotten. His fuel was low. He *really* started looking for the field.

The Iowa City crew had shut down when they got the word that Chicago was closing down for the night. Fortunately, Iowa City's night watchman heard Knight's engine, ran out, and lighted four flares, which Knight used to make a landing. He stayed in Iowa City until about an hour before dawn and then took off, hoping Chicago would be clear when he got there. It was, and he landed at 7:45 a.m. He had just flown from North Platte, Nebraska, to Chicago, Illinois—almost eight hundred miles at night with no lights, no markers, no instruments and in a surplus open-cockpit aeroplane! A pretty good "Knight's work!"

W. C. Hopson took off at 8:20 a.m. for Cleveland and landed at 11:45 a.m. E. M. Allison, one of the first pilots hired by the Post Office in 1919, flew the last leg. He took off at 12:30 p.m. for New York. Finally, at 4:50 p.m. on February 23, 1921, Hopson made aeroplane and aviation history by landing at Hazelhurst Field, Long Island, with eight bags of mail that had left San Francisco exactly thirty-three hours twenty minutes before. Each pilot received a telegram from Prager commending and thanking him for his great flying. Knight was singled out as a hero, but he maintained any aviator could have done it.

The next day, the House of Representatives passed the Air Mail appropriation bill 221 to 111. U.S. Representative Albert W. Jefferis, from Omaha, championed the conference report and took heavy verbal fire from U.S. Representative Tincher of Kansas. (Kansas City had tried to get the air-mail route through there.) The "Fat Boy" from Medicine Lodge (Tincher's *World-Herald* nickname) poked fun at Omaha. He said the bill was really an advertisement for Nebraska's Gate City, and he thought $1,200,000 was entirely too much money to pay for a useless service. Jefferis took the high ground and discussed the appropriation from a national standpoint: America was interested in air navigation, and the vast benefits accrued from this enthusiasm would bring the west and east closer together. He impressed his colleagues in the House, and the bill passed two to one. Here's a congressman fighting for the good of his constituents! Wouldn't it be great if all congressmen did the same? Omaha was not only in the middle of the flying, it was in the middle of the politicking.

Designing the Air Mail Service

The Air Mail Service still needed better aeroplanes, as well as to establish marked and lighted air routes and fields. The appropriation furnished some money to start designing a beacon system from Chicago to Cheyenne, the section that would be flown at night. Joseph V. Magee was given the job of studying the problem and making recommendations. It took him a year to come up with a plan, but it was an excellent one. Terminals at Chicago, Iowa City, Omaha, North Platte and Cheyenne would be equipped with thirty-six-inch revolving lights on fifty-foot towers. The lights would sweep the horizon three times a minute with a beam visible

220

for more than a hundred miles in clear weather. Each of the emergency fields, about twenty-five miles apart, had eighteen-inch rotating beacons blazing from fifty-foot towers. The course along the entire airway would be marked at three-mile intervals with gas lights flashing 150 times a minute. The 902-mile aerial boulevard would have flood lights and boundary lights at the landing fields, and the service's de Havillands would have powerful headlights mounted on the wings. This was the beginning of our sophisticated airway system.

Getting better aeroplanes was another matter. The DH-4s were continually improved with modifications and changes, but they were still old with limited payload and range. A test model of a Twin DH was constructed with two engines. It looked great, had more payload and flew faster. Fifteen of them were ordered. The first one came apart in the air on its first flight and crashed. Pilot Walter Stevens was flying one to Cleveland, when he made a turn, and the right wing snapped off. Stevens survived, but the plane was demolished. A few days later, pilot Oscar B. Santa Maria was cruising at two thousand feet when the left engine vibrated so badly it fell off, and the plane came down in an uncontrollable spiral. Santa Maria was not hurt badly, but that was the end of the Twin DHs.

Larsen—Part Two

Here come John M. Larsen and the German Junkers (his JL-6s) again. As mentioned earlier, three JL-6s flew to Omaha, and two of them went on to San Francisco. The two made it back to New York, and for the bargain price of $200,000 the Post Office acquired eight JL-6s and a bunch of parts. On an August 31 flight from Chicago to Cleveland, one crashed and burned in a cornfield. Pilot Wesley Smith survived but was badly burned. The next day pilot Max Miller and mechanic Gustav Reirson were killed in a crash near Morristown, New Jersey. The JL-6s were grounded but put back in service again on September 5. Eventually the JL-6s flew mail regularly from New York to Chicago and back.

The first flight of a Junkers-Larsen JL-6 to Omaha was made by J. Titus Christiansen from Blair, Nebraska. He landed at Ak-Sar-Ben Field at noon on September 13, 1920. When the Cheyenne mail arrived at 2:00 p.m., it was flown to Chicago by M. L. McMullen in the all-metal plane.

The next day, pilot Walter Stevens and mechanic Russell Thomas were killed near Pemberville, Ohio, when their JL-6 crashed in flames. On September 18, all JL-6s were permanently grounded, and DH-4s went back to flying the mail.

The benzol, which the JL-6s burned, ate rubber, so all its fuel lines were ridged tubing. This often cracked from vibration and caused great leaks which starved the engine and caused horrendous fires. Ashmusen seemed to have remedied this: The one he worked on and the two that were in Omaha in June didn't seem to have this problem. The JL-6s had all-metal construction, were faster and

would carry several times the payload at much less cost than a DH-4. It was too bad the fuel problem wasn't solved. Though Larsen should have been held accountable, he really was a promoter with obvious good government connections and not an aeroplane builder or pilot.

Even America's premier ace, Eddie Rickenbacker, took part in one day of the transcontinental flight. On May 26, 1921, he left Redwood City, California, but lousy weather turned him back an hour and twenty minutes after his first start at 5 a.m. He took off again at 8:32 a.m. and hoped to get to Washington, D.C., by the next night. Flying an Army de Havilland, Rickenbacker finally made it to Cheyenne thirteen hours and twenty minutes later. His flight of almost fourteen hundred miles was a world-record nonstop flight. Unfortunately, after a perfect landing, Rickenbacker hit a small ditch and flipped the DH-4 on its back. He spent the night and most of the next day in Cheyenne. His ship couldn't be repaired right away, so he hitched a ride with Omaha mail pilot C. V. Pickup. They arrived in Omaha at 11:53 p.m., and Rickenbacker left for Chicago at 1:15 a.m., riding in the mail compartment of W. C. Hopson's mail plane. He finally got another Army plane from Chanute Field, Illinois (General Mitchell made the arrangements), and landed at Bolling Field, Washington, D.C., at 6:15 p.m., Saturday, May 28, fifty-eight hours after he had left California. Now you can see what a great accomplishment the thirty-three hour twenty-minute flight in February really was.

The Pilots Were the Backbone

Somehow the Air Mail Service made it through 1921. In between inventing air-mail service, politics, aeroplane problems, more politics, learning to fly cross-country for three thousand miles day and night, more politics, getting the mail between the post offices and the airports, and just plain ignorance, it started to grow.

The most important asset the Air Mail Service had was its pilots. Omaha loved the pilots. They were on the front page of the *Bee*, the *World-Herald*, the *Omaha Daily News* and the weeklies, and not always for killing themselves. Almost every pilot that flew the mail, no matter where, was based in Omaha once or twice during his career.

Omahans often went out to the air-mail field to see the planes. I was six when I saw my first aeroplane, a beautiful de Havilland landing with the Chicago mail. I think my mom and dad were as hooked as I was because we drove out to Ak-Sar-Ben many times, in spite of the fact that driving a Model T from 24th and Manderson to 63rd and Center could be almost an all-day project. I remember my dad fixing a flat tire when the wheels had clincher rims. But I got to see the de Havillands, and once I saw a JL-6.

The early Air Mail Service had its tragedies, too. The May 5, 1921, *Bee* had front page stories on two. Walter Bunting had survived a bad crash in Omaha back in December, but was killed at Rock Springs, Wyoming, that morning. Blair, Nebraska, native J.

Titus Christiansen (the first pilot to fly a JL-6 into Omaha) was killed on May 1 as he tried to land at Cleveland Airport in a fog. He really was a hero. When Christensen broke out of the fog, he was headed directly into a row of houses. He booted rudder and stick and crashed his DH-4 into a street, to miss the houses. Eight thousand people—it must have been the whole town of Blair—along with a great fly-by of seven aeroplanes by his mail-pilot buddies and almost all of the pilots in Omaha attended his funeral. Members of the Stanley E. Hain American Legion Post headed the procession to the cemetery, and Reverend P. H. Vig of Dana College preached in Danish. At the command of Maj. Reed O'Hanlon, a squad fired three volleys over the grave and a musician blew taps. Christiansen's widow fainted at the grave just after she had dropped a silk American flag upon the casket. He sure went out in a blaze of glory.

On September 1, 1921, the Omaha Aircraft Company opened its doors at 1120 North 19th Street in a building built for Omaha Van & Storage Company (Bekins) by Peter Kiewit in 1897. The building is now owned by Micklin Lumber Company. This gave Omaha its second aircraft-manufacturing company and a foothold on becoming the air capital of the country.

This story started when Reed E. Davis saw an advertisement in the June issue of *Aerial Age* magazine. I have a 1921 copy of the magazine given to me by Dr. James W. "Doc" Wengert, psychiatrist at the Omaha Veterans' Hospital and historian. An advertisement had been placed in the Investors Wanted column by Giuseppi M. Bellanca, who had been born in Milan, Italy, in 1886, and had engineering and mathematics degrees from the Instituto Tecnico in Milan. Bellanca had been a United States citizen since 1912 and served as an engineer at the Wright Company and the Curtiss Company during World War I. Bellanca was seeking investors to build a revolutionary new "cabin" airplane.

Davis was an army flyer. He flew his own Jenny and worked on the *Bluebird* that the Ashmusen Company was building. Davis was also in the real estate business. He was always active in the Army Flying Reserve and recruited me into the Army Air Corps in 1941. I had a commercial license and an instructor's rating at the time. Davis had some very good contacts and found a number of people willing to invest in the aeroplane-building business. Bellanca came to Omaha and formed the Omaha Aircraft Company. James W. "Doc" Elwood (owner of the very successful Northwest Taxidermy School) was its President, and Dr. Charles W. Pollard, a prominent obstetrician, was secretary/treasurer. Bellanca told the investors that he could build his four-place cabin plane for $5,000. Construction started at the 19th Street plant with E. T. Richards and Davis in charge.

Bellanca had designed a great aeroplane—a four-place, high-wing, cabin-monoplane. It had a very innovative design and several new ideas built into it. The plane's wing struts were small airfoils that contributed lift. He incorporated some monocoque structures, and it was strong, light and clean. Powered by a ten-cylinder, 105-horsepower, radial French Anzani engine, Bellanca engineered it to carry four people at one hundred miles per hour and burn five gallons of gasoline. By January 1922, it was over half done. Then a major problem developed: The company ran out of money. The owners finally kicked in another $5,000. Now the Anzani could be ordered, and delivery was promised in thirty days. Work progressed, the Anzani was installed, and she was ready for doping and painting early in April. Around that time, they ran out of money again. This time "Doc" Elwood and his partners had had enough of this

The Roos-Bellanca CF-1, Ak-Sar-Ben Field, June 1922

venture and sold their interest to Victor Roos, who was involved with Nebraska aviation for many years afterwards.

The Roos-Bellanca Airplane Company moved the factory to Roos's building at 723 South 27th Street. Victor Roos had finished pilot training at Kelly Field, San Antonio, Texas, just as the war ended. He was a very successful motorcycle and bicycle dealer with a shop on the southwest corner of 27th and Leavenworth Streets. My dad knew Roos. I think they were Shriners together, and my dad bought me my first bicycle from him. Boy, was I thrilled with my Roos Flyer—it was sort of like having a Cadillac!

The Roos-Bellanca CF-1 was trucked out to the Ak-Sar-Ben Field for final assembly, checking and ground testing. On May 1, 1922, C. L. Bowen, chief pilot of the Ashmusen Company, flew the CF-1. The ship performed even better than hoped. It was probably the best in the United States at the time. It was flown at air meets and shows in Nebraska, Iowa, Missouri, Illinois and Indiana during the summer of 1922. They got orders for two and set up to build them at the 27th Street plant. Bellanca lived at the Brown Apartments at 508 North 21st Street and married Mrs. Brown's daughter.

Meanwhile, on September 6, 1921, Martha Grace Bushman died of injuries suffered in an aeroplane crash in a Jenny flown by her husband Francis, an ex-army pilot. The crash occurred August 6 at Ak-Sar-Ben Field.

Vern Treat's spectacular loops at Ak-Sar-Ben

She was the first woman to be killed in an aeroplane crash in Nebraska.

But this didn't kill interest in aviation in Nebraska. At the night show of the Ak-Sar-Ben Fall Festival on September 17, aviator Vern Treat of the famous Ruth Law Flyers attached flares to the wings of his plane and flew the five consecutive loops pictured above. He thrilled a crowd of fifteen thousand.

CHAPTER SEVEN:
THE INTERNATIONAL AIR CONGRESS

Probably the biggest and most important aviation event that Omaha has ever undertaken was the brain child of a small group of men. Earl Porter, Ralph Coad, Sam W. Reynolds, Arvid R. Almgren and W. R. Coates—all World War I veterans—were the officers of the new Aero Club of Omaha. Porter was their leader. Coad was a lawyer; Almgren worked for Milton Rogers & Sons wholesale hardware; Reynolds was in the Reynolds-Updike Coal Company, a whale of a tennis player, and became one of Nebraska's greatest golfers; and Coates sold insurance for New England Mutual. All of them loved aeroplanes and flying and thought Omaha should be the air capital of the United States.

They met one noon in February 1921 at the University Club on the north side of Harney Street between 19th and 20th (the building is still there) and made plans for an International Air Congress to be held in Omaha that fall. Then they took their idea to the Chamber of Commerce. Though the board listened, it decided that because of the post-war recession it was not the time to try such a large undertaking. Some of the older heads encouraged them though. Gould Dietz told them about the Pulitzer Race he watched the previous year and gave them Alan Hawley's address. Harley Conant was interested. Clarke Powell and W. A. Pixley told them to get some more information and get back to them. Arthur Fetters thought it was worth looking into. The young guys went to work anyway, even though they really didn't have much money and couldn't raise close to the estimated $30,000 necessary.

The Aero Club of America was very agreeable to such an event. Clara Ashmusen, Fetters and the brand new Omaha Aircraft Company contacted the Aircraft Manufacturers Association, who thought it was a good idea.

The Pulitzer Race was a much bigger problem. The 1921 race was already committed to Detroit. Hawley talked to the Detroit people about changing, and they said they would talk it over and let him know. Back to the Chamber they went, but the Board was still not willing to become involved in such a big project. Times never change: I recall several times when I was on the Board of the Omaha Junior Chamber of Commerce in the late thirties and early forties that the Senior Chamber turned us down, in spite of the fact

230

that Sam Reynolds, Willard D. Hosford and some of Omaha's most prominent were on that board.

Then things started looking up. Detroit said they would give up the Pulitzer Race if they could have it in 1922. The Aero Club of America gave its approval and donated $1,000. The national convention of the American Legion (formed by World War I veterans and fast becoming a major organization with a huge membership and political clout) was to be held in Kansas City. Omaha Post Number One had the largest membership in the United States. All the big guns of World War I would be at the convention, and they were going to have a huge air show. Sam Reynolds was on the convention committee and guaranteed that he would get most of the planes and flyers to Omaha for the Congress. Our old friend John M. Larsen said he would sponsor a race for passenger-carrying planes with a large money prize and trophy. He also got the Aircraft Manufacturers Association to donate $1,000.

Armed with all of this, the Aero Club tigers invited all of the big wheels and business leaders of Omaha to a dinner meeting at the Fontenelle Hotel on July 18, and they did a heck of a selling job. Twenty-one businesses and organizations endorsed the Congress.

An advisory board of two representatives from each organization was formed. The budget committee consisted of Willard Hosford, Clarke Powell, Everett Buckingham, Clarence E. Corey and William F. Bruett. A financial campaign to raise $35,000 would be conducted. All profits were to be put toward the purchase of a municipal landing field for Omaha. The Congress was off the ground!

The world received an official announcement of the International Air Congress to be held November 3-5 when full-page advertisements were published in both the *Bee* and *World-Herald* on Sunday, July 24, 1921.

On August 13, the Aero Club of America officially announced that the 1921 Pulitzer Race also would be held in Omaha. The winner would receive $3,000, second place, $2,000, and third place, $1,000. The winner also would receive the Mario Korbel-designed Pulitzer Trophy, which stood four feet high and was made of silver. Holding the Pulitzer Races in Omaha meant that the leading flyers of the world would enter the air meet. On August 18, the *Chamber of Commerce Journal* declared that every business and civic organization was working to bring the Congress to Omaha and that

An Omaha Bee advertisement for the 1921 Congress

the project had the unanimous support of its own Executive Committee.

In preparation for the Congress, Omaha's City Council unanimously voted a $21,000 appropriation to build a covered sewer through the flying field at 20th and Reed Streets. This Minne Lusa sewer never did get covered—it's still open today. The field consisted of 106 acres of flat land with the river on the east and the railroad on the west. It was just off the end of the 24th Street paving just south of where the Omaha Public Power District's North Omaha Power Plant is today. The Holt Manufacturing Company donated the services of ten- and twenty-ton tractors, and local Ford/Fordson agents gave the use of trucks and tractors for the grading project.

Entry blanks for exhibitors and race events were mailed out before September first, according to officials of the Aero Club. Many acceptances for the Congress were received from prominent birdmen. George

H. Joslyn's Western Newspaper Union, the business that made him a millionaire, printed editorial and feature stories on the event in practically every weekly newspaper in the West. The articles noted that Omaha claimed to be the most progressive aviation center in the country. Omahans had become very knowledgeable about the various types of airships and aeroplanes and knew the proper terms for each type.

It was a mighty effort which involved the entire business and industrial community and, it seemed, most of the citizens. Porter turned over his architectural business to his partner and devoted full time to the Congress. An office force mailed out thousands of invitations and pamphlets to airmen and celebrities such as President Warren G. Harding, Marshal Ferdinand Foch, Orville Wright, Glenn Curtiss and Judge Kenesaw Mountain Landis. The graduates of the Fort Omaha Balloon School, the center of America's wartime ballooning, were invited for their first reunion. Squadrons and escadrilles of flyers were asked to hold their first reunions in Omaha. The fifty-two American aces were invited. No one interested in aviation was left out. Kansas City's aid was enlisted to get the American Legion Convention delegates to come to Omaha after the convention. All of the railroads offered a fare-and-a-half rate from all parts of the country to Omaha. John Gutzon Borglum, world famous sculptor of Mount Rushmore in South Dakota, desinged a commemorative medal symbolizing the work of the American airmen during the war. Borglum's brother August M. Borglum, taught music and was Omaha's premier critic for years. His son, George Paul Borglum, taught French at Omaha University, and his wife, Imogene Borglum, was the prettiest teacher I ever had. She taught my fifth-grade class at Lothrop School.

Even with all of the tremendous enthusiasm, there was a lack of money. The total cost was estimated to be more than $50,000. International Air Congress stamps were sold, and posters were sent to cities everywhere. There seemed to be no end to promotions, publicity and beating the drums for the Congress. By October 1, there were almost a hundred entries in the various flying events, and the three-day program was released.

The Pulitzer Trophy Race

All pilots in this first event had to have an aviator's license issued by the Federation Aeronautics Internationale and be entered upon the competitor's register of the Aero Club of America. The race was a free-for-all for high-speed aeroplanes. One-hundred-fifty miles—five times around a closed course of thirty miles from North Field northwest to a captive balloon on a railroad track north of Fort Calhoun, then east to a captive balloon on the southern outskirts of Loveland, Iowa, returning to North Field. The Pulitzer Trophy would be the permanent possession of any flyer winning it two years in succession.

Event number two, set for 3 p.m. Thursday, was an aerobatic contest with cash prizes: first was $250; second, $150; and third, $100. It was a free-for-all contest for all types of aeroplanes. The contest was decided on points: Immelmann turns, fifteen points; barrel rolls, fifteen; "falling leaves," twenty; loops, twenty; vertical reversements, fifteen; tail spins, fifteen. I'd give anything to have gotten into that contest driving a Stearman, Waco or Fairchild!

Event number three, the Larsen Trophy Race, set for Friday at 10 a.m., was a derby for all types of commercial planes. The distance was 250 miles. Starting at Omaha Field, contestants flew to Des Moines, Iowa, landed on Curtiss Field, and returned to Omaha Field. Prizes were: first, $2,000; second, $1,000; third, $500. The contest was decided on points. The winner got a leg on the Larsen Trophy, said to be worth $5,000. Event number four, set for Friday at 10:45 a.m., a free-for-all race with $175 in prizes and was open to JN-4Ds, OX-5 Standards, Orioles with Curtiss OX-5s, Canucks and other planes with a speed of sixty to seventy-five miles per hour. The distance was ninety miles. Event number five, held Friday at 1:30 p.m., was a free-for-all race of ninety miles with $175 in prizes, was open to planes with speeds of seventy-five to ninety miles per hour. Event number six, set for Friday at 3:30 p.m., was a parachute-jumping contest with prizes amounting to $350. The jump was from one thousand feet or more, and the winner was the one landing closest to a circle on the field.

Event number seven, set for Saturday at noon, was a race for the Omaha Aero Club Trophy and cash amounting to $2,625. It was a closed handicap, open to all machines. The distance was 150 miles. Event number eight, set for Saturday at 2:30 p.m., was a bombing contest open to Army and Navy planes only. The first prize was a gold cup; second, a silver cup. The Congress also featured a half-mile row of all types of aeroplanes lined up in front of the Grandstand. The Grand Finale was "The Bombing of Courcelay," with one hundred costumed people in the cast and a model of a French village set up on the field.

The promotion was going great, but the money was not coming in. The preparation of the field was progressing very slowly, too. More than one hundred trees had been removed, but the grading was way behind schedule. The Minne Lusa sewer wasn't covered, and work on the stands hadn't started. It rained most of the first two weeks of October. Entries for all events were coming in faster than they could be processed. Still, tickets, stamps and pledges weren't filling the coffers.

On October 25 the young Turks of the Aero Club knew they weren't going to make it, so they crawled to the Chamber of Commerce Board again. Earl Porter resigned, hoping the Chamber would appoint a committee and take over. The Chamber Board scheduled a meeting at 10 a.m. the next day. Afterwards, the paper wrote,

> City Commissioner Joe Koutsky today declared, that if the Chamber of Commerce backs down in supporting the congress, he will resign his membership. "The men of the Aero club have devoted their time and money to promoting this event," he said. "It would be a decided black eye, not only to Omaha, but to them, if it fell through. . . . The congress has been advertised all over the United States and, in fact, all over the world.[4]

The old boys came through. The Executive Committee resolved at a special meeting Thursday, October 27, 1921, that it would be best to appoint a committee to work with a committee of the subscribers to help raise the additional funds needed for the Congress. It was thought that after all the advertising that had been done it would be an embarrassment to the city not to press forward with the effort to bring about the Congress.

They authorized a donation of $1,500, which reminds me a little of my dad bailing me out a few times. Did they ever put together committees! After all, the meet was only a week away. The Aero Club's committee was R.C. Tooke (Chairman), and members Ed Deeds, G.H. Sopp, Earl Porter and C.B.D. Collyer. The Chamber named Guy C. Kiddoo (of the M.E. Smith Company) General Chairman. The Executive Committee consisted of Randall K. Brown, Robert Trimble, Joseph Barker and Dr. E.C. Henry. From the original guarantors came Harley Conant, Gould Dietz, Nelson Updike, C.E. Berry, George Brandeis and Ward Burgess. Burgess was the chairman of the new Finance Committee. Charles Gardner and Charles Trimble of Ak-Sar-Ben were put in charge of running the meet, ticket sales and programs. Arthur G. Thomas (of the Stroud Little Red Wagon Company) was appointed director of publicity. Arthur Fetters of the Union Pacific was named

---

[4]*Omaha World-Herald,* October 26, 1921.

Transportation Chairman to supervise the running of shuttle trains from downtown to the field. Railroads would give cut-rates for the meet. Fetters, because of his involvement in aviation with Clark Powell, also served on the contest committee. The two worked with Maj. Ira A. Rader, representing the Aero Club of America, in finalizing the program of the Congress. Gould Dietz was in charge of entertainment. Harley Conant handled hotel reservations. Nelson Updike and T.P. Redman served as committees-of-one to invite Gen. John J. Pershing and Marshal Foch of France. They were at the American Legion Convention in Kansas City the first of the next week.

It seemed everyone who was somebody in Omaha—and a lot of just plain people, too—got in on the act, and things started coming together. Money came in; the field grading was completed; stands were erected; tickets were sold; and aeroplanes and airmen started arriving. C.C. Muentfering and his crew, assisted by engineers from Fort Omaha, surveyed the triangular course of exactly thirty miles for the Pulitzer Race. Omaha had "Air Congress Fever" and the fuel to keep it burning.

On Sunday, seven hundred Omaha Legionnaires detrained at Kansas City wearing International Air Congress badges. The committee in charge included Harry C. Delamatre, Jacob J. Isaacson, Hird Stryker, Sam Reynolds, Harry C. Hough, W. Pruett, John Kilmartin, Earle E. Kiplinger, P. Boyle, Walter S. Byrne, Claudio Delitale, J.C. Van Avery, Harry S. Byrne, George Carroll, C.C. Chapman, J. Milota, E. W. Sears and William S. Ritchie, Jr.

The program of the Congress was presented from the floor of the Legion Convention under a forty-foot banner by Hird Stryker, William Ritchie, Jr., and Sam Reynolds. The advertising worked, and many aeroplanes flown in Kansas City's Legion Air Meet came to Omaha afterwards, along with a bunch of Legionnaires.

By Monday of Congress Week, $49,000 of new backing had been raised, and Ward Burgess felt confident that $8,000 to $10,000 more would come in. That solved the money problem. Ed Peterson, a well-known Omaha railroad contractor, did the final grading of the field, and the pilots said it was just about the best they had ever flown from. Stands for five thousand were erected; two hangar tents were put up; and supplies of gasoline and oil were trucked to the field. The pylons were erected at Fort Calhoun and Loveland. The field and flying facilities were ready. Nine aeroplanes landed that day. S. S. Bradley, General Manager of the Aircraft Manufacturers Association, arrived, as did B. D. Thomas, designer of the Thomas-Morse airplanes, and Raymond Ware, Secretary/Treasurer of the company. The Curtiss Airplane Corporation group, led by Frank H. Russell (Vice President) and pilots Acosta and Clarence Coombs, along with several mechanics, were busy checking the field and making final adjustments to their bullet machines. A brand-new Curtiss Oriole that was donated to the Aero Club of Omaha for their disposition was also on the field.

Omaha's cowboy-mayor, James Dahlman (the same "Cowboy Jim" who had opened the 1910 air meet), proclaimed International Air Congress Week and asked all businesses to decorate their fronts with red, white and blue bunting and flags. Stores promoted the Congress in their advertisements, the *World-Herald* and the *Bee* gave away tickets as a promotion. Films were made by Harold F. Chenoweth, pioneer Omaha moviemaker. I had the pleasure of knowing "Chen." He filmed the Pulitzer Races and showed me the film once, but I don't know what happened to his film since.

On Tuesday, November 1, the Congress hit Omaha. Three Larsen all-metal monoplanes arrived. Reverend Edgar M. Brown (Omaha's "Flying Preacher" and pastor of Dietz Memorial Methodist Episcopal Church) and his old friend Augustus Post of New York (Secretary of the Aero Club of America) were passengers in one ship. Maj. Ira Rader, chairman of the contest committee, named the following officials: Referee was Caleb Bragg of New York City, Vice President of the Aero Club of America. Starter was R.W. Schroeder. Timers were Captain Labell of New York City, Sgt. George J. Emery and Karl McMullin of Omaha. On the technical committee were Arthur Fetters and Giuseppe Ballanca of Omaha, Glenn Martin of Baltimore (the same Glenn Martin who later would operate the Bomber Plant at Fort Crook during World War II) and Chance Vought of New York (the same man who later would build the Vought F4U Corsair fighters I had the thrill of flying several times during World War II). Judges were Reed G. Landis of Chicago (son of Judge Kenesaw Mountain Landis), Maj. J. J. D. Reardon of Fort Omaha, and Capt. R. A. Grear of Fort Crook, assisted by the following Omahans: Sam Reynolds, Frank Sheehan, Earle Kiplinger, Ralph Coad, D. T. Quigley, E. D. White, Lt. R. H. Finley, Clarke Powell, Guy L. Smith, T. W. Haynes, Victor Roos and my dad, George Adwers. Rader announced that a spotter plane, owned by

Fetters, would patrol in the center of the triangular course during each race, spotting for trouble and making sure the planes stayed on course and flew around the pylons.

By Wednesday, November 2, everything was ready to go. The four most famous speed-flyers in America were here with their planes: Bert Acosta in the Curtiss Navy plane, Capt. Harold H. Hartney (who had placed second in the previous year's Pulitzer Race) in the Thomas-Morse monoplane, Clarence Coombs in the *Cactus Kitten*, and Lt. John A. Macready in the Thomas-Morse biplane. Two Italian-made S.V.A.s with four-hundred-horsepower Curtiss engines were to be flown by James Curran of Chicago and Lloyd Bertaud for wealthy oilman C.B. Wrightsman of Tulsa, Oklahoma. Charles S. "Casey" Jones (the Casey Jones that started the QBs the, international "Quiet Birdmen" society) flew in his Curtiss Oriole; Longren Aircraft Company of Wichita flew in eleven ships; E. Matty Laird had

T. C. Leque's plane following its crash

his Swallow; William B. Robertson of the St. Louis Flying Club had his Canuck; Andy Nielsen flew his JN-4D; the Donaldson brothers of Milford, Iowa, had their Oriole. All in all, thirty-seven airplanes and more than fifty pilots entered the events. That day, T.C. Leque of Des Moines, Iowa, landed his Jenny on its nose. No one was hurt, though, and he planned to be in Friday's race.

A few days before the Congress, on September 28, Lieutenant Macready had set the world altitude record at McCook Field, Dayton, Ohio, by reaching 40,800 feet. When asked what it was like up there, he said it was cold (sixty degrees below zero) short on oxygen, and the sky was a very pale blue. Now almost everyone flies that high on any airline flight, and while I write this the space shuttle *Columbia* is flying about 150 miles above the Earth.

The field was a beehive of activity. Never before had so many aeroplanes been seen in Omaha, nor in many other places. M. V. Robbins, federal meteorologist, forecast a beautiful clear day, fifty-six degree temperature and a twelve-mile-per-hour southwest wind for Thursday. He also warned all pilots to be on the alert for the kites flying at the Elkhorn Upper Air Observation Kite Station. He said that some of the kites were a mile-and-a-half high in a radius of ten miles around the station. The station was located north of Elkhorn, Nebraska, almost exactly where the Air Force communications facilities are today. V. E. Jakl was in charge of the kite station, and he reported upper air conditions to the North Field every hour by telephone, just like today. On Wednesday, twenty-six businesses paid for full-page advertisements in the *Bee* and the *World-Herald* announcing Thursday's opening.

On opening day, Thursday, November 4, 1921, at 7 a.m., Lt. Gen. Baron Alphonse Jacques, commander of the Iron Division which seven years earlier to the day stemmed the German invasion of Belgium, arrived at Union Station. He was greeted by Governor Samuel R. McKelvie, Mayor James Dahlman, Adj. Gen. H. J. Paul of the Nebraska National Guard, T. J. Nolan (the local Belgian consul), William Ritchie, Jr. (State Commander of the American Legion), Walter Byrne (Commander of the Douglas County American Legion post), Gould Dietz (Chairman of the Reception Committee) and Louis Feys (president of the Belgian-American Society of South Omaha). A wonderful breakfast and reception was held at the Fontenelle Hotel.

It seemed that all of the residents of "Belgian Corner" in South Omaha were there to meet, greet and shake hands with the famous general, and did he enjoy it. They all spoke Belgian dialects—the general spoke very little English. National Guard Company K of the First Nebraska Infantry, under the command of Lt. C. F. Rainey, arrived at noon to assist and to present the colors. Twenty of Omaha's finest, under Sgt. George Emery, directed traffic and preserved law and order. Engine Company Number 15 from 22nd and Ames, under Capt. H. B. Howells, was on the field with chemical fire equipment. The people arrived at the field by special trains, streetcars, automobiles, bicycles, afoot and even by buggies.

Some of the local neighborhood people rented out parking and standing space for spectators. It was the biggest show in Omaha since the great Trans-Mississippi International Exposition of 1898 (located not too many blocks south of the flying field where Kountze Park is today).

At 1 p.m., General Jacques stood at attention and saluted the colors presented by Company K while Dan Desdunes's Band struck up "The Star Spangled Banner." Ten thousand voices in the packed stands and on the field and another five thousand outside of the fences joined in the great opening.

Errold Bahl of Lincoln, flying his brand-new tiny *Lark* monoplane, came diving out of the heavens, pulled up in a one-thousand-foot-high loop, shot a few rolls, "split-S'd" out and landed. The crowd went wild. Seven DH-4s from the Air Mail Service flew over in a wonderful fly-by, but they couldn't land because there was no room on the field, so they flew back to the air-mail field. The six entries of the Pulitzer Race were parked in front of the stands for all to see.

The judges and technical committee made their final inspection and the pilots—all Americans—drew for positions: Acosta first, then Bertaud, Hartney, Coombs, Curran and Macready. The crews manning the pylons at Fort Calhoun and Loveland reported that they were ready.

My Dad was there and told me in detail about what I will call "The Loveland Pylon Story." Lt. Arvid Almgren of Fort Omaha was the official judge of the pylon. Dr. N. Harry Attwood of Omaha was the physician in charge. Dr. John R. Nilsson of Omaha was his assistant. Guy Smith, Victor Roos and my dad were the other members of the group. Dad and Roos drove over to Loveland on Roos's Harley-Davidson sidecar-equipped motorcycle, and the rest of the team drove in Dr. Attwood's Stearns-Knight. In 1921 the next bridge north of Omaha across the Missouri River was at Sioux City, so it was about twenty-five miles to Loveland by road. The pylon, a Dempster windmill tower, was placed in a huge whitewashed circle on the Henderson farm, a quarter-mile south and east of Loveland on top of the highest bluff around. It's where the telephone company's relay tower is today, south of I-680 and south of the observation tower on the north side. Dad said they could see the flying field and the Fort Calhoun pylon easy from their location. They had an open telephone line from the Hendersons's house with the officials at the Omaha field, and their main job was to see that each plane went around their pylon. The pilots had to be under five hundred feet when they went around. They were also prepared to act in case of an emergency.

They were at their station before 1 p.m., and a great crowd had gathered and filled all of the roads with their autos and buggies. The bluffs seemed to be covered with people. Several planes flew over and around the pylon, evidently familiarizing themselves with the course. They were ready to go, then Omaha called and told them that the course would be flown counter-clockwise instead of

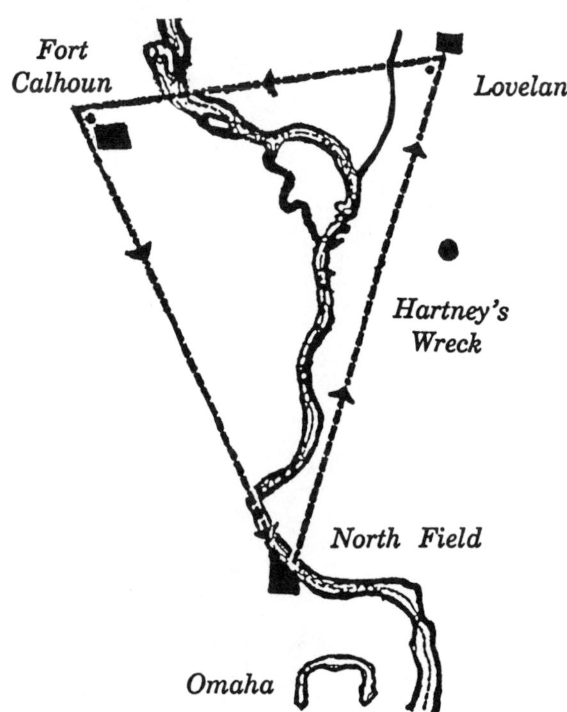

clockwise, which meant that the planes would fly around their pylon *first*, then go to Fort Calhoun. Dad heard later that the two Italian planes' props turned right, while the Americans always have turned left. I don't really know whether or not it would make that much difference going around a pylon; maybe it would with the wild airplanes they were flying. Some of the pilots objected to the change. But no matter—they flew counter-clockwise.

At 2:30, starter H.F. Wehrly waved the red and white flags, and Bert Acosta hit the throttle on his Curtiss. After a run of less than four hundred feet he was in the air. He circled once and crossed the starting line for his 150-mile dash. Flying less than five hundred feet up, he came past the starting pylon on the first lap in ten minutes, thirty-two seconds. He took the turns easily and without extreme banking. His motor hummed along perfectly, but it was noticed that one of the wings appeared slightly unsteady. As it turned out, a wire had snapped as he made his first turn at Fort

242

Calhoun. Acosta hit his stride during the very first lap, and he reeled in the succeeding ones. Clarke Powell's timers (Charles Hansen, Tinley L. Combs, C. B. Brown and H. E. Jacobson) showed he completed the second lap in ten minutes twenty-four seconds, the third in ten minutes twenty-four seconds, the fourth in ten minutes twenty-six seconds and the fifth in ten minutes twenty-three seconds. These times were all certified by Captain Labell's official timers.

The second pilot, Clarence Coombs, took off next in the *Cactus Kitten,* and he was flying wild and wide. He was burning gasoline and time on wide turns about the pylons and held to the outer edge of the course throughout the flight. After the race, the Texas pilot offered the information that he believed the craft was performing better and faster than if he had tried to pull it down to closer turns.

"The boat was wild and I let it have its head," was Coombs comment.

But if Coombs thought the boat was wild during the race, the thrill the ship gave the

Bert Acosta

crowd when Coombs essayed a landing at the finish was wilder. Just as the pilot took his dive for the ground his elevator mechanism stuck and the "Cactus Kitten" became an animated rubber ball.

For 100 yards the ship galloped across the field in an excellent imitation of some of the bucking broncos from its native state. [When a pilot bounces back into the air with a tail dragger, the plane is still flying, and Coombs was bringing her in hot.]

Coombs, despite his wild piloting, brought his triplane home less than two minutes behind the winner. That the craft has as much speed as the Curtiss-Navy was apparent when Coombs was on the straightaway part of the course.[5]

E. J. Cox, the Texas owner of the *Kitten,* was so elated he hosted a dinner for all the contestants at the Fontenelle Hotel that night.

Lieutenant Macready, in the Thomas-Morse, plainly did not have the speed to match the winner, but he flew perfectly. His best time for the thirty-mile lap was eleven minutes twenty-seven seconds. Lloyd Bertaud, in the S.V.A. Batilla, did the most spectacular driving of the race. Combined with the weird whistling made by his side radiators, the pilot's swoops for the ground as he came into the turns lent "daredeviltry" to the contest. From within twenty-five feet of the ground Bertaud would swing into an almost vertical bank and soar aloft as he rounded the pylon. But the ship didn't have the speed, so he finished ten minutes behind the winner.

---

[5]*Omaha Bee,* November 4, 1921.

Jimmy Curran flew the S.V.A. Diggens entry, but he dropped out after the first lap with engine trouble. *World-Herald* reporter William H. Graham covered the race from the air on board the

. . . Curtiss Oriole Plane 2,500 Feet Above Air Congress Field, Nov. 3—Leaving Air Congress flying field shortly after 2:30 o'clock with . . . "Casey" Jones at the stick of a 150-horse-power Curtiss ship with the *World-Herald* reporter aboard, armed with a portable typewriter strapped to his knees, we climbed to an altitude of about a half a mile to watch the progress of the aerial race of the age—the Pulitzer Trophy derby.

Like a giant hawk we circled high in the heavens above the pygmy air racers as they speed about the course almost twice as fast as we are going. . . .

. . . Suddenly as we wing our way towards Loveland, Casey, whose eyes seem to be better than mine, shuts off the motor and yells that one of the Curtiss navy planes is underneath us. Yes, there it goes. It is flying so low that the pilot must be skimming the leaves of the trees. . . .

. . . Casey tests out a few banks and uncanny movements, just to satisfy himself that my typewriter is fastened securely to my legs. Yep, it is, but my heart is capable of jumping around a bit.[6]

At first, Hartney was unable to get his Thomas-Morse monoplane started because of fuel line trouble. Because his ship was one of the best and because the captain was second the previous year, the judges gave him until 4:30 p.m. to get started. He taxied up to the starting line on time, but the big engine was missing as he got into the air. Missing or not, he headed for the Loveland pylon. He said that, even though the fuel pump was not working right, the MB-7 had gotten into the air as fast as any plane he had ever flown (remember, Hartney was an ace in World War I, second in the Pulitzer the previous year, and recognized as one of America's best pilots.) He leveled off at about three hundred feet, and she picked up speed like crazy. He was sure that, if she kept flying like that, he'd win the Pulitzer. He was about two miles from the Loveland pylon. He took her up a bit to clear it, and the big 320-horsepower

---

[6]*Omaha Morning World-Herald,* November 4, 1921.

V-8 gave a mighty "blurp," caught and spit a couple of times, and quit! It was the darned fuel pump. She started to stall at the same time. He shoved the stick to the fire wall and hit the ground in a cornfield almost immediately. He woke up in a farm yard a while later, and a lot of people gathered around. Hartney was just glad to be alive!

The Fetters patrol plane, flown by C. L. Bowen, saw Hartney go down and immediately flew back to North Field to let them know. It took him several minutes to get back and down.

My Dad, who was among the first to get to the site of Hartney's wreck, said the monoplane was about a mile and a half southwest of the pylon when it just nosed down and crashed in the cornfield. Lieutenant Almgren ran to the Henderson house to call the field to tell them what had happened and where. Dr. Attwood, Dr. Nilsson and Smith jumped into the doctor's car, while Roos and Dad took off in the Harley. The road was about a half-mile from where the plane crashed, so Dad got out, and the two physicians took a wild ride in the sidecar across the cornfield. Dad and Smith walked to the wreck. When the doctors got there, some farm folks had Hartney on the porch of their house. He was conscious, seemed to have a broken leg or hip, and was in great pain. The doctors

decided he had a fractured hip, and treated him as best they could at the time. Then Dr. Attwood sent Roos and Dad back to tell Almgren what was happening. They also told him to send an ambulance from Omaha to take Hartney to the Methodist Hospital at 36th and Cuming Street. Dad and Roos then drove back to the wreck.

They got there about the same time a *Bee* reporter arrived, riding behind Samuel F. Boord on his Omaha Cycle Company's Indian Motorcycle. Dad said Roos and Boord sort of eyed each other, but the Harley got there first. They went to look the wreck over. The ship was badly damaged. There was quite a crowd around it when some bright guy lit his cigar and threw the match on the gasoline-saturated ground. There was a "whomp!" and the $8,000 aeroplane was destroyed completely. Hartney was sick when he saw the ship burn because he considered her to be the most graceful, sweetest flyer he had ever flown.

In the end, Acosta came in the winner with a time a little over fifty-two minutes nine seconds, or 176.7 miles per hour. This wasn't quite as fast as C. C. Mosely's 178 miles per hour record time the previous year, but it won $3,000 and the trophy for a year. Clarence B. Coombs was second at a little over fifty-four minutes seven seconds. He won $2,000. John Macready was third at about fifty-seven minutes six seconds and won $1,000.

After the Pulitzer, stunt flyers put on an aerial circus such as was seldom seen anywhere. Loops, tailspins, nose dives, parachute drops and a myriad of other air-jockey tricks were performed almost continuously. N. D. Trinler in his Longren Aircraft Corporation biplane brought gasps, groans and moans from the throng as he turned topsy-turvy and tumbled through the air. He finally landed amidst rounds of applause.

The day's events came to a tragic end, though. Harry Eibe, parachute jumper for the Floyd Smith Aerial Equipment Company of Chicago, thrilled the crowd with a delayed-chute-opening jump. He didn't open his chute until he was well under five hundred feet and landed right in the middle of the Missouri River, which ran very fast at that location. Somehow Eibe got tangled up in the lines, and the chute pulled him under before the police river patrol could get to him.

On a happier note, General Jacques took his very first aeroplane ride in the Larsen monoplane, which was piloted by none other than Capt. Eddie Rickenbacker. The Larsen monoplane was the same that the Air Mail Service had just given up on. They'd finally worked out all of the bugs and were hoping to prove how good an aeroplane it really was. Rickenbacker had just arrived from the Legion Convention in Kansas City, where he was on the reception committee for the French general Marshal Foch. He was carrying a cane that had been made for him by a mechanic in his 94th Squadron from the propeller of the first German plane he had shot down. The famous birdman joked that it would be a good idea if the committee members all carried canes so the General would recognize them.

The *Bee* reported that the first day was very successful, especially considering that so many details had been worked out in so short a time. Contestants, officials and spectators all seemed pleased with the first day's events.

Day two was full of flying. All told, there were thirty-nine aeroplanes on the field, and it seemed like they were all in the air all of the time. Pilots were hopping passengers like crazy. They were getting $3 to $5 a ride, so the competition was fast and furious. Finally the Field Committee had to stop it. Imagine the wild flying around the field: no pattern, no control tower and every man for himself. They restricted passenger-hopping to before and after scheduled events, and everybody had to takeoff and land in the direction the "T" was pointing.

The main flying events on the second day were two ninety-mile free-for-all races for slower planes and a parachute contest for jumping accuracy. The owners of the nine slower Jennys on the field were up in arms that there was no race for them. Arthur Fetters saved the day by putting up a $100 prize for the fastest Jenny in the race. Ten were entered in the first ninety-mile race for ninety-mile-per-hour planes: Casey Jones of New York in a K-6 Curtiss Oriole; air-mail pilot J. H. Smith of Grand Island in a Laird Swallow; E. Matty Laird of Wichita in a Swallow; N. C. Tortenson of Milford, Iowa, in a K-6 Oriole; Jack Atkinson of Lincoln in a Lincoln Standard; Harry Buff of Topeka in a H-2 Longren; N. D. Trinler of Kansas City in an AK Longren; air mail pilot B. H. Pearson flying John M. Larsen's Avro; R. S. Miller of Minneapolis in a K-6 Oriole and R. Cambell of Des Moines in his K-6 Oriole.

At 1:30 p.m., Wehrly waved his red and white flags. Ten engines barked into life, and the pilots taxied to their starting positions. The planes were started in pairs, one minute apart. The positions were determined by draw, and the winner would be the pilot who took the shortest time to complete the course.

They took off, circled the starting pylon, and were given the starting flag. It made for some pretty wild turns around the pylons, since they practically flew chandelles around each one and sometimes flew neck-and-neck. The crowd, estimated at four thousand inside and 3,500 outside, loved it.

Casey Jones won the race with fifty-six minutes eight seconds and 100.39 miles per hour. N. C. Tortenson placed second at fifty-seven minutes two seconds, and R. S. Miller came in third at fifty-nine minutes nine seconds. All the winners flew Curtiss Orioles. Jones made his faster by taking the radiator off the top of the engine and bolting it onto the side of the fuselage like the Italian S.V.A. They were always learning new tricks in those days. Now they use computers—heck of a way to fly.

The second race was for seventy-five-mile-per-hour planes. The winner was F. A. Donaldson of Spirit Lake, Iowa, in sixty-nine minutes four seconds (80.08 miles per hour) in a Curtiss, and the prize was $300. Second place went to Casey Jones of New York at sixty-nine minutes twenty seconds in a Curtiss for $150. Third was Harry Buff of Topeka at seventy-two minutes thirty-three seconds in a Longren for $125. Andy Nielsen of Council Bluffs won the $100 prize with seventy-seven minutes thirty seconds. Four other Jennys started, and three of them finished.

Because of the fatal jump of Harry Eibe the previous day, the officials canceled the parachute jumping contest and donated the prize money of $350 to Eibe's widow who lives in Chicago. When this was announced to the crowd, they gave a great round of approval. His body had been recovered from the river by Omaha policemen just hours before.

John F. Kirk, Omaha's premier parachutist, made a beautiful jump from 2,500 feet and expertly guided his chute to almost the exact center of the field—really an accomplishment, considering the old round chute and cumbersome harness of those days. Our present-day sky divers make it look so easy. Because there weren't enough entries in the acrobatics contest, the committee used the contest money of $500 to pay for an air show of pilots and aerobats who were attending the Congress.

The show opened with a fly-by of a Curtiss Oriole owned by the Grand Island Aircraft Company, with J. H. Smith piloting and Pat McCarty of Vancouver, British Columbia, standing on the top wing. Smith flew east past the river, did a 180 and came back at about three hundred feet. This time McCarty was standing on his head! Smith flew west almost to Minne Lusa Boulevard, climbed, and did another 180. This time McCarty walked across the wing. Smith kept going east while climbing, did a 180 back, and chopped the throttle. McCarty climbed down from the top wing, straddled the fuselage just behind the engine, and stayed there while Smith made a perfect dead-stick landing smack in front of the crowd. There was another roar of approval.

Matty Laird of Wichita, flying his Laird Swallow, took off east and dipped down over the river. At that point, the crowd at the east end of the field rushed that way thinking he surely had gone in, but suddenly he zoomed up until he was at least five hundred feet high. Laird did a wing-over and roared back west, while daredevil Paul Duncan of North Platte was seen climbing out of the front cockpit and moving out on the right wing. The Laird Swallow flew west, then turned and came back at about three hundred feet. Duncan was hanging by his knees from the landing gear axle. On the next pass he hung by his hands, and the time after that he hung by his teeth from a rope. Duncan then climbed back into the cockpit, and Laird did a beautiful loop and landed. This was 1921, and the audience had seen the best flying that could be done.

Next, in a little Sport Farman plane, Robertson thrilled the crowd with graceful loops, tailspins, nose dives, "falling leaves," barrel rolls, Immelmann turns, banks and slips—all at no more than fifty feet above the ground. Upon landing he asked Pat Dozier, a reporter with the *World-Herald,* to ride with him. Dozier agreed and climbed in the little plane.

"Are you all set?" called Robertson. . . .

Before the answer could be spoken the little plane had dropped her tail and was well on the road to the blue. Not more than twenty feet had been traversed in the takeoff, and the reporter found himself looking down on the top of Cyle Horchen [*sic*] in his trick Laird Swallow. Or maybe he was looking up at the fellow. It's hard to tell about Horchen [*sic*]—he flies upside down so much. . . .

"Look to your left," ordered Robertson. . . , [and] each separate face [in the crowd] was passed as slow as though it were separated by the usual distance between telephone poles.

"Fifteen miles per hour," said the pilot. "If anything happens now, all you have to do is step out and walk along the side and fix it."[7]

This brings to mind a day in the thirties when I flew with L. D. "Dutch" Miller in Burnham Miller Flying Services' brand-new J-2 Taylor Cub with a forty-horsepower Franklin. It was the first time I had ever been in one. I was in the front seat, Miller in the back. He said, "Fly her." We took off straight east from the hangar (the first one at Omaha Municipal Airport, built by the American Legion in 1928). She was in the air in less than a hundred feet. We were about ten feet up when Miller hollered, "If you don't like it, step out and run along beside. You won't get hurt." I always wondered where Dutch got that neat little ditty. Now I know: from a guy named Robertson who flew a Farman a long time ago. Dutch was at the Pulitzer Races that year helping Andy Nielsen. They were both from Council Bluffs.

The show ended with the world champion upside-down flyer, pilot Clyde Horchem of Ransom, Kansas, zooming his Laird Swallow over the entire length of the field with his wheels pointing skyward. He would start diving, half-roll, fly inverted, then half-roll back, gain altitude, turn back and roll her over, all of this at about a hundred feet up. He flew his final landing approach inverted, rolled over and landed. This is a 1921 OX-5 Laird biplane, not a Pitts, Christian Eagle or Jungmiester!

I wrote the above account from the *Bee* and the *World-Herald* stories and from what my dad told me about the flying. He had come over from the Loveland pylon as soon as the races were over. I also relied on my own experiences flying all of those maneuvers (but at a heck of a lot higher than fifty feet!). I have never flown for a wing-walker. I wish I could have—it must have been a thrill. I have flown a Waco for a parachute jumper who climbed out on the lower wing to the farthest strut, and let me tell you she did drag and

---

[7]*Omaha Evening World-Herald,* November 5, 1921.

that wing went down. You have to fly her and bank away from the jumper after the jump.

I remember that we lived at 6754 Minne Lusa Boulevard at the time of the Omaha Aero Congress. I was in the first grade at Minne Lusa School, which was a two-room annex then. After school, my mom walked my friends Don Ralston, Albert Joy and me over to 24th Street (which ended at Vane Street) to watch the planes. I remember the flying because they were flipping and flopping. We probably saw the end of the acrobatics. No matter what, I can say I saw the Aero Congress of 1921.

Friday night, the big Aero Congress Banquet was held at the Fontenelle Hotel. More than two hundred airmen, dignitaries, officials, businessmen and their wives attended the affair in their beautiful ballroom. Eugene Eppley's people orchestrated it, and it was a scrumptious affair. Eppley owned the hotel and subsequently had Omaha's airport named after him. At the main table were Toastmaster William Ritchey, Jr., State Commander of the American Legion; Mayor Jim Dahlman; Eddie Rickenbacker; Augustus Post, Secretary of the Aero Club of America; Howard S. Coffin, Governor of the Aero Club; Maj. Ira Rader, Chairman of the contest committee; Guy C. Kiddoo, Chairman of the Chamber of Commerce Air Congress Committee; and S. S. Bradley, President of the Aircraft Manufacturers Association. Also present were Rev. Edgar M. Brown, who gave the invocation, and Arthur H. Fetters.

The program started with presenting the Pulitzer Trophy to Acosta. Just before the presentation, Coffin read a telegram from Caleb Bragg, Chairman of the Contest Committee of the Aero Club of America, which said that Acosta's time of 176.7 miles per hour was the new world record for a competitive flight. It bettered the previous record of 173 miles per hour of Lt. George Kirsch at the race for the Meurthe Cup in France the previous year.

Those at the banquet cheered as Acosta came forward to receive the beautiful statuette. After receiving the trophy and the $3,000 winner's check, Acosta said that the record had to be broken and he did it. He congratulated the other contestants for their sportsmanship and thanked Omaha for working to make the Congress possible. Since the world's fastest aeroplane flight took place in Omaha and over the Loveland and Fort Calhoun bluffs, the Air Capital of America was certainly in Omaha that day in 1921. Coombs, who took second with the Texas *Cactus Kitten,* accepted his

$2,000 check and noted that Omaha could be proud that the record was set here. Macready, accepting his $1,000 check, thanked the organizers and declared the race better all around than last year's race.

Major Rader presented the checks to all the other winners of the meet races. He announced that the next day's only official event would be the Larsen Trophy Race, which would start at 1:30 and would be flown over the Omaha-Loveland-Fort Calhoun course. After the presentations, Coffin spoke on the need for federal laws to govern flying and urged the Chamber of Commerce and the Aero Club of Omaha to stand firmly behind the Wadsworth Bill, which had been drafted to effect this purpose. Frank H. Russell, Vice President of the Curtiss Airplane Corporation, presented the city a Curtiss Oriole. Mayor Dahlman accepted it and said that it would be used to further aviation in Omaha. I have never been able to find out what happened to the Oriole.

Eddie Rickenbacker was the main banquet speaker. He prefaced his address with a toast to Harold H. Hartney. This was quite a tribute. Hartney had been Rickenbacker's commanding officer in France, and on this occasion Rickenbacker was following the tradition in which flyers toasted their fallen comrades. He then drew a tremendous round of applause when he announced that today he had joined Omaha Legion Post Number One in his adopted home town. During his talk, Rickenbacker reviewed flying in the last eleven years, starting from the time Curtiss, Mars and Ely became the first to fly in Omaha and Charles Baysdorfer flew north of Waterloo. His talk was a good one, with plenty of bravado and excitement. The flyer concluded by remarking that stunt flying was the greatest enemy of aviation. On the other hand, flying in a safe manner would inspire confidence in the businessmen whose backing was needed for the success of commercial aviation.

Here was America's number-one World War I ace, speaking in Omaha just three years after the war. Rickenbacker went on to help found Eastern Airlines and was President from its organization to 1952. Reverend Brown said a prayer for Harry Eibe and asked for a blessing on the group as they adjourned.

There is no way you could get a bunch of aviators together without a party, and a great one was held later at the Brandeis roof garden. All the flyers and everyone that had anything to do with the Congress were there. They ate, drank, danced and lived over the days when they were flying. Rickenbacker was loud in his praise of Colonel Hartney. He also told of memories he had of the East Omaha Speedway.

On day six, the only scheduled event was the Larsen Trophy Race, an efficiency race designed to prove the load-carrying capability of an aeroplane over a measured distance in the shortest time with the least amount of fuel. From the start, Larsen had been mysterious, exciting, different, enterprising, motivated strictly by ulterior motives, and sometimes a scoundrel, but in spite of it all likeable. He made a substantial contribution to American flying and qualifies as an aeronaut in my book. I believe that Larsen announced the Larsen Trophy and put up $6,000 prize money in order to promote the acceptance of his all-metal cabin monoplane, which had a very controversial record. He planned to win at Omaha, and he made a big splash.

After looking over the planes that entered the race, the committee decided to have two classes. The first was 240 miles (for the heavier ships), and the second was 150 miles (for lighter ships). They put up another $2,625 prize money for the new class. They didn't change the efficiency formula that was in Larsen's original announcement: Efficiency equals weight minus time multiplied by speed divided by gas (where E is efficiency, T the weight of the lighter planes, S the average speed, G the gasoline consumed, and W the weight of gasoline, oil and food carried). Now, Larsen claimed anybody could figure out this formula; the only trouble was the committee couldn't find "anybody." I don't think the technical committee concerned itself much with the formula, though, because it was pretty obvious that the three Larsen planes were going to walk off with the prize money. The committee did reserve the right to verify all gasoline and water figures.

All the contestants wanted to do was get to flying. The Larsen Trophy Race was scheduled to start at 1:30 p.m., but low-test gasoline had been delivered instead of high-test, so they had to wait until Ray C. and Edward C. "Ned" Goddard could get another tankload delivered from their fuel yard located at 13th and Webster Streets.

The entries included three JL-6 monoplanes entered by Larsen. Their pilots were Max Goodnough, Eddie Stinson and B. H. Pearson. Stinson had learned to fly at the Moisant School; he was the birdman for whom the Scout was built; and he later became a foremost aeroplane designer and builder. I once had the pleasure of flying his great gull-winged SR-9. Major Rader's committee also allowed two Italian-made S.V.A.s to enter after receiving Larsen's approval. They were smaller planes with capacity for only the pilot and one passenger, but much faster than the planes in the 150-mile-race class, so people thought Larsen was a real sport to approve their entry. The eighth plane to fly in the Pulitzer Race was owned by C. B. Wrightsman of Tulsa, Oklahoma, and piloted by E. F. White. The other plane, owned by the Ralph Diggens Company of Chicago, was flown by H. G. Lewis, an air-mail pilot from Omaha.

Because of the delay, the 150-mile race for lighter planes was run first. The entries were three Laird Swallows owned by the E. M. Laird Company of Wichita, Kansas, with pilots Walter Beech (the founder of the Travel Air Manufacturing Company and later the Beech Aircraft Corporation), Buck Weaver and Matty Laird himself. Also entered was an OX-5 Oriole owned by Curtiss Airplane Corporation of New York, piloted by Casey Jones; a Lark monoplane entered by Zook, Harding and Bahl Company (from Lincoln, Nebraska), with pilot Errold Bahl; one from the New Longren Company of Topeka, Kansas, piloted by Tom Trinler; and an OX-5 Oriole entered by the Donaldson brothers of Spirit Lake, Iowa, piloted by F. A. Donaldson.

At 1:45, the 150-mile race got started when Walter Beech took off in his Laird Swallow. He was followed at one-minute intervals by Jones's Oriole, Weaver's Swallow, Donaldson's Oriole, Laird's Swallow, Bahl's *Lark* and Trinler's Longren. Beech was still in the lead at the end of the first lap, but Donaldson was gaining. Bahl dropped out and landed with motor trouble. Donaldson was leading at the second lap. Weaver was gaining, but Beech's engine was missing as he rounded the home pylon. Jones and Trinler were neck-and-neck and rounded the pylon together (much to the crowd's delight), and Laird was right on their tails. Donaldson still led at the end of the third lap; Weaver was second; Trinler was third; Jones was fourth; Laird was fifth. Beech gave up and landed because he was so far behind. The fourth lap ended with Donaldson well in the lead, Trinler second, Weaver and Jones battling for third, and Laird right behind. Donaldson won, coming in almost a half-mile ahead of Trinler. Weaver nosed out Jones for third, and Laird was just behind. The official results had to wait for the point tabulations of Major Rader's committee. They were announced within twenty-four hours.

At 4:15, starter Wehrly finally waved the beginning of the big plane race, and B. H. Pearson hit the throttle of the Larsen JL-6. The big, orange, all-metal monoplane rumbled across the field loaded with Larsen, John Markel, Grace O'Brien, J. F. Armstrong and ten-year-old Eddie Rice, all of Omaha. Ernst Buhl was aboard as mechanic. One minute later, Stinson took off in his gray JL-6 with Mrs. C. E. Dailey, wife of an Omaha dentist, Richard W. Coad of the Dodge Car Agency, Mark McGregor, John Nilsson (son of Dr. Nilsson) and Mr. and Mrs. Jack Summers, all of Omaha. Next, pilot Max Goodnough took off in his silver JL-6. His passengers were Harry Larsen (John Larsen's brother), R. A. Findly, Herbert J. Connell and Jack Nourse, all of Omaha. White took off eighth with six hundred pounds of Lawrence-shot as dead weight. The last was pilot Lewis with E. D. Barnes of Grand Island as a passenger. I suppose the passengers and dead weights were figured out for each plane, but it sure was a loose system. Anyway, I think Larsen was so sure his JL-6s would win that it didn't bother him.

The race was a humdinger. Those 1921 planes were going a staggering one hundred miles per hour, so it took eighteen to twenty minutes to fly the thirty-mile lap. There were six ships, so somebody was going by the home pylon most of the time. Pearson led the first five laps, with Stinson second, and Goodnough third. White followed right on Goodnough's tail. Lewis was last; his engine was acting up. The crowd loved the way White and Lewis rounded the pylon, almost in a vertical bank. The JN-6s rounded in a wide turn.

Then all heck broke loose on the sixth lap. Pearson's engine burped and quit just before he got to the Fort Calhoun pylon. He put her down in a field east of where the DeSoto Bend of the Missouri River started south in those days, not too far from where the riverboat *Bertrand* recently was found. None of his seven passengers suffered a scratch, and the ship wasn't damaged. Since Pearson had been leading by a substantial margin, the engine failure had great repercussions. That evening, Larsen attributed the failure to a

> broken washer, the size of a silver dollar, [while speaking] at a country grocery store at Loveland, where his party was stranded for several hours.

"It was for the intake valve. Probably could be bought for 5 cents, but it might have caused loss of life and the wrecking of a very valuable plane."[8]

It was getting dark. On the same lap, Lewis dropped out and landed at the field. When Pearson didn't appear, it caused a temporary uproar, but Stinson dropped a note telling where Pearson had landed. Robertson then jumped in his Sport Farman and took off to look for the downed plane.

In the long run, Stinson appeared to be the winner, having crossed the line first. Goodnough looked like second, and White appeared to be third. Stinson landed with no difficulty and gave the officials the first real word about the location of Pearson's downed plane. Stinson commented that Pearson's orange ship was faster and had a lead over him of about five hundred yards. When Stinson approached the Fort Calhoun pylon during the sixth lap, he watched Pearson turn back in the direction of Iowa. Stinson then slowed and watched until he saw Pearson sail safely into a field. On his last lap he saw the plane in a field south of Fort Calhoun.

At the end of the race, White landed right behind Stinson, but Goodnough headed back to where Pearson's plane was down. *World-Herald* reporter Cecil W. Grange talked C. L. Bowen into flying him up to the downed plane in the Ashmusen *Bluebird*. He reported:

Darkness made it difficult to see clearly, besides which Bowen kept his ship high.

Soon it became difficult to distinguish much on the ground. We saw a fire on a hilltop ahead. The river was just under us. No sight of the missing planes. A train, creeping compared to our speed, was visible to our right. . . .

As we got to the river a great thrill came. It was necessary for Bowen to get below the level of the bank to see the surface of the field clearly. He did. The wheels of the Bluebird skimmed the river, a slight spray was thrown up before I over the edge again Bowen had put his ship onto the ground in a perfect landing. He taxied to the south side where he parked for the night as a crowd of persons rushed from the sidelines to see who were [*sic*] were.[9]

---

[8]*Sunday World-Herald,* November 6, 1921.

[9]*Ibid.*

Goodnough never found the downed plane and almost ran into the crowd who rushed onto the field when he landed. Somebody had enough sense to stop any more flying, and about that time Major Rader was called to the field telephone. It was Larsen calling from Loveland explaining that everyone was safe and to please send a car to get them. With that the day ended, and the crowd went home. T. J. O'Brien sent two of his Dodge Cars up to Loveland, and all of the flyers, officials and guests left to prepare for the Aviators Ball to be held at the Auditorium.

Robertson walked into Fort Calhoun and got a ride to Omaha in plenty of time to get to the ball, but the downed flyers weren't rescued at Loveland until almost 10 p.m. O'Brien regretted missing the ball but thought the fine dinner that Larsen and Markel gave all of them at the Fontenelle Hotel was a very pleasant affair.

The first International Air Congress concluded with style at the Grand Aviators Ball at the Auditorium Saturday night. More than five hundred aviators, Army officers, officials and everyday Omahans attended the gala affair. At exactly 8 p.m., Ralph William's famous dance orchestra played the Grand March, which Rickenbacker and Gertrude Stout, president of the Junior League, led. Major and Mrs. Rader led the Flyers' section, while Major Walsh and Marion Hamilton led the Army section. The Officials' section was led by Gould Dietz and his wife. Hundreds of colored balloons decorated the scene. It was a ball right up there with the best—the women in beautiful gowns, the men in formal attire, and the military in dress uniform. It proved a fitting finale to a landmark event in America's aviation history—the first International Air Congress ever held in the United States. Omaha seemed to be the Air Capital of the United States that November night, and the old Red-tailed Hawk was perched on top of the highest flag pole on the auditorium.

Major Rader's contest committee met Sunday afternoon at the Congress headquarters at the Fontenelle Hotel to determine the winners of the Larsen Race. They had no problem with the 150-mile race, the results were the same as they appeared to finish: Donaldson first with 3,783 points for $1,500 in prize money, Trinler second with 3,437 points for $750, and Weaver third with 2,871 points for $375. But the 240-mile race was a different story. First, Larsen kept bugging them for the results. Second, one of the judges at the Fort Calhoun pylon filed a late report that White's plane had cut inside on two laps. Third, they really needed an Einstein to figure out the Larsen Formula.

Rader said to go with the committee's findings come heck or high water. They told Larsen to quit calling them and promised to advise him before releasing the winners' names. They questioned Capt. R. A. Grear of Fort Crook, who was in charge of the Fort Calhoun pylon, and he stated that there had been no inside violation on the part of the contestants. Applying Larsen's formula wasn't so easy. They finally set their own method of applying the data and agreed to unanimously accept the results.

The committee concluded that White was first with 4,671 points, Goodnough second with 4,640 points, and Stinson third with 4,212 points. They gave Larsen the results, contacted C. B. Wrightsman and White, gave the press the release, and then made themselves scarce before Larsen could find them.

Meanwhile, on Sunday afternoon, Gene Eppley had arranged with Clara Ashmusen, owner of the five passenger *Bluebird,* to give the four women members of the Fontenelle Orchestra their first aeroplane ride. A little before 4 p.m., C. L. Bowen took off to the east with his four excited passengers. The big ship rose gracefully off the ground and was about a hundred feet up when the engine quit. Bowen knew he was headed for the river, so he attempted to turn back to the field. He wasn't high enough—she stalled and crashed. Bowen later commented:

> After we had crashed, I climbed out, although hurt in the back myself, and by this time about fifty persons had gathered. I called on them to help me get the young ladies out as I tugged away at one, but no one would come. It was quite a good little while before I finally got aid. The two girls in the front seat were unconscious, but the two behind did not seem to be hurt so badly.[10]

The injured were Lucy Atkinson of Detroit, Michigan, with a crushed right knee; Margaret Haggerty of Cleveland, Ohio, who dislocated her hip and tore ligaments; Thelma Fisher of Fort Wayne, Indiana, who suffered body bruises; and Vera Rasche of Oakland, Maryland, who broke her nose and suffered bruises. Bowen escaped with a bruised back. They were all rushed to the Swedish Mission

---

[10]*Omaha Evening World-Herald,* November 7, 1921.

Hospital at 24th and Pratt Streets. There, Drs. Rudolph Rix (who delivered me) and A. S. Pinto (who administered my early smallpox vaccinations) treated them. Bowen was heartbroken. "I've been flying for seven years and those girls are the first I have ever hurt. ... If I had only been up 100 feet I might have brought it out of that tail spin."[11] Clara Ashmusen said that they would build another *Bluebird.* Haggerty's hip was set, and she was expected to recover with no problems. Fisher and Rasche were released the next day, but Atkinson's leg couldn't be saved.

Getting back to the races, on Monday, Larsen filed an official protest with Rader's committee. Roscoe G. Conklin, member of the Omaha Aero Club, testified that a three-gallon gasoline tank in White's aeroplane had not been taken into account in computing the scores. So the committee went out to the field and checked White's plane again. Rader then issued the following statement: "The decision of the Contest Committee has been made, awarding first prize to the S. V. A. That is final and was made after the consideration of all of Mr. Larsen's charges and protests."[12]

Larsen retained the law firm of Smith, Schall and Howell, and District Judge Charles A. Goss issued a temporary restraining order on Guy Kiddoo, Rader, Fetters and White to keep them from paying or collecting the $3,000 first prize in the Efficiency Race.

The judge set the hearing for November 14 and placed Larsen's bond at $500. Larsen alleged that White's victory was due to fraud and deception. The S.V.A. plane carried a "concealed gasoline tank" which was not figured in the gasoline consumption computation of the race results. Interviewed by both the *Bee* and the *World-Herald,* Larsen said that he was a champion of good sportsmanship, not a "welsher," and that he had been urged by the Aero Club of America to enter his planes in the race in order that competition might be stimulated. Larsen was asked what it had cost him and how much Borglum's fee was. His guess was $25,000.

"The trophy won't go to filthy hands," Larsen answered to one question. "It isn't the prize money I care about. That's just cigaret [sic] money, but I object to having White's name placed on the trophy as winner of the race

---

[11]*Ibid.*

[12]*Omaha Evening Bee,* November 8, 1921.

260

when I know that he was not the winner." . . . "It's a plain steal, that's all. I won hands down." . . . .

Larsen denied that his planes were the German "junker" planes, but stated they held a resemblance to them. He denied that Gutzon Borglum designed the trophy.

I hate to take away from Omaha's sons, but Borglum elaborated on the design after I had given him the basic idea. It is now being finished in New York city [sic]," he said.[13]

According to the *Omaha Bee,*

Wrightsman, owner of the S. V. A., retained Paul L. Martin as counsel.   He refused an offer on the part of Larsen's attorneys to compromise on second place, and posted a bond of $1,200 pending the court proceedings. Wrightsman said, "I don't need the money. . . . If my plane is not entitled to first place it is not entitled to any prize. We'll fight this thing out."[14]

While Larsen was filing his injunction suit that morning, deputy sheriffs were looking for him, armed with an order for attachment against his property or that of Capt. Harold Hartney, formerly his pilot, in connection with an accident at the air-mail field in August, 1920. Remember the Jensen house on West Center Street? Well, nobody ever paid Jensen, so he filed suit for $2,639 against both men for the damage. When the sheriff finally caught up with Larsen, he claimed the government owned the aeroplane, which was on a test flight. He had no property in Omaha. The three JL-6 planes then on the north airfield belonged to the Larsen Aero Company. So the deputies had the attachment order but found nothing to attach it to. They did find Captain Hartney—he was still in the Methodist Hospital recovering from his crash in the Pulitzer Race. I never have found any record that Jensen ever was paid for any damages to his house.

On Monday morning, November 14, Captain White, pilot of the S. V. A. biplane, the declared winner, strongly denied any fraud or deception when he took the stand in Judge Arthur C. Wakeley's division of district court in the injunction case brought by Larsen. Expert testimony was presented by both sides showing that the 220-

---

[13]*Omaha World-Herald,* November 8, 1921.

[14]*Omaha Evening Bee,* November 8, 1921.

horsepower SPA engine could or could not make the distance on the amount of gasoline used. An affidavit by Hartney maintained that White's plane couldn't have traveled on nine gallons of gasoline an hour. Giuseppe Bellanca, Omaha aviation expert, asserted in an affidavit that the S. V. A. plane could have made the distance on the fuel shown on the official records and on even less. He declared that White's performance was nothing unusual for the plane. White was the only witness called and the session was devoted entirely to his examination and cross-examination.

On Tuesday morning, Larsen took the stand and testified that White had a hidden gasoline tank in the upper wing and used gasoline from it that was not figured in the efficiency calculations. Defense counsel Martin questioned Larsen in a technical manner that challenged calculations Larsen offered. Rader testified that every member of his committee had read or heard the complete text of Larsen's protest and that Larsen's contention that White used gasoline which was not charged against him was absolutely preposterous.

In the afternoon session Larsen was again placed on the stand. As in the morning, he parried words with Martin and kept the court in an uproar. Judge Wakeley was forced on several occasions to warn Larsen that he must answer the questions of the defense counsel. A decision was expected the next day.

Judge Wakeley's decision dissolved the restraining order. White and owner Wrightsman won the race. On Thursday, November 16, 1921, Larsen left Omaha for New York City in one of his JL-6 Junkers with Eddie Stinson piloting.

On November 16, Guy Kiddoo, chairman of the general committee of the Aero Congress, stated that the total cost of the meet was $55,000 and all bills were or were to be paid. Admission receipts totaled $16,000. That indicated a paid attendance of fifteen to twenty thousand. There were probably that many people who watched the meet for free. Omaha businessmen and several prominent individuals paid the $39,000 balance through pledges and underwriting.

What did it accomplish? Most of the world heard about Omaha, Nebraska, that week. Acosta flew an aeroplane faster than anyone had previously—176 miles per hour. The Midwest saw many different aeroplanes and some great flying. City officials, businessmen and others probably learned that Omaha should acquire land for an airfield. And it proved that the aeroplane was here to stay.

In the summer of 1921 the Air Mail Service grew up. The safety of the pilots became the first priority now that the pioneers had proved they could deliver the mail. From July 1, 1921, to September 1, 1922, not a single pilot died on duty. There were several reasons: The maintenance and repair of the aeroplanes was tremendously improved, a radio network was established with continuous weather data, and the pilots now wore parachutes even though they were considered "sort of sissy." The Air Mail Service head, Paul Henderson, told the pilots that he would rather have twenty-five percent delivery without any pilots killed, than have ninety-six percent delivery and have pilots get killed.

By August 1923, the 902-mile illuminated aerial boulevard was completed between Chicago and Cheyenne, so a full dress rehearsal of round-the-clock operations was performed over four days. On August 20, Dean Smith flew the beacon-lit route from Chicago to Omaha. When he landed at 11 p.m. to the cheers of at least a thousand people, he said that it was duck soup.

The eight transcontinental day-and-night trips went off without a hitch. The best west-east trip was twenty-six hours fourteen minutes in eleven relays. The fastest east-west time, flown against the prevailing winds, was twenty-nine hours thirty-eight minutes in twelve relays. If such a schedule could be maintained, it would cut the transcontinental time of coast-to-coast mail by three days.

Henderson and the Post Office Department were elated, but not the nearsighted Washington politicians. The lights weren't turned on until the summer of 1924 because the dubious Congress declined to appropriate enough money for further night flights. They did manage to improve most of the ground facilities and to extend the lighted airway to Rock Springs in the west and Cleveland in the east.

The Omaha operation was moved from the Ak-Sar-Ben Field to Fort Crook in April, 1924. Offutt Field, which had just been named, was much bigger, almost a mile square, and it was located on a military base. A new large hangar with complete shop facilities was built. Several buildings were given to the service for headquarters and offices. Another building housed supplies and

radio equipment. Carl F. Egge, a pilot himself, became the first Superintendent of the Air Mail Service and had his office outside of Washington, D.C.

Congress finally allocated $2.75 million to the Air Mail Service for the fiscal year beginning July 1, 1924. Round-the-clock service started on August 1, and "via air mail" became a household term as people began filling the red, white and blue air-mail mailboxes that were placed in strategic locations in cities served by air-mail. Omaha was one of them, and I remember that sending or receiving an airmail letter was a big occasion. West-to-east air-mail regularly crossed the country in about thirty hours, east-to-west in about thirty-five hours. "Airmail" was here to stay, and Omaha was in the middle of it from the start.

CHAPTER NINE:
THE AROUND-THE-WORLD FLYERS

In 1923 Gen. Mason M. Patrick, head of the U.S. Army Air Service, announced that the Army was going to make the "ultimate test" of aeroplanes and pilots with the first flight around the world. Both England and France were planning flights, but General Patrick wanted the United States to be first. Donald Douglas, an up-and-coming young aeroplane designer and builder, built and delivered four Douglas World Cruisers—single engine, four hundred-horsepower biplanes convertible to floats or wheels. The Army picked five of its best pilots to fly them: Maj. Frederick L. Martin, Maj. John Harding, and Lts. Leslie P. Arnold, Leigh Wade, Erik Nelson. Each pilot picked his own mechanic. The planes were christened the *Boston,* the *Chicago,* the *New Orleans* and the *Seattle.*

They took off from Lake Washington, near Seattle, on April 6, 1924. On September 23, 175 days and 26,000 miles later, they landed on the same Lake Washington. Two of the four planes finished, as did Lieutenants Smith, Arnold, Wade, and Nelson. Major Harding and Sergeants Ogden and Alva Harvey also completed the trip. Major Harding and the *Seattle* dropped out after a crash in Alaska that wrecked the plane. Smith became the group commander. Lieutenant Wade in the *Boston* was forced down in the Atlantic Ocean by engine failure between the Orkney Islands and Iceland after 19,000 miles. The sea was very rough. The *Boston* turned over and sank just after Wade and Ogden were rescued by the Navy escort, the U.S.S. *Richmond.* They continued the flight in the two remaining ships until they reached Boston, where the prototype had been flown. Then they flew to the finish.

On September 9 they landed at Bolling Field in Washington, D.C., and were welcomed by President Calvin Coolidge (who broke all precedents by smiling and looking happy as he shook hands with each man). The flyers were greeted by huge and tremendously enthusiastic crowds in every city they visited as they circled the globe. The United States was no different. Every city in the country wanted them to stop on their way to Seattle. Omaha was chosen as one of the stops, probably because of the Air Mail Service at Fort Crook. I really was excited because since the start I had been following their flight in the *World-Herald* and the *Bee.* The

Bee was so wound up that they scooped the *World-Herald* and reported the fliers' arrival a day before they got here.

They did arrive at 1 p.m. Wednesday, September 17, escorted by several Army planes. The *Chicago,* the *New Orleans* and the *Boston II* flew majestically over 24th Street and Ames Avenue, where I was watching. The whistles in town were blowing to announce their arrival. Every factory had whistles and used them to announce unusual events. The Saratoga Laundry steam whistle at 24th and Laird Streets was shrilling at several second intervals.

The around-the-world fliers in Omaha (l. to r.): Lieutenants Ogden, Harding, Arnold, Nelson, Wade, and Smith

I could hear the Florence Water Works whistle, and the Missouri Pacific Railroad steam engines on the belt line added their great notes. Way off in the distance you could hear others—I knew one of them was the Jones Street electric power plant, also the Omaha Council Bluffs Street Railway Company's power plant and the Union Pacific Railroad shops. It sounded like fifty locomotives. Everyone honked the horns on their cars, and many church and school bells were ringing.

Schools were all out for the great event. Boys ran around waving their arms and flying just like the heros. I was privileged to see the first fliers ever to fly around this world as they flew over Omaha on the final leg of their epochal journey. The fliers landed at Fort Crook, where a crowd of several thousand had gathered. The ships were parked on the hangar apron, and the formal reception was held there. Gen. Halsted Dorey officially welcomed them. Omaha Mayor Jim Dahlman said that the world was rejoicing over the fliers' achievement, and he gave them silver cigarette cases as gifts from the people of Omaha.

Big Charles Gardner then megaphoned introductions of the airmen to the crowd. After the reception, Gardner announced the selection of Omaha for the 1925 convention city of the American Legion. Lieutenant Wade was greeted by his sister Mrs. E. A. Francis of Denver. Jack Knight, the same of airmail fame, greeted and visited with his old friend Lieutenant Nelson. Several girls slipped past the soldier guards, and plied Lieutenant Smith with questions before the guards were aware of their presence.

The fliers were then driven to the Fontenelle Hotel where they were the guests of Gould Dietz and Eugene Eppley. Lieutenant Smith was met at the hotel by his cousin, Mrs. Charles M. Williamson of Indianapolis, Indiana. The fliers were exhausted and greatly appreciated a shower and rest. They left at 10:30 a.m. for St. Joseph, Missouri, and Muskogee, Oklahoma.

The DH-4 was the backbone of mail flying. De Havilland had designed a darn good aeroplane back in World War I. The DH-4 flew all of the mail until the service was turned over to private contractors in 1927. The Mail Service constantly modified it—six hundred structural alterations were made in its career. The four hundred-horsepower Liberty engine proved to be a great one, too. After mechanics replaced the original camshaft gear with a new stronger one, engine failures practically stopped. Many of the changes and modifications were contributions of the Omaha maintenance shop.

The pilots (who were mostly seat-of-the-pants flyers in those days) came up with some pretty ingenious navigational and flight instruments themselves. The first instrument they discovered was a good reliable clock, and clocks were installed in all of the planes. Holding the plane level without an artificial horizon was another problem they solved when "Dinty" Moore, an Omaha-Cheyenne pilot, taped a half-empty pint whiskey bottle on the dash so he could see the amber fluid and try to hold it level.

Wesley Smith, who flew the original cross-country flight in 1922, and Howard Salisbury, the Omaha instrument specialist, came up with what was probably the grandfather to the turn-and-bank indicator. First they mounted a military turn indicator's air tubes on the exhaust pipe to eliminate icing. Then they put a steel ball in a curved glass from a carpenter's level and mounted it on the dash. It sure did help when a pilot flew into the soup.

Jack Knight flew to Cheyenne once with a 150-pound General Electric radio in a compartment just behind the cockpit. He had a receiver in his helmet and a transmitter on his chest. There was a 200-foot wire trailing behind as the antenna. He gave periodic reports on his position that were sporadically received by the ground stations. E. Hamilton Lee also made tests flying on the eastern runs. Both Knight and Lee flew many years for United, so they watched radio progress to the sophisticated navigational instrument and traffic-control system it is today.

The government's airmail success paradoxically contributed to its demise. Postal officials and most congressmen long maintained that they had no desire to fly the mail forever. As soon as it became successful and profitable, they wanted to put the routes up for bid by private contractors. America's civil aviation was growing by leaps and bounds. The Air Mail Act passed by Congress authorized the Postmaster General to contract out airmail routes. It was signed by President Coolidge in February 1925. In July the Post Office advertised for bids on eight feeder routes that ran laterally to the transcontinental airway. That started the great airline businesses we have today.

BOOK FIVE:

The Aeroplane Becomes the Airplane
(1925 to 1938)

After Ak-Sar-Ben advised the Chamber of Commerce that the airfield would have to stop operating by July 1, 1925, the Chamber convinced the City Council that Omaha needed an airport. On May 5, 1925, the city of Omaha bought 198 acres of land east of Carter Lake and north of Locust Street. The Council didn't think it could buy land for an airport without a vote of the people, so the land was bought as park property. There wasn't much private flying and practically no commercial flying going on in Omaha in 1925, but thank goodness for our city officials. Much credit goes to Mayor Jim Dahlman and Commissioner Dean Noyes, plus the Chamber's Aerial Transportation Committee—Gould Dietz, Clarke Powell, Arthur Fetters and Eugene Eppley—who had some input.

So, when a June 4 tornado demolished the hangar and four private planes at Ak-Sar-Ben, Omaha had an airfield at the new location, though it was really just a cow pasture. The field was put under jurisdiction of Street Department Commissioner Dean Noyes, an aviation enthusiast. He put his crews to work, and they cleared and leveled the field almost a half-mile in every direction.

By July, the half-dozen private airplanes still left in Omaha were flying from the new field near Carter Lake, although there were no buildings, fences, hangars, fuel facilities or roads. Carter Lake Boulevard ran by the field's west side, and aviators had to drive cross-country from there to get to the field.

My dad would drive us over to the field on nice Sunday afternoons. Sometimes there would be a crowd of people watching Andy Nielsen, Charles Kenwood, Arthur Fetters, Victor Roos, Reed Davis and a few airmail pilots flying their planes. Sometimes there would be two or three in the air at the same time, and every now and then some brave soul would pay five dollars for a ride.

Roy Furstenberg's Overland Airways, the biggest operation on Omaha's new field, had six airplanes. I don't suppose that at that time there were fifty pilots in the whole countryside. Of course, airmail flying was booming down at Fort Crook, but flying in and out of Omaha's air field wasn't booming for a number of reasons: Airplanes still weren't completely reliable; pilot comfort, let alone passenger comfort, wasn't too hot; almost none of the planes could carry a big payload; and cross-country flights could be real experiences. The facilities in other cities in the Plains and Midwest weren't much better, if there were any.

In May the Army flew their largest dirigible, the *TC-6,* to Omaha from Scott Field, Illinois. It flew over town for a couple of days. It was huge, almost a block long, but it couldn't beat one of our wild May storms and was wrecked landing at Fort Crook.

Captain H. C. Gray, the pilot, said,

> We'll bring back the ship, with the same crew, . . . to show the people of Omaha that an airship can weather the storms you have out here. Omaha is a bad place for dirigibles—it's entirely too cyclonic. But we'll bring the TC-6 back again.[1]

But the Army got out of the airship business and let the Navy sink with them.

There was a lot of flying activity in Lincoln, mainly because of Ray Page and his Lincoln Standard Aircraft Company. His plant and school were located at 2409 "O" Street, and his field ran south of Van Dorn to High Street, west of 20th Street. They built three darn good airplanes, the LS-5 Commercial, the Lincoln Standard Tourabout and a great single-seater, the Lincoln Sport, which could be bought as a kit. Page was ahead of his time with the kit.

On June 28, Augey Peddler flew the Sport off of St. Mary Avenue between 24th and 22nd Streets in downtown Omaha (I was there). Everybody thought it was great. They even had extra cops to handle the crowd.

By 1925 the automobile was becoming the primary people-mover. Good roads were being built everywhere; the Lincoln Highway was paved all the way to Ames, west of Fremont, and people could go all the way out to San Francisco or back to New York on it. Autos were getting better, and good trucks and buses were being built. They were starting to make the mighty railroads sweat a little.

With very little money and guidance from the new Airport Advisory Board, Commissioner Dean Noyes kept working on the Omaha Airport. The field was smoothed, leveled and put in good shape. A road was built to the flight line (using the term loosely) from Carter Lake Boulevard. Underground fuel-storage facilities

---

[1]*Omaha World-Herald,* May 5, 1925.

Construction Details of Lincoln Sport Plane

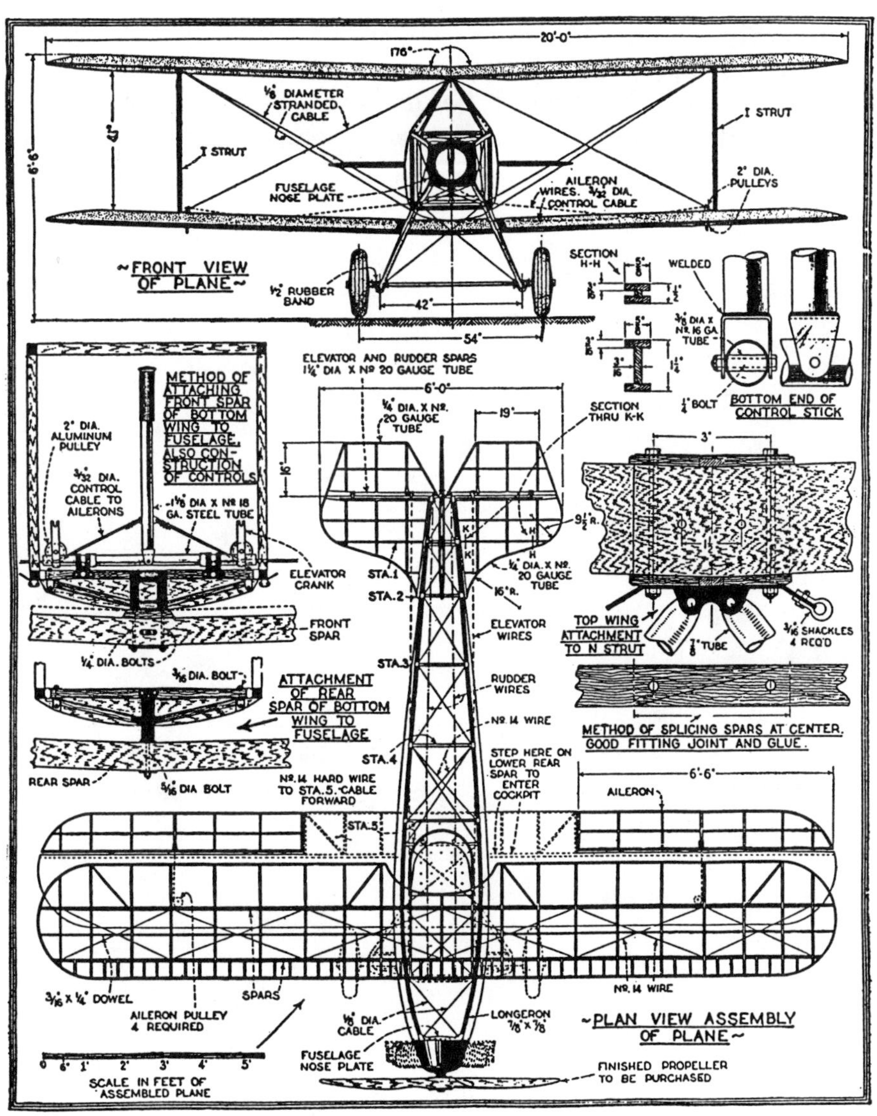

LINCOLN AIRCRAFT CO.

2409 O STREET   LINCOLN, NEBRASKA

were installed. It was hoped that boundary lights could be put around the field, and that the big cottonwood trees on the north and south landing approaches could be cut down. (The trees weren't really that bad—they could be dodged easily.) There were no runways in 1925, but it was the start of today's Eppley Airfield.

CHAPTER TWO:
PREPARATIONS FOR FLYING THE ATLANTIC
(1926 and 1927)

1926 was almost a forgotten year in Omaha flying. The same few guys with less than a dozen planes were still the only flyers in town. Every now and then some barnstormers or locals would put on an air show, but they never broke any attendance records. The airmail had a good home at Fort Crook. One good thing happened: The airfield was put officially under the Street Department and Commissioner Noyes.

By 1926 airplanes were designed and built pretty well, but there was still the lack of a really good airplane engine. Charles L. Lawrence, former racing-car engine builder, designed the remarkable Wright Whirlwind J-5 engine, the engine that in my opinion made aviation. It was built by the Wright Aeronautical Corporation (Orville, the surviving brother, was no longer connected with the company). This was the first engine ever built that was almost 100% reliable—a nine-cylinder, radial, air-cooled 200-horsepower engine that ran and ran on about ten gallons of gasoline an hour. It had the lowest weight per horsepower of any engine built to date.

The company decided to build a plane to prove and demonstrate their great engine. One of Wright's test pilots, Clarence D. Chamberlin, was born and raised in Denison, Iowa, about sixty-five miles northeast of Omaha. Chamberlin had flown a Bellanca CF when he was barnstorming in Illinois (it had to be one of the four built in Omaha). He thought it was the best airplane he had ever flown, so he highly recommended that the Wright Aeronautical Corporation get Giuseppe Bellanca to design their airplane, which they did. The Wright Bellanca high-winged monoplane weighed 1,850 pounds empty and carried four passengers at one hundred miles per hour. It was full of Giuseppe's innovations. From the day Chamberlin test flew it, it was generally believed to be the best airplane in the world. It looked a lot like the Bellanca CF built at 1902 South 27th Street, Omaha.

The whole world seemed to want someone to fly nonstop across the Atlantic Ocean from America to Europe or vice versa, and it looked like every flyer and his brother were willing to give it a try: World War I aces from every country, army, navy, government, just plain flyers, screwballs, rich and poor. Organized and unorganized trans-Atlantic attempts were made, both east-to-west and west-to-east. Newspapers promoted the flights, and publicity was everywhere, even in the brand new media—radio. The *Bee* and the *World-Herald* ran front-page stories almost every day about a new or planned attempt. Attempts that did get off the ground almost always ended in a disaster with people killed or injured, and many flights ended with the disappearance of the plane and those in it. Though some trans-Atlantic pilots were rescued, none had ever been successful.

Raymond Orteig, a French-born New York hotel owner, put up a $25,000 prize under the auspices of the Aero Club of America to the first aviator of an Allied country to fly from Paris to New York or New York to Paris. People started getting serious about such a flight.

Igor Sikorsky, a great Russian airplane designer (to become the father of the helicopter), was the first to build a plane specifically for flying the Atlantic. His huge S-35 tri-motor (British Jupiter engines) biplane was to be flown by René Fonck, France's leading World War I ace. On September 20, 1926, it was loaded with 2,400 gallons of gasoline and a crew of four when Fonck hit the throttles. It lumbered down Roosevelt Field's soft runway, never got into the air, crashed in a marsh and burned. Fonck and his copilot got out, the other two were killed.

The Omaha Chamber of Commerce promoted Aviation Day at Omaha's new Municipal Airport. Twenty-five thousand of us watched the planes on the National Air Tour and the local pilots as they put on a great air show. The star of the show was the huge Ford Tri-Motor, sixty-eight-foot wingspan, nine-passenger, radio-equipped, upholstered, and equipped with wicker chairs and an electric refrigerator (which the passengers said was stocked with "Near Beer." Ha!). Boeing put on a Pageant of Mail Service featuring their brand new 40B mail plane and a horse-drawn stagecoach.

By spring 1927, there were five teams in the United States and France getting ready for a shot at the Atlantic. On April 26 the Keystone Pathfinder tri-motor (Wright Whirlwinds) biplane *American Legion,* piloted by Lt. Cmdr. Noel Davis and Lt. Stanton Wooster, crashed on a maximum-load takeoff test flight at Langley Field, Virginia. Both Davis and Wooster were killed. The four teams left kept right on working, and a legend started.

CHAPTER THREE:
LINDBERGH,
THE SECOND DAY THE WORLD CHANGED,
AND THE OMAHA AIRPORT

Twenty-three years after Kitty Hawk, one night in November 1926, a twenty-five-year-old airmail pilot flying a DH-4 on the St. Louis-Chicago run for Robertson Airways (owned by W. D. Robertson, who flew his Canuck at the Omaha Aero Congress in 1921) decided that he was going to fly from New York to Paris by himself. Charles A. Lindbergh's dad was the Congressman from Minnesota who co-authored the Eighteenth Amendment to the Constitution; his mom was a school teacher in Little Falls, Minnesota. Charles dropped out of the University of Wisconsin after one year, hopped on his Indian motorcycle and sped off to Lincoln, Nebraska. On April 1, 1922, he paid Ray Page (President of the Nebraska Aircraft Corporation) $500 to learn how to fly, maintain, and repair an airplane, and started at the East "O" shop.

This was before Page started building the Lincoln Standards, so their airplane fleet consisted of two original Standards with a lot of their modifications, one-half owned by Errold Bahl. Their field was still on the south side of town. I. O. Biffel (who taught Thor Bronderslev) was Lindbergh's instructor. Harlan "Bud" Gurney, who also was a student and became Lindbergh's life-long friend, gave him the nickname "Slim." It took Lindbergh a month-and-a-half to get the ten hours of dual the school required before solo flying. Biffel was struggling with flying; he had watched his long-time buddy, Ed Gardiner, stall out at the top of a loop, whipstall, lose a wing off his old Standard, and die in the crash.

When Biffel finally certified that Lindbergh was ready to solo, Page wouldn't allow it because Lindbergh didn't have the $500 damage deposit. The real reason was that the flying school was down to one airplane—the one he owned with Errold Bahl—and Bahl had just bought him out for $1,000 and was coming to get the plane. The Harding, Zook and Bahl Airplane Company had just broken up. There just wasn't enough demand for the great *Lark*. The same story was true for almost all of the airplane builders in 1922. The same story's true in 1994. There are no small private-type airplanes being built by airplane companies, and "fun-flying" is practically gone.

280

Zook and Harding took the one *Lark* left. They later sold it to a fellow in Fredonia, Kansas, where it just sort of disappeared. Bahl, probably the best acrobatic pilot around, bought the Standard to go back into the barnstorming business. He took the Standard up over the field and "wrung her out." Lindbergh was really impressed. He decided then and there that he was going to fly with Bahl.

Lindbergh talked Bahl into letting him go along as his helper, mechanic and general flunky at no pay while paying his own keep. Though Lindbergh didn't solo, he got a lot of experience, even learned to "wing walk," and made several parachute jumps that summer with Bahl. Bahl later recalled,

> One stunt Lindbergh did with me was to stand on top of the wing of the plane while I looped the loop. He had read about others doing this stunt so decided to try it. We rigged up wires to hold his feet and then tried it one afternoon at Capitol Beach [on the west side of Lincoln]. I made the loop all right but came out of it in rather a hurry. The jolt caused Slim to fall to his knees and tore two holes in the top of the plane that nearly broke his heart.[2]

Lindbergh told a story about his wing-walking while flying with Bahl, a story which could have changed history:

> The day was warm and clear, but the air was rough, yet the Air Show had to go on. At noon Errold flew over the main street while I walked the wings and waved to the people a hundred feet below. Next I let myself down on the tubular wing skid, hung below the wing spars to protect the wing tips during the landings. I hung by my knees during the first "pass" above the town. On the next pass I hung by my hand and waved to the folks with my left hand. My arm grip was so well developed by gymnastics I had no fear as to my strength. Suddenly the plane must have flown into the vortex of one of the many "Dust Devils" that were so numerous that day. A tremendous jerk snatched the wing skid out of my hand and left me looking at the hard packed street in the center of Calloway on the western Loop River of Nebraska. I gasped one breath and then suddenly felt the

---

[2]Gladys M. Bahl, "The Forgotten Pilot: Errold G. Bahl, Early Instructor and Friend of Charles A. Lindbergh," photocopied manuscript, San Bernardino, California, 1982. p. 19.

skid slam back into the palm of my right hand, fingers closed and quickly the left hand had the grip doubled. The worst part came next, oh that street looked hard, although I knew I was safe, the scare was so severe I had difficulty climbing back atop the lower wing.[3]

Another time, on June 19, 1922, Bahl, with Lindbergh in the front seat, landed his Standard on the sixth fairway at the Auburn, Nebraska, Country Club, when his engine quit. He wanted to talk to the Secretary of the Nemaha County Fair and Livestock Association about performing a flying exhibition at the fair in September and figured a "forced landing" would get some attention. It sure did!

Rev. R. Edgar Elliott, who served as pastor of Auburn's Avenue Methodist Episcopal Church until 1929, and his usual foursome (C. A. Nordland, International Harvester dealer; Col. Herman Ernst, Secretary of the Agricultural Association; and R. F. Neal, one of the Board of Governors of the Country Club) were approaching the sixth green when Bahl landed. Reverend Elliott had had Bahl's family in his flock when he served the Methodist Episcopal Church in Humboldt, Nebraska. The *Nemaha County Herald's* June 22, 1922, story said:

> Lieut. Bahl then expressed a desire to secure permission to leave his plane where it was for the night. . . . the desired permission for parking the plane over night was secured. . . . Lieut. Bahl saw active service as an aviator during the war and is now engaged in giving exhibition flights. He and his mechanic were the guests of Rev. Elliott and his family during their stay here.[4]

Nordland's daughter, Frances Holmgren, told her father's story of Lindbergh's visit to Auburn. He told her that Bahl had introduced his mechanic as "Slim" and that he never got his last name. Bahl didn't get a deal to fly at the fair. When Lindbergh became famous in 1927, Mrs. Elliott remembered the tall, polite, quiet young aviator who spent the night as their guest in 1922. From then on she called his room "The Lindbergh Room." The

---

[3]*Ibid.,* 13.

[4]*Nemaha County Herald,* June 22, 1922.

house he slept in is still there, and I had my picture taken standing in front of it.

In 1923 Lindbergh finally bought an old Jenny in Georgia, and barnstormed for a year all over the countryside—Texas, Oklahoma, Kansas, Nebraska, Iowa, Minnesota, Wyoming and Colorado—where he wrecked it beyond repair. He joined the United States Army Air Service and graduated from flying school at Brooks Field, Texas, in March, 1925. (I flew at Brooks during World War II.) He flew in the Missouri National Guard, then went to work for Robertson. That's when he decided to fly alone across the ocean.

When Lindbergh got back to St. Louis, he talked to Robertson about his idea. Robertson said he himself would invest money and introduced Lindbergh to other St. Louis people. Lindbergh raised $10,000 in a month, and, with his own $3,000, figured he had enough money for the task. He told himself if he could get the Wright-Bellanca, he could fly across the Atlantic. He went to New York and talked to the Wright Aeronautical Corporation. They told him that Charles A. Levine bought the Wright-Bellanca WB2 and was setting up a company to build them. So he talked to Levine and Bellanca. Bellanca thought his idea was great and said with some modifications and extra gasoline tanks the WB2 could fly the ocean easily. But Levine wanted $25,000 for the plane, and Lindbergh said he didn't have that kind of money. Levine said that he'd donate $10,000, so the plane would only cost $15,000. Lindbergh said he'd take it. Then Levine dropped the bomb: He said *he'd* name the pilot and rider. Lindbergh told him to forget it and didn't get over being mad until he got back to St. Louis the next day.

Robertson introduced Lindbergh on the telephone to Claude Ryan, builder of the high-wing monoplane that flew the airmail West Coast routes. Lindbergh told Ryan that he wanted a Whirlwind-powered plane that would fly nonstop across the Atlantic. Ryan's company had a severe cash shortage, and he was in the process of selling out, but he did some figuring and said $7,500 plus the engine (it actually cost $10,580) and three months' delivery time. Lindbergh wired Ryan $3,000 and arrived in San Diego on February 23, 1927. Ryan's people, with Lindbergh's help, modified their Model M-2 for the man who would fly it.

On April 28, the silver-painted plane was rigged, and Lindbergh test flew her at the Dutch Flats Airport in San Diego, California. She flew like a hawk, a little wild, but he was completely satisfied. The *Spirit of St. Louis* was painted on her burnished aluminum cowling, the official Department of Commerce *NX211* was painted on the top of the right wing and the bottom of the left, *NX211* on the top of each side of the rudder, *RYAN* and *NYP* on the bottom, and she was ready for destiny.

By May 1, 1927 the trans-Atlantic competition boiled down to the final four. The Lavesseur PL-8 *L'Oiseau Blanc*—an open-cockpit, water-cooled V-12, 450-horsepower, French biplane flown by France's most popular World War I ace, Charles Nungesser, with François Coli as navigator—was going to fly from Paris to New York. The Wright-Bellanca WB2 *Columbia*—a Wright Whirlwind-powered, high-wing, cabin monoplane—was piloted by Clarence Chamberlin. The Fokker Tri-motor C-2 (Wright Whirlwinds), *America*—a full-cantilever, high-wing monoplane, modified version of the plane Richard Byrd flew over the North Pole—had pilots Bert Acosta and Bernt Balchen, Byrd as navigator and George O. Noville as radio operator. The *Spirit of St. Louis* was the fourth contestant.

Lindbergh advised the Aero Club of America on May 6 that he was officially entering the competition for the Orteig Prize. He was waiting for the weather to clear so he could fly back east when the crew at Ryan's got the news that Nungesser and Coli in the *L'Oiseau Blanc* had left Paris on May 8. Nungesser was sighted west of Ireland five hours after takeoff by a ship in the Atlantic. They had enough fuel for forty-three hours of flying and would have to average seventy-five miles per hour to make it. Nobody ever saw Nungesser or Coli again.

Then there were three. It was really down to two because on April 18, Anthony Fokker flew the *America* on its maiden flight at Teterboro Airport, New Jersey, with Byrd, Floyd Bennett and a radioman as crew. The flight went well, but they hit a soft spot on the field on landing and flipped the plane over on its back. Byrd fractured an arm, Bennett broke a leg, but Fokker and the radioman weren't hurt. The *America* wasn't too badly damaged. Fokker said he'd have her flying in two or three weeks.

On May 10, Lindbergh left San Diego at 3:55 p.m. and arrived at St. Louis a little after 6 a.m. He shot a 160 mile per hour fly-by and pulled her up 1,500 feet in a climbing turn. It got his Lambert Field friends' attention. He had breakfast with them and spent the day showing off the *Spirit* to his sponsors and backers. He left for New York the next day and arrived at Curtiss Field, Long Island, in a rainstorm. The Aero Club had a hangar for him, and the Curtiss-Wright people met him and offered to help in any way. He had some bad carburetor icing on his flight from California, so the engine people installed a carburetor air-heater (which probably saved his neck on the Atlantic flight).

The day after Lindbergh landed at Roosevelt Field, Errold Bahl drove out to field, and the two aviators had a great reunion. Errold was going to graduate school at Columbia University and flying off of Teterboro Field at the time.

The press, who were living at the fields and feeding copy to their papers, mentioned that a "Charles Lindenberg" had entered the race, but they didn't give him much of a chance. The *Bee* printed that a "Charles Lindenberg" had entered. People in Omaha called Lindbergh "Lindenberg" right up to the day he flew the ocean. We read every word about the flyers, felt we knew them, and had our favorites. Mine was Byrd, but I didn't know Chamberlin was from Iowa, or I'm sure he would have ranked number one. Anticipating Lindbergh's flight across the Atlantic Ocean has to be one of my strongest memories. We lived it and talked about it; our teachers let us talk about it at school; we ran home to read the paper; our moms and dads talked about it. What an exciting time it was!

Lindbergh never did get used to the press. They finally spelled his name right and dubbed him "Lucky Lindy, the Lone Eagle" and infuriated him with a headline "Flying Fool Adopts a Mystery Air, Indicating a Quick Takeoff." They also called him the "All American Boy," a nickname that stayed with him.

That whole week the weather here and over the ocean was windy, cold, rainy, no ceiling, and blustery. Byrd offered both Lindbergh and Chamberlin the use of adjoining Roosevelt Field with its mile-long runway. Byrd had it leased for his own flight. Lindbergh and Chamberlin became pretty good friends. They had a lot in common—they were both Midwesterners, barnstormers and each other's biggest competition.

Lindbergh really worried about the Wright-Bellanca because he knew it was a good airplane, and he recognized Chamberlin as a darn good pilot. In fact, just two weeks earlier, Chamberlin and copilot Bert Acosta had broken the world endurance record. They stayed up fifty-one hours eleven minutes—more than enough time to fly to Paris.

Then Charles Levine decided that he was going to ride with Chamberlin, so he fired Lloyd Bertaud whom he had hired to be Chamberlin's navigator. Bertaud persuaded a New York judge to issue a restraining order which held up the flight, temporarily grounding the *Columbia.*

On May 19, Lindbergh checked with the weather bureau and was told that a high-pressure system would displace the storms from the North Atlantic to Europe. That did it. He decided to go. On May 20, 1927, at 7:45 a.m. Eastern Time, the *Spirit of St. Louis* got into the air.

And it wasn't all luck—he had a fine airplane with a superb engine (which, by the way, was named Wright). He was completely dedicated to flying. People who knew him said that all he ever thought about was flying and airplanes. His flying friends, like Gurney from Lincoln and airmail flying days, said Lindbergh didn't even think about girls. Lindbergh was one heck of a pilot. He flew everything built—from a Standard to a Jenny, a Canuck, a DH-4, a Ryan, a Lockheed, a Me-109, a Ju-88, a FW 190, a P-35, a B-24, a P-38 to an F4F, an F-80 and a barn door. Lindbergh also knew he and the *Spirit* would make it because he had checked and watched every nut, bolt, rivet, cable, spar, thread and rib, every piece of it being put together. He was the only one that ever flew her.

When "Lucky Lindy" and his *Spirit of St. Louis* landed at Le Bourget Airfield, near Paris, France, on May 21, 1927, Omaha and the rest of the world went crazy. Yes, Lindbergh changed the world when he landed at Le Bourget. He was on the Wrights's right hand in the airplane and aviation story. After that, everywhere Lindbergh went thousands gathered. Lindbergh never got used to it, and it haunted him and his family all his life.

Others flew for Europe after Lindy. On June 4, 1927, Clarence Chamberlin finally got the *Columbia* off for Europe with Charles Levine as a passenger. The Wright-Bellanca performed perfectly. Three thousand nine hundred and eleven miles later they landed in a wheat field, out of gas, about a hundred miles west of Berlin. It was the longest nonstop flight ever made. Some two weeks later, on June 29, Richard Byrd and his crew took off in the *America.* They got to Paris, but the weather was so bad they couldn't land. Instead, they started back west to where they last had visibility. Running out of gas, Acosta and Balchen ditched the *America* in the surf near Courseulles-sur-Mer, France. They weren't hurt, however, and rowed to shore in their rubber raft. The *America* wasn't too badly damaged and was saved as well.

In May and June of 1927, the world witnessed the first three successful flights made across the Atlantic Ocean. Today there are probably several hundred trans-oceanic flights going on everyday. In January 1991, our grandson Pfc. Ryan James Adwers left Forbes Field, Topeka, Kansas, and arrived in Saudi Arabia in some twenty to twenty-four hours—a distance of 6500 miles and spanning several continents. The three airplanes that made it across the Atlantic had a total of five Wright Whirlwind engines—the best aircraft engine ever made to that date—and they all operated until they landed or ran out of gasoline.

And the flights were all connected to Nebraska: Charles Lindbergh learned how to fly in Lincoln. Guiseppe Bellanca built his first airplane in Omaha. Clarence Chamberlin was born and raised in Denison, Iowa, sixty-five miles from Omaha, and had flown Bellanca's C-4 from Omaha's Ashmusen Field. Bert Acosta set the world speed record at 176 miles per hour in the Pulitzer Race in Omaha in 1921.

The Omaha Airport

Lindbergh's flight got Omaha and its Municipal Airport off dead center. As of May 1927, it still had no hangar, fences, roads, lights, runways, fuel or oil facilities. It was just a 195-acre field, really not much more than a pasture. This was partly because of some legal problems that seemingly couldn't be solved: The city feared legal liability for airplanes flying there, and the voters would have to approve bonds to finance any building. In addition, no one in the aviation business seemed willing to make any investment. This all changed with Lindbergh's flight. The people wanted action.

On July 26, Omaha's American Legion Post picked up the ball. A group of Legion leaders, Chamber of Commerce people and other civic leaders met at the Fontenelle Hotel and laid plans for building a hangar. The Street Department couldn't afford to spend much on the field, and no funds had been included in the general budget. It was decided at the meeting that the Legion would start a popular subscription drive to raise the $30,000 needed to build a hangar. The Legion would own it; the city would let them use the ground; and the hangar would be available to the city and other interested parties.

The next day they kicked off the campaign—Gould Dietz, Clarke Powell, Everett Buckingham, Arthur Fetters, Gene Eppley, George Brandeis and others, mostly World War I veterans and Legionnaires. Allen Tukey was the Chairman, and eight teams of over a hundred men (headed by Gene Holland, Hird Stryker, Sam W. Reynolds, Amos Thomas, Frank R. Landers, T. J. McGuire, Anan Raymond and Walter Byrne) hit the streets. Tukey said that they could start building the hangar about ten days after they got the money. They raised $5,026 the first day.

Then on Friday, July 1, 1927, at 2 a.m., the Post Office Department terminated their operation of airmail service. Ruben Wagner landed his Douglas at Ft. Crook with the Cheyenne mail and turned it over to Boeing Air Transport Inc. The mail sacks were put into a brand-new $26,000 Boeing 40B, and pilot I. O. Biffel (who else?) hopped off a few minutes later with the first private-carrier load of mail for Chicago. Government and Boeing officials were in Omaha for the historic changeover. Later that day, six hundred pounds of mail arrived from Chicago in the last flight of a DH-4, piloted by E. M. Allison, one of the original pilots in 1920. Thus the Post Office airmail service flew into the "Wild Blue Yonder" almost seven years to the day it started, and an old Red-tailed Hawk dipped his wings. Meanwhile, with just under $11,000 in funds, the Omaha Municipal Airport hangar drive was slowing down. The Chamber and Legion decided to have a big "Aviation Day" celebration at the field on Sunday, July 10, to raise money and to get Boeing involved at the Municipal Field. Fetters was named Field Marshal. He was assisted by Victor Roos, Eugene Eppley, Guy Smith, Gould Dietz, Clarke Powell, J. W. Elwood and George Brandeis.

Aviation Day was a huge success. Omahans swarmed all across the field. Company K, the Union Pacific's National Guard unit, the police, Chamber and Legion officials and volunteers couldn't keep the crowd of twenty-five thousand away from the planes. I know because I was there. C. E. "Bob" Steele in his Alexander-Eaglerock and Eddie Stinson and family in his brand-new Stinson-Detroiter showed some great flying. Charles Kenwood "fancy-flew" his Jenny and Lawrence Enzminger his Travel Air. Capt. B. F. Giles of Ft. Riley flew an Army Scout. Boeing put on a show called "The History of the Mail Service" and flew their new 40B mail plane. The star of the show, though, was Clyde Ice's Ford Tri-motor.

Two weeks later, interest in aviation grew to a frenzy when, on July 15, 1927, Omaha was advised that Lindbergh and the *Spirit of St. Louis* would fly to town on August 30 and stay overnight. Lindbergh was coming! This was bigger than the President of the United States, the King of England, Mary Pickford, Clara Bow, Tom Mix, Babe Ruth and Red Grange all coming to Omaha on the same day.

Then, two days later William P. McCracken, Assistant Secretary of Commerce for Aeronautics, inspected the field with Commissioner Dean Noyes. He said that unless the field was graded, filled and marked, Lindbergh might not be able to land here. Noyes doubled his efforts and with much volunteer help got the pasture in good enough shape to be called an airfield.

Still, the hangar fund drive had bogged down at $13,500, so the Committee had a Second Kickoff Breakfast at the Fontenelle on August 10. They got $4,000 more in pledges that morning. That afternoon at the City Council meeting, the three commercial flyers at the airport vigorously protested the new ordinance that required that each flyer post a $10,000 bond for one passenger and $5,000 for each additional. Cecil "Bob" Steele, Charles Kenwood, A. H. Glaven and Andrew A. Risser all said that the prohibitive premiums would mean the end of commercial flying. To solve the airport problem, however, the Airport Boosters suggested that they and the City Council get together with the insurance people and work it out.

Andy Risser's folks farmed near Wisner, Nebraska, and he bought a Travel Air OX-5. The Travel Air, OX-5 powered biplane was a darn good airplane. Its ailerons were only on the top wing and extended beyond the wing tips. They gave the ship a distinctive look. Later models had Whirlwind and Challenger engines. Travel Air Manufacturing Company was owned by Clyde Cessna, Lloyd Stearman, and Walter Beech. Need I say more? These three giants in the airplane building business had all worked and flown for E. M. Laird's Wichita Aircraft Company in Wichita, Kansas, in 1921 and all had flown Laird Swallows at the Omaha Air Congress and Pulitzer Races. Cessna had always favored monoplanes, so later he pulled out of Travel Air and started building full cantilever, high-wing cabin planes. Stearman also went out on his own and built one of the fastest biplanes ever, which evolved into the great PT-13 and PT-17 World War II Primary Trainer. The first time I flew one was at Randolph Field, and I am flying one on the back cover of this book! Beech stayed with Travel Air, which was purchased by the Curtiss-Wright Corporation in 1929. Along with acquiring ten percent of the commercial airplane business in the country, they got a bright young engineer—Ted Wells.

Ted Wells had a long and illustrious career with Beech from 1929 to the '70s. Wells was an excellent pilot who flew and won some great races with the Travel Air "Mystery Ship," a low narrow chord wing with a big cowled radial engine that was one of the world's fastest planes. He also was a county fair snipe sailor and won many national titles. He had worked with Herbert Rawdeeno and Walter Burnham, designers of the highly successful two-hundred mile per hour Mystery Racers. Beech could only stand his fancy desk job for so long, and when he pulled out of Travel Air to build biplanes he took Wells with him. They designed and built the immortal Beech Staggerwing.

"First Family" of the Air

Three Flying Kenwoods Still Go Up—Eldest Learned Four Years Ago.

Jack, "Slim," and Charles Kenwood

As a result of the new board ordinance Omaha passed in 1927, Risser went to Norfolk, Nebraska, where he started a flying school. In 1928 he taught Ethel Tillotson how to fly. Tillotson was a good student and soloed the Lincoln Page Trainer after nine hours. She appears to have been Nebraska's second woman pilot. Tillotson made application to the Department of Commerce in Sioux City for a private license. This was never issued because she "spun in" on June 22 and was the first woman killed while flying herself.

Risser ran his service for many years, and a book, *Tower, This is Andy,* has been written about it (1992) by Robert L. Carlisle.

Steele kept his plane at the North Field, so Kenwood was the only one left who flew out of Muni. In mid-August, 1927, though, John Kenwood wrecked his brother Charles's Jenny when he crashed into a clump of oaks near Rainbow Point in the bluffs, just east of the field. This temporarily put the Kenwood Flying Service out of business and alleviated the city's bonding problem.

The Kenwood brothers flew in Omaha for many years. Charles was the oldest and started flying in 1921. He was killed flying the mail in Wyoming in 1929. John flew until he died in 1988. I knew John. He "slow timed" and delivered B-24s built at the Detroit Ford Plant during World War II. Manuel, better known as "Slim," was my instructor when I got my license in 1938. He could fly a J-3 better than anyone I ever rode with. He originated the saying, "I could land one on a tennis court if you take the net down." He flew for the Northern Natural Gas Company and could drive a Cessna 195 like he owned it—he could fly Northern's pipeline from Omaha to Ft. Stockton, Texas, without gauges or looking at the ground. I was flying with him one day over southwest Kansas when he got tired and asked me to take her, then promptly went to sleep. I thought I could lose him, so I got off course about twenty miles and changed the heading about twenty degrees, then let him sleep about twenty minutes. I woke him up and told him I was lost. He made a ninety-degree turn to the left, turned back and said, "Garden City is back about fifteen miles, Dodge [City] is south and a little east over there, and fly the damn airplane straight."

The Kenwoods also played a role in starting Ted Wells's flying career. Wells got hooked on flying when his dad took him over to watch the mail planes and to see an air show at Ashmusen Flying School in 1921. Fetters, who knew Wells's dad, was building a new plane with Clara Ashmusen, so he arranged for Wells to help. Charles Kenwood was flying at Ashmusen Field and told Wells that when he turned sixteen he would teach him how to fly, which he did in his Jenny. Wells was Kenwood's silent partner, so when John Kenwood wrecked the Jenny, Wells talked his dad into buying him a brand-new Travel Air (which didn't get finished until after he went back to school at Princeton).

On Friday, August 17, 1927, the whole town and countryside was getting ready for Lindbergh, when suddenly they realized that Clarence Chamberlin had come home to Denison, Iowa. On that day, Chamberlin's hometown gave the second man to fly the Atlantic a parade that rivaled New York City's. Over ten thousand people lined the streets and cheered their boy. They gave him a hero's reception and a huge banquet. Iowa Governor John Hammill made him a colonel in the Iowa Guard and presented a bronze plaque from the state. Many Omaha people were there, including some of Chamberlin's old flying buddies, Andy Nielsen, Victor Roos and Maj. B. Q. Jones, a Ft. Crook air officer. The day ended with a grand dance at the beautiful Uwanna Ballroom on the Lincoln Highway just south of town.

The next Tuesday, August 23, Omaha honored Chamberlin. He arrived on the 6 a.m. Northwestern with his father, E. C. Chamberlin. He was driven to Ft. Crook, where he and an old Army friend, Sgt. O. W. Hayes, flew an Army DH-4 over town and then landed at the Municipal Field. There he was met by a reception committee of acting Mayor John H. Hopkins, Leo Bozell, Frank Landers (Commander of Post Number One), Arthur Fetters and W. A. Ellis. The group drove to the Fontenelle Hotel, where almost two hundred people had gathered for breakfast. It was an informal meeting, and Chamberlin impressed and entertained everyone with his aviation knowledge and wit. When asked about Charles Levine's sometimes unusual behavior, he grinned and said that Levine was a little different and difficult, not a very good pilot, and he also owed him money. Chamberlin was very generous with his time and was given a tour of Omaha with many stops where crowds gathered to meet him. He was greeted with great cheers by the hundred boys at Father Flanagan's Boy's Home.

He was honored at a noon luncheon attended by nearly five hundred men and women at the Chamber of Commerce, where Victor Roos introduced him. He told in detail the story of his transatlantic flight, which was remarkably free of problems. He had hoped to make Berlin, but ran out of gas, and made a dead-stick landing in a hayfield. He said the locals just couldn't believe he came from America. According to Chamberlin, the flights made across the Atlantic had made it plain that passenger ships used for commercial aviation must be larger, stronger, safer and equipped with radio apparatus. He also noted that Omaha needed a finished, lighted landing field and larger hangar facilities to stay on the aerial map. He concluded that young people, the pilots of the future, would make better pilots because they will have grown up with the planes as a part of their lives. He was given a standing ovation.

That evening the entire Chamberlin family was honored at a dinner in the Palm Room of the Fontenelle. More than a hundred invited guests attended. He thanked Omaha for such a great day and said he hoped he had helped a little in the hangar drive. He did, with Chairman Tukey reporting that the drive had hit $21,000 that day.

The number two aviator in the world had just been here. Now Omaha was getting ready for number one. It seemed that everybody was in the act. Everywhere were committees, church groups, women's clubs, civic clubs, lodges and fraternal organizations. Lindbergh was coming to Omaha. Arthur Fetters was in charge of all the arrangements at the field. The Street and Park Departments had men and machines manicuring the field, and they painted a huge circle, three hundred feet in diameter, in the middle of it. Col. Amos Thomas had 250 Nebraska National Guard Troops safeguarding the entire area. A fifteen-mile parade route was laid out from the field to Florence Boulevard to downtown and Ak-Sar-Ben Field where Lindbergh would speak. The police department, fire department and hundreds of Boy Scouts lined the parade route controlling the crowds.

Acting Mayor John Hopkins and Governor Adam McMullen headed the reception committee of councilmen, dignitaries, American Legion and Chamber of Commerce leaders. Lindbergh stayed at the Fontenelle, where a huge dinner was held. One thousand tickets were sold at six dollars each. Downtown stores were decorated with bunting and flags. Everyone in Omaha, Council Bluffs, Elkhorn, Valley, Waterloo, Bennington and the whole countryside seemed to be planning to be there, including Mayor Jim Dahlman, who had been laid up for eight weeks with a broken hip—he had been hit by a streetcar in front of City Hall.

On the day Lindbergh was to arrive (Tuesday, August 30) traffic was horrendous on Locust Street, which was a dirt road. It seemed every car and truck in the county was there—and dust was everywhere. I met Don "Tiny" Ryan and Ralph Dickerson at 24th and Pratt Streets, and we pushed our bikes most of the time. When we got to Carter Lake Boulevard, the police were stopping traffic. We pushed and rode north to the field, chained our bikes to a cottonwood tree, and went the rest of the way on foot.

They had fenced in an area about a hundred feet square, and soldiers were standing around the outside with a big crowd. There were quite a few people out on the field—pilots, somebody said. Inside the fence with a lot of people and a gasoline truck was Frohardt with his camera, and he told us to climb over and act like we were his helpers. Mayor Dahlman and Governor McMullen were sitting in a big Packard touring car parked just outside the fence.

At about 1:30 p.m. the advance plane, a Fairchild, flew from the east over the field, circled twice, then landed. It taxied to about a hundred yards from the fenced area; the pilot cut the switch; and a man got out. Some officials ran out to the plane, and they had some kind of a meeting. Lindbergh would not land or taxi if people were on the field. Remember that the crowds surrounded him at Le Bourget before he had cut the switch, and he had to be rescued, and they tore all of the fabric from the *Spirit.* The officials came back; police drove out on the field and chased people off; the army kept people back; they fired up the Fairchild and taxied it into the fenced area; and they closed the gate. We were still inside, waiting excitedly.

Then it happened: Straight out of the east over Rainbow Point at about eight hundred feet, the *Spirit of St. Louis* flew right over the big circle and turned ninety-degrees south. I'll never forget that big *NX-211* on the bottom of the left wing. He turned ninety-degrees east, ninety-degrees north, flew about to the river letting down, ninety-degrees west on base leg, then ninety-degrees south on final. He greased the landing. They had taken the gas tank out so he could see. He taxied very slowly towards the enclosure. The crowd was held back; the gates were opened; and he taxied in. The Packard was driven in, and the gates were closed. The *Spirit of St. Louis,* Lindbergh, dignitaries, politicians, newspeople, police and Bob Adwers (keeping very low and small) filled the area. I almost got away with it, but one of the men in blue spied me and said, "Hey kid. What the heck are you doing in here?" and deposited me over the fence. Tiny and Ralph had suffered the same fate. I was going to get in Frohardt's picture come heck or high water, and I did. I'm the kid with the cap to the left of the two soldiers just above the right wing of the Fairchild. Charles Lindbergh is standing under the wing of the *Spirit of St. Louis.* An old Red-tailed Hawk watched from a cottonwood tree on the south side of the field.

William T. "Buck" O'Hanlon, a friend of mine who played alongside me on Central High School's freshman football team in 1929 and who lived down the street from Mayor Dahlman, gave the following account of Lindbergh's day in Omaha:

Lindbergh's arrival at Omaha

We knew that the Mayor was going to ride with Lindbergh. Maybe he would bring him home with him, so we hung around his house. Unbelievably, about 3:30, we heard sirens over towards Center Street. They came along Hanscom Park, turned east on Woolworth, and here came four police motorcycles and several cars, and the big Packard stopped in front of the Mayor's house. Lindbergh was sitting in the back seat with him. We crowded around, and he said, "Hello boys." The cops helped the mayor into his house, and the procession headed towards downtown. I stood right beside Charles Lindbergh.

O'Hanlon has the rare distinction of having served as a flying officer in the Royal Canadian Air Force, the British Royal Air Force and the United States Army Air Corps during World War II. He also is the only man I've ever known who soloed in a Gypsy Moth.

After Lindbergh left, Omaha returned to just being Omaha. The hangar drive went over $30,000, and they started building in November. Aviation was in Omaha to stay.

CHAPTER FOUR:
AVIATION CONTINUES TO GROW (1928)

Aviation took off in Omaha in 1928. The new hangar was dedicated April 1. Andy Nielsen and Charles Kenwood were the first operators to use it. They had a shop in the back, put in a gasoline pump, carried a supply of oil and other materials, and gave lessons, rides and an occasional charter flight. People still loved to drive out to the airport and watch what flying there was.

Arthur Fetters, one of the most important men in Omaha's aviation history, issued the following challenge in a March 1928 *Omaha Chamber of Commerce Journal*:

> Is Omaha an air-minded community? Not yet. While there has been certain activity along this line by private citizens, clubs, committees and associations, much remains to be accomplished before Omaha can be called an air-minded community. Lack of vision for the future is largely responsible for our present status.
>
> 'Omaha as a Center of Air Industry by 1940.' Let us adopt this as a slogan.
>
> It is fortunate indeed, that Omaha is located on the Trans-continental airway, and this has given the proper stimulus toward making Omaha an important airport.
>
> After a great deal of hard work, the Chamber of Commerce committee on Aerial Transportation, ably assisted by some of the city commissioners, has established the nucleus of a Municipal Airport, but a great deal still remains to be done to develop this field into a first-class airport that will attract fliers here from all parts of the country.
>
> The Boeing company, operating air mail between Chicago and the coast, have developed a splendid organization. . . .
>
> The Department of Commerce is doing its best in establishing and lighting air routes. . . . The Weather Bureau is likewise . . . establishing special and more frequent weather service. . . .

The airplane has shown remarkable possibilities as a means of business transportation.[5]

Early in 1927, the Post Office had announced that it would take bids on the United States transcontinental trunk route. The office had already contracted out the feeder routes, but it wanted bids for the New York to Chicago and Chicago to San Francisco routes. William E. Boeing, a very successful Seattle airplane builder, had just designed a new plane for flying airmail. His young assistants and designers talked him into bidding for the western route. He got the route, having responded with a much lower bid than his competition.

The new Boeing Air Transport company was a huge success from the start, mainly because of the new Boeing Model 40-A, the best mail-carrying airplane built to date. The 40-A was a fabric-covered biplane. The pilot sat in an open cockpit, but it had a closed cabin for two passengers and still could carry four hundred pounds of mail. It was powered with the new air-cooled, radial, nine-cylinder, 525-horsepower Pratt & Whitney Hornet. The 40-A would cruise at one hundred miles per hour and operate at high enough altitudes to get over the western mountains. Altogether, it was one heck of an airplane. Boeing hired former mail-service pilots, and they flew the 40-A like crazy. The new airmail company made money from day one (and still does today as United Airlines).

The *Bee* and the *World-Herald* ran airplane stories and pictures almost every day in 1928, like the one on C. E. "Bob" Steele's Flying School it featured. We kids used to ride our "wheels" (the 1928 name for a bicycle) up to Steele Field and hang around watching the planes and hoping a miracle would happen and we'd get a ride. Steele was always friendly, and every now and then he would ask us to do something for him. Ralph Dickerson and I spent a lot of time at the field and became well-known to the "Birdmen Gods," but still we got no rides.

Meanwhile, the Gibson Plantation Coffee Company's big Lincoln Standard had become a very familiar sight as it flew over town, which brings us to my first airplane ride. That year my big brother Ardon promised me an airplane ride for my birthday. We were to go on Sunday. The big day finally arrived, and we drove to

---

[5]A.H. Fetters, "Omaha, Air Capital of the Country by '40?" *Omaha Chamber of Commerce Journal,* March 24, 1928, p. 12.

For a Want Ad and 59 Cents
The Bee-News will give
A Ticket for an
AIRPLANE RIDE
in this fine machine

Each ride will last 7½ minutes and will make a round trip in the air covering ten miles. It will rise from and alight on the Municipal air field in East Omaha.

The Bee-News has made arrangements with the Gibson Plantation Coffee Co. for the use of its airplane to give these Want Ad rides. The Gibson company uses this machine regularly to distribute coffee in the territory around Omaha. The airplane is equipped with a 150-horse power Hispano-Suiza motor, has a wing spread of 44 feet 7 inches, and travels 100 miles an hour. The pilot is Fred Kelly. The ride may be taken next Saturday afternoon or Sunday; if weather conditions prevent flying, then the following Saturday or Sunday.

Candy

If you don't want an airplane ride.

To those preferring, The Bee-News will give free a pound of Cobb's Candy (from Kilpatrick's) with every 3-time Want Ad to start in Sunday's Bee-News, cash in advance, at the regular rate of 23 cents a line per insertion, presented at the office of The Bee-News.

ALSO

The Bee-News will give an additional insertion of the ad free of charge—four insertions for the price of the three insertions you pay for.

All Day Saturday
At The Bee-News Office

The Bee-News will give a ticket for a ride for 59 cents and a 3-time Want ad to start in next Sunday's Bee-News, cash in advance, at the regular rate of 23 cents a line per insertion if presented at the office of The Bee-News. The advertiser must be of legal age.

An Airplane Ride

For every Want Ad and 59 Cents Extra

ALL DAY SATURDAY!
At The Bee-News Office Only

the airport in my brothers' brand-new '28 Chevy. The Lincoln Standard could seat four, so my brothers Ardon and Jack and I hopped in. Pilot Kelly took off north. We flew over the Missouri River, turned west, climbed to six or seven hundred feet, turned south, turned east, throttled back, turned in and landed. I thought I was Lindbergh! The flight probably lasted about five minutes, but I had flown in an airplane. I really was hooked and knew I had to learn to fly. On August 4, 1929, Howard G. Gibson (owner of the company) was killed flying the same

Bob Adwers holding a model plane

plane when he hit some turbulent air near De Witt, Iowa.

A number of Omahans continued designing and building planes in 1928. By the summer of that year, Omaha Airways, Inc., designed and built a two-place plane, called the Sport, in a factory located at Commercial Avenue and Sprague Street. I used to go down and watch the workmen building the plane. The president of the company was Roy Furstenberg, who had flown and operated a flying service in Council Bluffs, Iowa, for years. The Sport was designed by master mechanic Harold Phillips, and L. D. "Dutch" Miller flew its first test flight.

At that time, the Department of Commerce certification requirements included fast-spin recovery, but the Sport was unstable and spun pretty wildly. One day that spring, Miller was testing her when she went into a fearfully flat spin. He couldn't get the nose down to break the stall, so he left her and pulled his chute. Afterwards, he and the Sport were hauled back to the field in the same truck.

Overland then hired an aeronautical engineer, William C. "Chet" Cummings (who later taught William Durand stress analysis) to make necessary modifications. Sport number two passed the tests and was certified. Charles Kenwood flew the Omaha-built plane and beat Andy Nielsen flying his Travel Air in a big race covered in a front-page story in the Sunday, May 20, 1928, *Omaha Bee.* The paper reported that

> they raced their planes at more than 100 miles an hour between the municipal airfield and Steele airport, four miles northwest of the field.
>
> Kenwood . . . won the race by a small margin. His plane, made in Nebraska, outdistanced a Wichita made Travel-Air ship, owned by Omaha Airways. . . .
>
> . . . Nielsen said. . . , "I've challenged Kenwood to race again a week from Sunday, at 11 a.m. . . . [, and] it will be a wonder if the result is the same."[6]

They were putting on a show hoping to get a bunch of people to come out to the airport and ride in their planes, and they succeeded.

On July 2, Walter and Arthur Myers, who operated the Omaha Grain Exchange's radio station WOAW, flew their homemade plane at Steele Field. The brothers were students at Steele's Flying School and had built their plane in their father's garage at 1451 South 15th Street. I didn't see their first flight, but I later saw them fly it many times.

After Omaha Muni began taking shape, Clyde Ice, a real trail blazer who ran the only flying service in South Dakota from Rapid City, started flying his Ford Tri-Motor into the field. His Ford, which he named *Wamblee Ohanko* (Lakota for "Swift Eagle"), really attracted attention. The biggest airplane around, it could carry twelve passengers. He often took passengers up on short flights. Ice flew "Ina & Al," whoever they were, in the *Wamblee Ohanko* on

---

[6]*Omaha Sunday Bee,* May 20, 1928.

September 9, 1928. I found the certificate of flight pictured below in a Louisville, Nebraska, antique shop a couple of years ago. Clyde Ice gave me a ride in the *Wamblee Ohanko* once, and I didn't have to pay. By the way, on May 28, 1992, he celebrated his 103rd birthday and was nominated to the National Aviation Hall of Fame.

In June 1928, the Army told Boeing they wanted them off of the field at Ft. Crook. No specific time was set, but they suggested Boeing look for another field. Boeing talked about using a proposed field near Bellevue, but by the end of 1928 Omaha city fathers had met all of Boeing's demands, and the company settled on Omaha Municipal Field. All those who had worked so hard for so many years to have a first class airport in Omaha had victory in sight. Boeing Air Transport was the key to aviation progress in Omaha because, as the carrier for transcontinental passenger and mail service, it was on the ground floor of the airline business. The day of the daredevil pilot operating a flying service was on the wane.

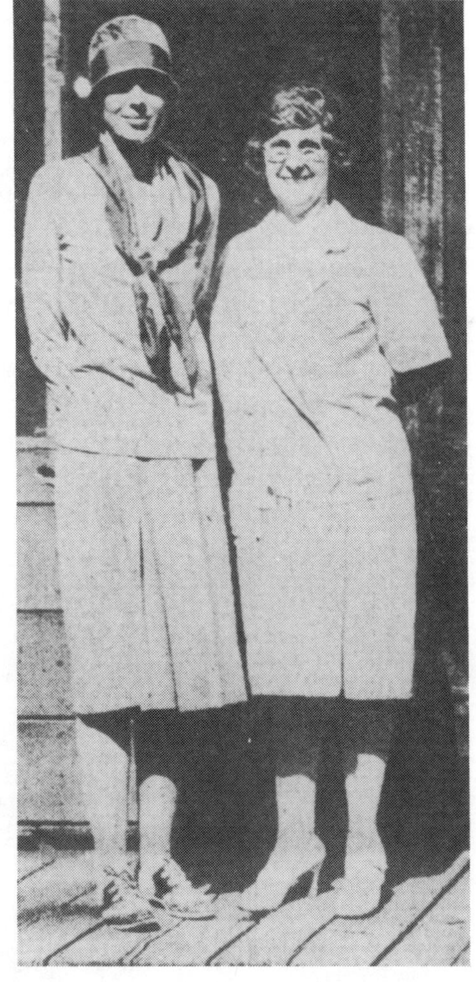

In October 1928 Amelia Earhart (or "Lady Lindy" as she often was called) visited Omaha and stayed at Cora Nelson's boarding house, as pictured above. The Atchison, Kansas, native was the first woman to fly the Atlantic. She stopped here overnight during a flying vacation.

One of the most memorable flying events I ever saw happened at a University of Omaha (24th and Pratt Streets) homecoming football game in October 1928. Frank J. Grace was President of the

Overland Tire and Rubber Company, which owned a Waco biplane they used for promotions. His twin daughters were University of Omaha cheerleaders and thought that dropping the game ball from an airplane would be a great first in college football. So did everybody else—President Whetstone, Professors Vartanian and Borglum, Coach Lloyd M. Bradfield and the team. Grace was agreeable, so that's how the game started.

At 1:30 p.m., the stands were full; the players were on the

Dropping the ball at Omaha University

field; and the Waco was coming in from the east about two hundred feet up. Pilot John Kirk flew over Meredith Street to about 27th, turned south, and flew east above the Beltline to 16th and came back to the field. The players from both teams were on the forty-yard line. There was supposed to be a big prize for the guy who caught the ball. Kirk slowed the Waco down when he was about a hundred feet high, and Grace threw the ball out. It came down like a shot. The players ran like heck, but didn't come close, and fortunately nobody caught it. The ball hit on about the twenty-yard line and didn't even bounce—it broke flat. Neither Harvard, Notre Dame, Miami, Texas, Colorado, nor the mighty Cornhuskers ever topped Omaha University's 1928 Homecoming kickoff.

With the new bond money, and advice and guidance from the new Airport Advisory Board, Commissioner Dean Noyes got the Omaha Municipal Flying Field smoothed, leveled and put in good shape in 1928. A road was built to the hangars from Carter Lake Boulevard. Underground storage fuel facilities were installed. Boundary lights were placed around the field, and the big cottonwood trees on the north and south landing approaches were cut down.

Boeing had behaved rather obstructively while the Chamber of Commerce, Legion, city and other aviation enthusiasts were working to get an airport for Omaha. While Boeing had been using Ft. Crook facilities, for probably a nominal fee, so they put stumbling blocks in the way of a move to the new Muni field. Some (like saying it was too foggy too often) were dodges, but by the end of 1928 the field definitely was taking shape.

Aviation continued its growth in Omaha. The American Legion sold their hangar to the brand-new Rapid Airlines, Inc., formed by Clyde Ice and others. They started flying passenger service between Omaha and Kansas City, and Omaha and Sioux City. Walter Klopp, treasurer of the pioneer A. T. Klopp Printing Company, was on the board of directors. He was learning how to fly, and the printing company had just purchased a three-place Waco. F. Robert Shields was Rapid's instructor.

Lawrence Enzminger, a Union Pacific locomotive-engineer-turned-airplane-pilot, formed the Midwest Aviation Corporation and started building a second hangar just north of Rapid's at Omaha Municipal. Midwest became the distributor for Travel Air planes, and they also ran a flying school and charter service. Most of the old Omaha pilots went to work for Midwest, and I think some of them were in the corporation. By summer 1929, they were a busy outfit. Clifford E. Burnham, Dutch Miller and Barney Burnham were the chief instructors.

Overland Airways was third at Omaha Municipal. Their plane, named *Miss Omaha,* flew very well. My research indicates that this was the sixth successful flying airplane built in Omaha and Douglas County, the first being the Baysdorfer *Hawk,* then the Loch-Coleman *Number One,* the Ashmusen *Number One,* Fetters's *Flyer,* the Ashmusen-Fetters *Bluebird* and the *Overland Sport.*

Other corporate airplanes flew off of Muni quite often. Enzminger sold the Bankers Reserve Life Company of Lincoln the latest Travel Air with the brand-new Curtiss Challenger engine in it. It was flown mainly by W. G. Preston, vice president and treasurer of Bankers. Andy Nielsen had a brand-new Curtiss Robin with a Challenger engine. The Robertson Division of Universal Air Lines, Inc., started flying mail and passengers back and forth from St. Louis to Kansas City to Omaha on May 1. They flew new six-passenger Fokkers and landed at Ft. Crook. This new service connected Omaha with the South and the lines which extended to Cuba, Central and South America. Boeing, of course, was flying both passengers and mail to Chicago and Cheyenne with connections to both the East and West Coasts.

On September 9, at 8:30 a.m., the First Annual Nebraska Air Tour, sponsored by the Chamber of Commerce to promote aviation across the entire state of Nebraska, left Omaha Muni Airport. The "good will caravan" consisted of a fleet of sixteen airplanes and thirty-seven men. Maj. H. J. Houghland, the air officer for the Army's Seventh Corps Area, was the flight commander. He led the flight in his Army Douglas. Lawrence Enzminger was the assistant flight commander, flying the new Travel Air cabin plane with four passengers. Andy Nielsen was the other assistant flight commander. He had Steve Spitsnagle, Ak-Sar-Ben's publicity director, with him. The star was Clyde Ice's Ford Tri-motor. Planes were sponsored by the Junior Chambers of Commerce of Omaha, Lincoln and Hastings. Other planes in the tour were from Holdrege, Grand Island, McCook and Fairbury. They made quite a sight as they took off and headed for Auburn and Beatrice. It was quite a tour, all in all very successful, and the people in the towns loved it. They got to see airplanes close up, and naturally the pilots had to hot dog and flew some great acrobatics.

One tragedy occurred when Lincoln pilot Frank Cropsey, flying a Lincoln-built Arrow, lost a wing pulling out of a "split-S" in a show at Columbus. He went out in style in front of a crowd of more than four thousand. There was a lot of flap about stunt flying afterward. Eyewitness Andy Nielsen told me that Cropsey just ran out of air and was hauling her back too hard. Incidentally, the Arrow Aircraft Company made darn good airplanes. One is hanging in the Lincoln Airport terminal lobby today.

Without a doubt the biggest aviation event that ever happened in Omaha (probably even in Douglas County, Nebraska, the United States, or on the earth), was that the kid who had said, "Hey Mom. See Bird? I Can Fly," flew an airplane. I carved airplanes out of wood, made flying models with rubber-band engines from the big kite-like pushers to Baby ROGs (rise off ground), folded paper planes, and read everything I could find about airplanes. I had a small test tube of Veedol oil that was from one of the engines of the *Graf Zeppelin* when it made its first trip to the United States, and I was pretty good at launching the paper Fourth of July hot-air balloons. I was making six dollars a week from my paper route and was putting a lot of it in my "learn-how-to-fly bank."

One day in the spring, Ralph Dickerson (a good friend of mine at Lothrop School) and I wheeled up to Steele Field to hang around.

FIRST
All-Nebraska Air Tour
September 9 to 14

*Organized by the air enthusiasts of Nebraska
to promote aviation in the state*

Itinerary of Tour

		Arrive	Leave
Monday Sept. 9th	Omaha		8:30 a. m.
	Auburn	9:15 a. m.	1:30 p. m.
	Falls City	2:30 p. m.	for night stop
Tuesday Sept. 10th	Falls City		8:30 a. m.
	Lincoln	9:00 a. m.	11:30 a. m.
	Fremont	12:15 p. m.	2:30 p. m.
	Norfolk	3:30 p. m.	for night stop
Wednesday Sept. 11th	Norfolk		8:30 a. m.
	Columbus	9:15 a. m.	1:30 p. m.
	York	2:15 p. m.	for night stop
Thursday Sept. 12th	York		8:30 a. m.
	Grand Island	9:15 a. m.	11:15 a. m.
	Kearney	12:00 Noon	2:30 p. m.
	Broken Bow	3:30 p. m.	for night stop
Friday Sept. 13th	Broken Bow		8:30 a. m.
	North Platte	9:30 a. m.	1:00 p. m.
	McCook	2:00 p. m.	for night stop
Saturday Sept. 14th	McCook		8:30 a. m.
	Holdrege	9:30 a. m.	1:00 p. m.
	Hastings	2:00 p. m.	4:30 p. m.
	Omaha	6:30 p. m.	

Program from Nebraska's first air tour, 1929

Steele was working on his Eaglerock with his chief mechanic, Robert Shields. We hung around watching and staying out of the way when Steele said, "Hey, Rover boys (as he always called us), you wanna fly?" (We had been waiting forever.) "This hangar needs cleaning, bad! Clean it up." The hangar was two corrugated metal buildings about thirty or forty feet square; the floor was pea-gravel. Before long, we had that hangar up to military standards. Then he had us wash the ship—we found out what it was like to wash an elephant with buckets of water carried from Spangard's dairy barn about a block away. Steele said he was pleased with our work, and he owed us a ride.

He was true to his word. The next day he took the stick out of the front seat, and Ralph and I climbed in. We flew over to Muni, landed and looked at the planes there while Steele was doing whatever he was doing. Then we flew back, and he gave us a little ride west to 24th Street (we could see Lothrop School), then north (I believe I saw my house). After we landed, he told us we were good kids and to come back; maybe he'd have more work.

That ride started it—I was going to be an airplane pilot. I was at the field every weekend until school was out. I was a big fourteen; I had just graduated from the eighth grade, and I had sort of a job at Steele Airways. I never did get paid money, and every now and then we would drive Steele's Dodge Touring car to the filling station; sometimes I paid for the gas. But if I was at the field, he always let me ride with him, and most of the time he left the stick in the front seat. Pilots—they'll start teaching you how to fly.

I was getting pretty good at being a hangar cleaner, and I was learning what made the Eaglerock tick. She was a beauty with her OX-5 (Eaglerock valve and rocker-arm modification) really cowled back over both cockpits. The aluminum was burnished with those circles like Lindbergh's NX 211. The fuselage was painted blue and the wings silver. She didn't have brakes or a tail wheel. The gas tank was in front of the cockpit, and the radiator was completely cowled. The front seat was big and could hold two people. It had a half-door on the left side that made it easy to get in. She was clean and pretty, would cruise between ninety and a hundred miles per hour, and flew like a hawk. She was mostly aileron and landed a little hot, but the wide gear with the bungee-cord shock made her roll straight. I can still see her sixty-five years later.

By the middle of the summer Steele had me flying pretty well. I was taking off and landing. I could handle her in a stall and could make a two-turn spin and come out on the road. He taught me how to loop and how to snap her. But I wasn't tall enough to get the stick far enough forward, inverted, so he didn't let me roll her. He had a neat way he entered any maneuver needing excess speed. He would pull her up into a hammer-head stall, get his speed in the dive, then loop, roll or do an Immelmann. I never forgot it, and I taught it to Bob Alf with his Stearman a couple of years ago. Like all of those guys Steele could fly acrobatics. He went over to Muni almost every Sunday and holiday and hopped passengers. Sometimes there would be a big crowd watching, and some Sundays he'd take in a couple hundred dollars.

Bob Steele

There never were a lot of people at Steele Field. He always had a few students, flew a few charters, and made a few passenger hops. He flew for Nat L. Dewell, the photographer, quite often. Since he was good and knew how to fly for a jumper, he flew for almost all of the parachute jumpers that came to Omaha for shows, amusement parks entertainment, celebrations, other occasions or just for the heck of it. Also, the half-door made it easier to get out of the front seat. I was with him the Sunday he flew for a jumper named Chet Vienot.

I think he sold one or two planes while I was spending time at the field. He was just a twenties' pilot who learned how to fly in a Jenny in World War I. I thought he was great—not quite up to Lindbergh, but he did fly with the old Red-tailed Hawk.

One of the most memorable days of my life happened on a Labor Day weekend. Steele and a pilot named Jim Goggins, who had a Travel Air, had made arrangements to barnstorm to a Labor Day show at Dodge, Nebraska, about fifty air-miles from Omaha. The towns of Snyder, Dodge, Howells and Clarkson are right together on a line on State Road 91. Those people were starved to see an airplane. Steele had promised me I could go with them. We left the field at about 8 a.m. I was in the front seat with two full five-gallon cans of gas, two four-by-four chocks, and a wire milk basket with eight quarts of oil in it, sitting on the jumper's spare chute. Chet Vienot was riding with Jim Goggins. You can't imagine how thrilled I was.

They flew pretty close formation. We were faster and never got over three hundred feet high. We got to Dodge in thirty or forty minutes and spotted a good alfalfa field a little west and south of town. Steele dragged it and then landed to the south. He taxied up as close as he could get to the house, where there were a bunch of men and boys. They were waving and looking friendly. They were agreeable to our using their field for a free ride.

Steele signalled Goggins, who was circling. He started climbing, went back over town, buzzed them a couple of times, got their attention, then he started climbing for altitude so Vienot, the jumper, could jump. The field had a graveled road on one side and a dirt road on the end. I'll bet half the cars in north Dodge and Colfax counties were there. Steele told them to watch the Travel Air.

Goggins got her up about two thousand feet, slowed down a little south and east of the field, and Vienot jumped. The people loved it. Vienot was good. He made it to the field and landed pretty close to a crowd that must have numbered five hundred by then. Jim waited until Vienot landed, did a two-turn spin, looped out of his recovery, came in and landed. Steele said to me, "Adwers, we've got 'em in our pockets." He gave a fellow ten dollars to bring gas from town. I went with him several times, but most of the time I helped Vienot keep the two ships gassed, oiled and watered.

Everybody wanted to ride in an airplane. We charged three dollars for the regular flight (about five minutes), five dollars for the long flight (not much more than seven or eight minutes), and ten dollars for the brave soul that wanted a thrill. I don't think Steele made a flight without two in the front seat (well—maybe a couple of times because some of those farm guys were big). They flew constantly from a little after 10 a.m. to almost 7 p.m. Vienot and I filled the tanks and added oil and water. Steele and Goggins would get out when we were doing it. They had a hamburger and a glass of water, then headed back into the cockpit. Somebody in the crowd probably had a jug of liquor because I noticed Steele was getting pretty jovial by 7 p.m., and I began worrying about flying back to Omaha in the dark.

Steele finally announced, "One more load!" After that, I gathered up the cans and oil basket, and Vienot took his chute. I climbed in the front seat. Steele kicked the chocks out, came around, and gave me the stick. He looked at me and said, "Throw those damn cans out. We're rich!" He told me later that he had over $600 in his pockets. I put the stick into the socket and slipped the locking pin in. He climbed in the back seat and said, "Take me home." Goggins was a little ahead, so I followed him. He was going on to Muni. It was dusk when I got to the field. I figured that we would land to the south. I expected Steele to take over, but, no, the stick was still mine. I turned and looked back at him—he was sound asleep, at least I thought he was. That's all I needed. I flew a three hundred foot pattern straight in, pulled the throttle back, set a glide, held her off to the ground and had her rolling straight. The ground woke him up, but he didn't do a thing except help a little parking her. Then he said, "By God, Adwers, you're an aviator!" I was, and I still am.

My flying career went into a holding pattern when I started at Central High School. I would get out to Steele Field every now

The Arrow plant in Lincoln, Nebraska

and then, and, if Steele could, he would take me with him and let me fly. Thanks, Bob Steele. May the yonder always be blue!

The airplane "came of age," at least in Omaha and the rest of Nebraska, in 1929, and many aviation events took place. Cliff Burnham flew his Travel Air over downtown Omaha on August 10; the Oakland to Cleveland Air Derby landed at our airport on August 27; the joint Army and Post Office Department Endurance Transcontinent Shuttle refueled over Omaha, with Capt. Ira Eaker as its chief pilot on September 1; on the same day, Ted Wells won the Portland to Cleveland Race, beating "Speed" Holman; the Graf Zeppelin was flying fast over Texas on the last stretch of its globe-circling flight on August 28. In addition, the great Arrow Sports were being built in Lincoln, and Lt. Lester J. Maitland, the famous San Francisco to Honolulu flyer, was running an aviation column called *Sky Roads* in the newspapers. I remember I could hardly wait to get to my paper station "B" at 24th and Lake Streets in Omaha to read it each day.

In 1929 Ed McCord and some of his fellow birdmen began flying off of his farm just south of Nebraska City, too. They are still flying at the same field today. I've landed there off and on for fifty years.

Memorial Stadium, Lincoln, Nebraska

I guess no more fitting a climax to 1929 could be told than when Chief Bowen, flying Midwest's Cessna, flew the *World-Herald*'s photographer over Memorial Stadium, where he took the first picture of a Big Red game from the air (shown on the opposite page). He brought it back to Omaha in twenty-one minutes, so it could be printed in the Sunday paper.

CHAPTER SIX:
THE MUNICIPAL AIRPORT (1930)

Aviation in Omaha was well on its way in 1930. On June 10 Northwest Airlines, Inc., of Minneapolis, Minnesota, started service between the Twin Cities and Omaha via Sioux City. Chief Pilot Chadwick B. Smith landed their 550-horsepower, Pratt-and-Whitney-powered, seven-passenger all-metal Hamilton on its first flight to Omaha later that day. Passenger travel still was not all that luxurious, however—the pilot sat in an open cockpit in front, and the passengers sat in a closed cabin. The one-day-a-week service flew both ways on Thursdays.

In August 1930, my sister Franny and her friend, Louise Mallinson, went to their sorority's national convention in Minneapolis. My sister rarely did anything without style, so they flew to Minneapolis on Northwest Airlines. None other than "Speed" Holman (then the operations manager for the airline) was their pilot.

On September 5, Walgreen Drug brought their big Sikorsky amphibian to town and flew it around for several days. We all rode our wheels to Carter Lake because it was said they were going to fly her off of the water. They never did, though, because the lake wasn't high enough. It was sort of a mudhole in those days.

The Omaha Municipal Airport also began to operate that year. In June the Chamber's Aerial Transportation Committee announced that the city and Boeing had made an agreement for a fifty-year lease at the Municipal Airfield. Boeing would start immediately to build a hangar and would move their operation from Fort Crook as soon as the new facilities were ready. This concluded ten years' work by the Chamber of Commerce, starting back in 1919 with Harley Conant's Aerial Navigation Committee. It was a good agreement for both, and it is still in effect today (1994).

On November 4, 1930, Omaha voters approved the issuance of $100,000 airfield improvement bonds a year for five years in order to build runways, a terminal building (which we always called the "Ad Building"), parking and roads. Jay Dudley, Sr., became the first airport manager. He was my brother-in-law's father.

Work on the airfield moved quickly that year, and on November 30 Boeing moved into their new $60,000 hangar-terminal building at the field. Several thousand people attended the dedication, the main attraction of which was their brand-new fourteen-passenger Boeing Trimotor Transport. They made a number of flights, carrying dignitaries in the luxurious ship. It was the biggest airplane I had ever seen.

That year as well, Marion Nelson, who would become the last of Omaha's flying service operators, left his family farm east of Harlan, Iowa, with $300 in his pocket and headed for Omaha to learn to fly. He signed up with Midwest Aviation and had Cliff Burnham as his instructor. Nelson soloed in the OX-5 Travel Air in four hours. Burnham told him, "The reason I soloed you in such a short time was because you came right off the farm and all I had to do was tape a plow handle on the stick."

"Dutch" Miller with Burnam-Miller Travel Air plane

Nelson worked for Midwest as a mechanic and did odd-jobs while he was building up time for his license. In 1931 Midwest became Burnham-Miller Flying Service, and Nelson began flying for them when he got his license (No. 23221). The first pilots' licenses were issued by the Aero Club of America; then government licensing started with the passing of the Air Commerce Act in 1926. Charles Baysdorfer's license was No. 183 and was issued by the Aero Club of America; Dutch Miller's, No. 464, was government issued, and mine, issued in 1938, is No. 72706.

Nelson was one of the best pilots around. He was in demand as an instructor, flew a lot of charters, and test-hopped Overland Airways's last plane. In 1933 he had made enough money to buy a Stinson Reliant, and it was about the classiest airplane on the Omaha field. He cornered almost all of the charter business and still instructed. In 1934 Nelson Flying Service joined Burnham-Miller Flying Service, Krantz Aviation and Enzminger Aircraft as the operators of Omaha Municipal Airport.

Errold Bahl (third from left) and McFadden's Vega

Here comes Errold Bahl again. In 1928 he was the pilot for McFadden Publications of New York and flew their brand-new Lockheed Vega on the first Goodwill Trip to Mexico. Bahl's "swan song" was sung in 1930: He had just quit flying Ford Tri-motors in the Caribbean for Pan American and became the pilot for Union Electric Light Company in St. Louis; they owned a Ford Tri-motor Club plane. He really loved the job and told his wife that this was it—no more moving. On October 26, 1930, he was driving his Model A to pick up some friends. He ran through a stop sign and was killed instantly in a collision with another car. He packed a lot of living and flying into his thirty-six years. I know he is flying with the Hawk. Six months later his son Errold (who assisted in my research) was born.

Omaha had a handful of women pilots by 1930 as well, all inspired by Amelia Earhart. Louise Tinsley, the secretary at Midwest Aviation, became the first woman to hold a private license in Nebraska. She soloed after only a few hours of instruction.

The biggest aviation event ever held in Omaha, in fact in the entire Midwest, was inspired by the 1930 National Air Races held in Chicago. The Chicago races had been promoter Cliff Henderson's most successful to date. The primary reason for the success was the addition of the Thompson Trophy Race, which ran over a 150-mile, three-pylon, closed course and was for planes of unlimited horsepower. By the day of the Chicago race, September 2, 1930, there were seven entries—all of them the fastest airplanes in the world. Charles W. "Speed" Holman won in a Laird Solution with an average speed of 201.95 miles per hour and set a world record. The plane was built by the same Matty Laird who had the Swallows at the Aero Congress in 1921. James Haizlip placed second, flying a Travel Air Mystery Ship, which was designed in part by Ted Wells. Ben Howard came in third in a plane he had built himself, the revolutionarily-designed *Pete.*

Charles Holman was an outstanding figure in national aviation. Today the Twin Cities's St. Paul airport is named Holman Field, after their most famous aviator. He got the nickname "Speed" from his motorcycle-racing days. It stayed with him when he learned how to fly, and it certainly described his flying. Holman won the 1926 New York to Spokane, Washington, National Air Derby. In both the 1927 and 1928 Transcontinental Air Race, he was in the lead when his planes quit, requiring him to make sensational forced landings. He holds the world record for consecutive loops (1,433 of them), made over the St. Paul airport. In 1929 he flew a Laird to first place in the Gardner Cup Race from St. Louis to Indianapolis and back at an average speed of 157 miles per hour. Holman also created a sensation at the National Air Races at Cleveland when he flew acrobatics with a Ford Tri-motor. Dutch Miller and Barney Burnham were there, and I heard from them many times how he slow-rolled, looped, did hammer-head stalls, and spun the plane. Miller and Burnham swore it was true—and it was. Holman had been flying for Northwest Airlines since 1926.

Bernie Wickham, son of a prominent Council Bluffs road-building contractor and member of the Omaha Junior Chamber of Commerce, piloted his father's company's five-place Travel Air cabin plane, which was kept at the Midwest Aviation hangar. In 1930

Wickham flew his good friends Marvin M. Myers, Crawford Follmer, Rudy C. Mueller and Lawrence Shaw to the air races in Chicago. Impressed with the meet, they all said that they should promote an air meet in Omaha. Sound familiar? When they returned to Omaha, the first man they talked to was Gould Dietz, the dean of Omaha aviation, who still was active in the Aero Club of America. He was highly in favor but advised them to go slowly, to do it right, and to talk to Arthur Fetters.

Fetters encouraged them. They talked to the city officials. Noyes thought having air races in Omaha was a great idea. Wickham and his friends contacted Henderson, director of the National Air Races, who encouraged them and suggested that his brother might be available to manage a meet for them. They then put together a detailed proposal and presented it to the Chamber's board. They impressed the board, which appointed an advisory committee to work with them. What a committee! It was composed of W. Dale Clark (President of the Omaha National Bank), James E. Davidson (President of the Nebraska Power Company), A. W. Gordon (financier), and W. H. Schellberg (President of the Union Stock Yards Company). Gordon would be a non-flying Colonel in the Army Air Corps in World War II. I rode his son Tinner Gordon's final check ride in Primary Flight Training, and, yes, the younger Gordon passed. I was a first lieutenant at the time, and Colonel Gordon was there that day.

The sponsor of the air races was the Omaha Air Races Association, a non-profit corporation. Its President was Marvin Myers; Vice President, Verne W. Vance; Secretary Treasurer, Crawford Follmer; Assistant Secretary Treasurer, W. W. Mayer, and Directors were Myers, Follmer, Vance, H. W. Peterson, A. L. Jacoberger, Mueller, Donald M. Haley and Hayden Ahmanson.

The $25,000 budget was underwritten by the following: Willard Hosford; Omaha and Council Bluffs Street Railway Co.; Nebraska Power Co.; Omaha National Bank; Union Stock Yards Co.; the *Omaha World-Herald*; J. L. Brandeis & Sons; Eppley Hotels Co.; Sam A. Houser; A. W. Gordon; Nebraska Clothing Co.; Peter Kiewit Sons' Co.; Northwestern Bell Telephone Co.; Orchard-Wilhelm Co.; Omaha Printing Co.; John McGuirk; Standard Oil Co.; National Construction Co.; Fairmont Creamery Co.; Stockyards National Bank; United States National Bank; First National Bank of Omaha; O. E. Engler; Kimball Laundry Co.; Harry A. Koch Co.; the

Panitorium; the Refinite Co.; Henshaw Cafeteria; Omaha Steel Co.; Uncle Sam Breakfast Food Co.; Eggerss-O'Flyng Co.; Klopp Lithographing and Printing Co.; Union Outfitting Co.; Andrew Murphy and Sons; Harding Cream Co.; Thomas Kilpatrick & Co.; Carpenter Paper Co.; George & Co.; Livestock National Bank; Boeing Air Transport; Haas Brothers Co.; H. C. Noll Co.; Western Air Service Corporation; and Rees Printing Co. Their financial participation guaranteed the success of the air races.

The Junior Chamber said they would provide all the manpower, and the show was on the road. On March 1, Phil Henderson was appointed General Manager of the races. He immediately started organizing events, talent, participants, rules, and so forth. Henderson knew what he was doing and soon had everything in place for a very successful meet. In April, Henderson's old high school friend and world-famous flyer, James C. "Jimmy" Doolittle, stopped in Omaha to publicize the races. Doolittle was the first person to land an airplane blind (using only instruments and radio) and either set or held most speed and distance records. As an Army Air Corps General, Doolittle later led the B-25 Tokyo raid, the first bombing of Japan during World War II.

On April 10, Henderson released the flying events and cash prize list (which totaled $7,500) and announced that some of the country's best flyers had already entered or planned to enter the Omaha race. Omaha's best were all entered, too. Henderson also appointed W. E. Cleveland (who had been the official referee of the national races) Omaha's referee. Both the Army and Navy planned to attend and perform flying maneuvers.

By May 10, the stands, which were located just east of the two south hangars at Omaha Municipal Airport and seated fifteen thousand, were almost completed. The stands were situated so that spectators could see the planes fly around all three pylons. The one-hundred-foot-high home pylon was located directly east of the stands and just east of the north-south runway. The other two pylons were placed approximately three-and-a-half miles from the field—one just west of Crescent, Iowa, and the other almost straight north of the field. A two-hundred-foot-diameter circle was laid out just north and east of the home pylon for the bombing event, dead-stick landing and parachute-jumping. The official referee stand was built in front of the grandstands, and an excellent public address system was installed. Six large concession tents were located on the field in the stands area. Dean Noyes's people had the field in excellent shape. Fences were erected, and parking for ten thousand cars was leveled and cleared. The Midwest Aviation Hangar was the official headquarters for all pilots, and the servicing of all the planes was handled through Midwest.

Advance ticket sales were very good. All of the box seats were sold for the three days; reserved seat and general admission cost one and two dollars; and regular seats were fifty cents. All tickets included parking. Carter Lake Boulevard was closed off north of the field, and cars were issued tickets to get in from Locust Street. Gould Dietz, when shown preliminary arrangements, made sure that dead beats wouldn't watch the air meet for free; thus, all that could be seen from off of the field was the high flying. He remembered 1910 and 1921.

The town was revving up for the show. Downtown stores decorated their fronts with flags and bunting. Model-plane building contests were held. Companies promoted the races in their ads. Mrs. H. C. Noll (a popular Omaha woman flyer) was named Chairman of the Women's Committee, and Mrs. Louis Bock (also a pilot) was her assistant.

Both the *Bee* and the *World-Herald* and radio WOAW and KOIL wrote and read stories. Mayor Richard L. Metcalfe, who was off to an International Meeting of Mayors in Paris, proclaimed the week of May 11 "Aviation Week" in Omaha. This was the first aviation event in Omaha's history that "Cowboy Jim" Dahlman didn't proclaim or kick off, as he had passed away in January 1930. He had been Omaha's mayor for twenty-one years (the longest of any mayor to date) and had been a staunch supporter of aviation.

Then, on Tuesday afternoon, May 12, fifty-six Boeing P-12 fighting planes of the United States Army Air Corps Twentieth Pursuit Group landed at the field after circling over downtown. I saw them fly over and even watched them land from the east entrance of Central High School. Principal J. G. Masters let us out of classes to watch, and it sure was great. I didn't see that many airplanes flying together again until I was in the Army in 1941. Commanded by Maj. Clarence L. Tinker, the group was based at March Field, California, and headed for Wright Field, at Dayton, Ohio. While in Omaha, Tinker renewed some old acquaintances: He had been stationed at Ft. Crook while serving as air officer for the Seventh Service Command in 1923 and also had led a flight of nineteen planes in the Offutt Field dedication in 1924. This aviator later became Major General Tinker and commanded the Seventh Air Force at Hickham Field, Hawaii.

That evening the flyers were honored at a Chamber of Commerce dinner. Tinker said that the purpose of the flight was to assemble all 760 airplanes of the air division, the first time such an effort ever had been made.

The next morning they took off for Chanute Field in Rantoul, Illinois.

> At 7:53 a.m. Major C. L. Tinker, commanding officer, flanked on left by his aide, Captain W. E. Lind, roared down the north-south runway, pulled back on the stick, and climbed high into the air. . . .
>
> Wings reflecting the sun's rays, motors rending the calm of the ether, flying smooth and level as though they were on a floor, the 56 ships swept across the city in a far-flung line. Nearing the airport they began to lengthen out in a single file of elements, dipped a thousand feet toward the field and were off eastward for their next stop. . . .[1]

I watched from Central, with my mechanical drawing teacher, Oscar J. Franklin. We saw one-tenth of the United States's military flying force. It sure was impressive. The final touches were in place, and airplanes from all over the country were landing almost one behind another. Most of the famous aviators were here.

Commissioner Noyes miffed some of the Junior Chamber of Commerce when he put in a section of four hundred reserve seats for his friends, just east of the Boeing hangar. Noyes defended his action in arranging his section by recounting the role he had played in the success of the races (and he really had played a key part in building the Municipal Airport). Marvin Myers said the Chamber had been hoping to use the chairs, and Noyes took them. Noyes, being an elected official, said, "Anytime I can't take care of my friends, it's too bad."[2] My dad was his guest on Saturday.

What a great week it was for me! My best friend George Seemann lived two doors west of us at 52nd and Chicago Streets.

---

[1]*Omaha Evening World-Herald,* May 13, 1931.

[2]*Omaha Morning World-Herald,* May 16, 1931.

His dad owned the Seemann Chevrolet Company in Benson, one of the Chevrolet dealers furnishing cars and drivers for the pilots, officials, and dignitaries during the meet.

On Saturday, May 16, I turned sixteen, so "Big George" (the elder) Seemann said I could drive one of his cars at the field—a red roadster. I was excused from school Friday, rode over to the garage with Seemann, picked up that beautiful roadster, and began an unforgettable three days. I followed Seemann to the airport, and the police let us onto the field. We parked in front of the Rapid Airlines Hangar, the meeting place for the pilots and everyone who was someone. My job was to drive them out to the planes, to deliver things and to be available.

Opening Day—Friday, May 15

According to the *World-Herald,*

Throngs began pouring into the municipal airport shortly after noon today to watch the opening of the three-day program of air races and stunts. Eighty planes were on the field, many of them from far states.

The first event, the formal dedication of the airport, began promptly at 1:30. The races were to follow quickly.

The sky was cloudless and the sun was bright, but a 20-mile wind from the south caused pilots of some of the more temperamental racing ships to cast anxious eyes at the flapping flags. . . .

The dedication ceremony was begun with the firing of two fireworks bombs. Then G. Crawford Follmer, president of the Junior Chamber of Commerce, which is sponsoring the races, raised the flag as the Technical High School Band played the Star Spangled Banner.

After Follmer delivered the address of welcome, M. M. Meyers, president of the Air Race association, Acting Mayor Hopkins, Gould Dietz, Miss Marcelle Folda, and C. M. Wilhelm, were introduced.

As the field was being dedicated three army reserve planes from Kansas City flew in formation overhead. Soon they were joined by three naval reserve planes, all painted bright yellow from St. Louis.[3]

Harold Neuman of Elmhurst, Illinois, won the bombing contest. His sack of flour almost hit dead-center in the circle in its three-hundred-foot fall. Ray Liggett of Rapid City, South Dakota, was second; Bill Reedholm of Boxholm, Iowa, third; and Omaha's own Cliff Burnham was fourth.

The first race was for open- and cabin-ships and went three times around the course for a total of fifteen miles in length. A closed-course air race around three pylons is about the most breath-taking and wild flying imaginable. The wind was out of the south at about twenty-five miles per hour. The four ships entered, took off north (downwind) in a line from the grass, flew to the north pylon, rounded it, headed back to the home pylon, rounded it, and flew toward the northeast pylon. Their time started when they rounded the home pylon. Was it ever wild! They bunched as they went around and then spread out between pylons as the faster ships pulled away. John Livingston, from Moline, Illinois, flying a beautiful clipped-winged, 110-horsepower Warner Monocoupe, won in eight minutes twenty seconds. J. H. Bridges of Wichita, Kansas, was second in a Cessna. Barton Stevenson of Kansas City was third in a Travel Air.

The dead-stick landing was the next event. In it, the pilots had to fly right over the circle, chop the throttle and land, actually stopping inside the circle. The wind was blowing about twenty-five miles per hour from the south. Harold Neuman stopped his Travel Air eleven feet from the center. Tex Rankin of Portland, Oregon, stopped his Great Lakes Trainer twenty-one feet away. Tex LaGrone of Kansas City was third at thirty-four feet. They knew how to fly a 360 overhead approach. Nobody else came close.

Then the parachuting began. Don Charles Rae of Chicago (the national parachute-jumping champion) really got the crowd on their feet when he reefed in the lines of his parachute until it was a small part of its normal size and went shooting towards the earth in one of the most daring leaps ever seen here. His tactics landed him in the circle in spite of the high wind.

There was something going on every minute; the next event was a fifteen-mile race for light planes. The three winners, all

---

[3]*Omaha Evening World-Herald,* May 15, 1931.

flying Buhl Bull Pups, were Chester Loose of Moline, Illinois (first, at 75.73 miles per hour), Richard Williams of Omaha (second) and Art Davis of East Lansing, Michigan (third). Did they ever thrill the crowd as they bobbed around the pylons! Dorothy Hester, a nineteen-year-old stunt flyer from Portland, Oregon, flew some acrobatics in a Great Lakes Trainer that equalled almost anything flown today. The next race was for cabin planes, and John Livingston won it easily at 115.63 miles per hour. He held his Monocoupe back in winning.

In the last race, a free-for-all for five-hundred-cubic-inch engines over a distance of twenty miles,

> Livingston's victory . . . proved the biggest upset of the day's speed events. His . . . Monocoupe, powered with a 110 horsepower engine, was expected by pilots to finish only third against Bill Ong's Howard Special racer and Stub Quinby's special racer, built by the makers of Livingston's ship.
>
> But Ong was never a contender for the lead. His motor was acting badly from the start, and the two Monocoupes were away ahead of him. Livingston took the lead, and the greatest speed contest of the day followed as Quinby sought to wrest it from him. Wide open they flew on the straightaway, and fairly clipped the corners of the pylons as they flattened out in vertical banks on the turns.
>
> About the time Quinby passed Livingston, at the far pylon, Ong's motor gave out entirely. It knocked and hammered, and emitted so much smoke that he thought the plane must be on fire. He barely got to the field for a landing.
>
> Quinby seemed a sure winner as he finished the fourth of the five five-mile laps. But he had miscalculated, and thought it was the finish. He zoomed high into the air, to waste his speed. Livingston kept on going, and by the time Quinby realized his mistake, it was too late for him to overtake the other.
>
> Livingston made the five laps in 11 minutes 32 seconds, Quinby in 11 minutes 42 seconds.[4]

One of the most popular planes on the field was the customized Ford Tri-motor. Powered by three Pratt-and-Whitney

---

[4]*Omaha World-Herald,* May 16, 1931.

420-horsepower Wasp engines, the two-pilot, nine-passenger craft had a washroom, clothes closet, table, kitchenette and luxurious upholstery. The plane was open for inspection, and the line to it was always full.

The Sioux City Glider Club expertly flew their glider. Towed by an auto along the runway, when the glider reached about two hundred feet, it was cut loose to turn and land on the grass. The craft was made by the club members. Roy Kline and Howard and Harry Pence were the pilots.

In addition to all the events, Boeing Air Transport had two flights arriving each afternoon. The racing activities were scheduled around them. The crowd loved seeing the big tri-motored 80-A's land and takeoff on the runway right in front of them, and, when it was announced that Jack Knight was the pilot of the first one, the crowd gave a huge cheer. Jim Ewing of Kansas City was the announcer, and he kept the crowd completely informed of all of the activities using the finest public address system Omaha had ever heard. Maj. Al Williams (a good pilot who impressed us all at Randolph Field during World War II by flying an inverted chandelle that started on the deck and ended up about fifteen-hundred-feet high in a Grumman Gulfhawk) was at the microphone when the Army and Navy fliers performed. Tex Rankin was the announcer when his protege Dorothy Hester flew.

Now to Dorothy Hester. At the time, I didn't have any idea who Dorothy Hester was, but I had heard that she was a heck of a flyer, and she was only three years older than I. First, I got to drive her out in the red Chevy to her Great Lakes Trainer when she flew her record-breaking inverted-barrel-roll flight. We called them snap rolls, and that's what they were. She took-off at about 2 p.m. and flew fifty-six inverted snap rolls in about two hours. She climbed to about four thousand feet right above the field, half-rolled, dove, then did a complete rudder and stick-roll, stopped, and then another. She would do three or four until she lost altitude, then she'd climb back up and do the whole thing over again. She stopped at fifty-six and no gas—both the Great Lakes and her, I think. I was impressed.

Over five thousand fans had seen a great opening day and were eagerly looking forward to the next day, when the Autogiro and "Speed" Holman would be there and hopefully Bill Ong would get *Pete* running. That night Cliff Burnham flew his famous "falling leaf" with red flares on the wing tips of his Travel Air. He was the best at it that I ever saw. The pilot really has to know when to hit opposite stick and rudder before she stalls completely out. It was a fitting finale to a great day of flying.

The Second Day

A crowd of more than ten thousand crowded the stands as the Army and Navy planes opened the show with a formation fly-by. There was a high overcast, but the wind was below ten miles per hour. It was a good day for flying.

Ewing announced that the Autogiro would arrive about 3 p.m. Dorothy Hester then put on her show, which included flying across the field inverted at about a thousand feet up. She also flew two outside loops. It was announced that she was going to try to break the world record for loops the third day. Hester ended by flying across the field on her side as close to a vertical bank as she could keep it flying. She was good.

Jerry Wessling of Rapid City won the parachute jump, landing almost dead center, nosing out Don Charles Rae by less than four feet. Johnny Livingston, in his Monocoupe, won the dead-stick landing, stopping his ship five feet three inches from the line. Art Davis was second in his Waco at six feet nine inches, and Harold Neuman was third in his Travel Air at eleven feet four inches.

The first race was for OX-5-powered cabin or open-cockpit planes. With ten participants, this race had the most entries of any during the meet. It turned out to be a wild, woolly thriller. All the ships had about the same speed (ninety to ninety-five miles per hour), so they stayed close together for the entire race. At the start, two ships touched wing tips on the takeoff, but they all got into the air. When they flew back to the home pylon, it was wild. Four ships started around almost together. E. L. Neibeck of Grinnell, Iowa, flying a Curtiss Robin, was practically in a vertical bank when he hit someone's prop wash. He levelled her out but hit the ground with the left gear and tore the wheel loose. He dropped out of the race and climbed and circled the field.

Meanwhile, the race continued for three laps with Harold Neuman barely nosing out Arley K. "Barney" Burnham of Omaha. Each was flying a Travel Air. When the sky cleared and all the racers were on the ground, Neibeck made three slow passes before he worked up enough nerve to land the Robin. The fire truck and ambulance were in the big circle, and ten thousand breaths were held. Neibeck kept the left wing high, brought her in on the right wheel and almost stopped before the left wing touched the ground. He climbed out to a mighty roar from the crowd.

Next, John Livingston, flying a Monocoupe cabin plane, won the balloon-busting with a phenomenal time of ten seconds. Steve Wittman of Byron, Wisconsin, was second, and Harold Neuman was third. Balloon-busting takes a lot of skill and luck: The pilot flies into the wind, releases a balloon, does a 180, finds it, then flies through it with his or her prop. I got pretty good at it in a Fairchild PT-19, but it had no top wing to blank out the pilot's view. Livingston's time (ten seconds) was unbelievable, but I saw it. It was possibly a world record time.

In the second race, Art Chester of Joliet, Illinois, battled Bart Stevenson of Kansas City and Frank M. Transue of Alliance, Ohio. Rounding the first pylon, the three planes were almost together, but Chester pulled ahead. His time was ten minutes thirteen seconds or 111.58 miles per hour.

Tex Rankin's stunt flying won the praise of pilots, especially an inverted "falling leaf" that he flew down to about a thousand feet before he rolled over and landed. Next the "windmill plane" stole the crowd's attention:

> The auto-giro flown by Johnny Miller, came out of the Iowa haze at 3:40 p.m.. from Paughkeepsie, N.Y. [sic]. . . . It looked like an old Dutch windmill bewitched.
>
> The four rotors whirled in an apparently leisurely fashion. The auto-giro flew over the grandstand, turned completely around slowly without banking, and then dropped down to the exact center of the chalk circle in front of the grandstand. When it lit, it stayed.[5]

Miller's Autogiro distracted the crowd from watching the Navy ships. It was the first time an autogiro had flown west of the Mississippi River. The ship rises and descends with power applied to the blades, but no power is applied when cruising. W. Dale Clark was the first of many Omaha passengers to ride in the whirlybird.

The third race was won by Arthur Davis of East Lansing, Michigan, in his Taper-wing Waco, after a close fight all the way on

---

[5]*Omaha Sunday Bee-News,* May 17, 1931.

330

the fifteen-mile course with George Harte of Wichita, Kansas, flying a Travel Air. Davis's speed was 135.44 miles per hour.

The final event of the day was a twenty-five mile free-for-all race for a plane of any piston displacement for the Junior Chamber Trophy. It was supposed to start at 4:45 p.m., but it appeared that "Speed" Holman would miss the

The autogiro in Omaha, 1931

race. Finally, at about 4:30 p.m., Holman's black Speedwing Laird came roaring out of the northeast. He seemed to be going two hundred miles per hour when he pulled up at the pylon. He was a good one thousand feet high when he rolled her, then split-S'd into a 360-overhead and landed. He had left Minneapolis one hour and forty-five minutes before. As he taxied to the parking line, Myers and Rudy Mueller jumped into my Chevy, and I took them out to meet him. Holman needed gas and oil, then he fired up and taxied out for the start of the race, which they had been holding up for him. He didn't know the course, so he followed the other five ships on the first lap.

Art Davis was leading in his Taper-winged Waco. Holman poured on the coal and caught up to Davis on the third lap. Then he hot-dogged around the north pylon at about a thousand feet, dove for the home turn, and went around it in a screaming vertical bank. He lapped everybody but Davis, whom he beat by more than a half a lap. He seemed to have been going two hundred miles per hour when he pulled up at the home pylon, but his speed was 147.87 miles per hour.

The crowd loved it, and so did I. Some of the pilots thought he could have behaved a little better, but it's hard to contain a tornado. Anyway, he taxied back to where I was parked, got his bag out of the side compartment, talked to the Rapid Airlines mechanic who was in charge of fueling the planes, met the press, had his picture taken, and jumped into my Chevy with Myers and Mueller. I heard them inviting him to some big affair that night.

Again the huge crowd was completely happy with the day's events and looked forward to the next day's finale. Burnham did his superb night-flying, and there was a lot of passenger-hopping for those who stayed late. Of the many parties held Saturday evening, the biggest was given by Bernie Wickham at the Omaha Country Club.

The Third Day

The weather was perfect—clear with only a ten mile per hour wind—and

Charles "Speed" Holman

by 1:00 p.m. there were more than twenty thousand people in the stands and on the field. A salvo of aerial bombs opened the show, and Old Glory was raised to "The Star Spangled Banner" played by the Central High Cadet Band.

Dorothy Hester was already in the air three thousand feet above the field. She rolled the Great Lakes on its back, dove for speed, and shoved the nose up in an inverted loop, the first of her world-record- breaking flight. She stayed at it for slightly more than two hours and made sixty-two successful loops out of sixty-nine attempts. Her instructor, Tex Rankin, who held the men's record with seventy-eight loops, told the crowd he thought Hester might break his record, too. At the end of her last loop, she split-S'd out of it, rolled both right and left, did a wing-over, came in and landed. She stopped right in front of the starter's stand, got out, and was given the loudest ovation I had ever heard. I had had the pleasure of driving her out to her plane, and this time I drove her to the

Boeing hangar, where she rested. Tex Rankin and Mrs. Noll were with her. I was really in hawk heaven.

Hester was flying with the Red-tailed Hawk. She flew the record in a good airplane, a Great Lakes Trainer biplane with a one hundred-horsepower Menasco or Cirrus engine and a fixed-pitch wood prop. The top wing was swept back. She had to pump the wobble pump to keep the engine running when she was inverted, hang onto the stick, keep her feet on the rudders and fly the maneuver through some strong negative "G" forces.

Dorothy Hester

I never flew an inverted loop. I tried, but I always fell out at the top. I've spun a Stearman and Fairchild inverted and snapped and rolled each of them inverted, but I've never flown a loop. I thought enough was enough.

On July 13, 1989, Dorothy Hester Stenzel watched Joann Osterud fly 208 consecutive outside loops in two hours four minutes above North Bend, Oregon. Osterud flies for United, owns Osterud Aviation Airshows and performs aerobatics all over the country. She flies a high-tech Supernova aerobatic plane with a 230-horsepower Lycoming engine with a nitrous-oxide injection system. The only resemblance between Hester's and Osterud's planes is that they both have propellers. Hester talked Osterud into trying to break her record. My hat's off to Joann Osterud, but she'll never take the place of Dorothy Hester and those sixty-two loops that long ago May day in Omaha when forty thousand Nebraska eyes watched or the thrill that a sixteen-year-old boy had driving Hester across the field in the red Chevy.

In the next event, Art Chester, flying the Davis Special, won the first race for 275-cubic-inch-displacement engines at 110.27 miles per hour. Bart Stevenson barely nosed out Frank Transue for second. Both were flying Monocoupes.

Omaha's Cliff Burnham again won the bomb-dropping. His sack of flour landed forty-three feet nine inches from the center. Tex Rankin was second, and Art Chester third. The three Navy ships then bombed in formation. Dropping from about fifteen hundred feet, the bombs trailed white smoke, then a parachute opened with an American flag attached. They all hit the field. It was a real crowd-pleaser.

Next, the Autogiro put on a great demonstration, taking off and landing with a run of less than fifty feet, but its turning 180-degrees practically standing still really thrilled the crowd. It also landed once coming almost straight down. It was unbelievable. The word "chopper" hadn't been invented yet.

John Livingston won the dead-stick landing for the third day in a row. He really knew how to drive that beautiful clipped-wing Monocoupe and stopped seven feet eight inches from the line. Art Schench came in second, Harold Neuman third, and Steve Whitman fourth.

At that point Dorothy Hester was scheduled to fly her acrobatics, but, after those sixty-two loops, she was sound asleep at the Boeing hangar. Henderson and the Junior Chamber officials were debating who should fly next. They chose Holman. He was holding court over at the Midwest hangar. He said, "I'll fly it." I drove him, Mueller, Myers and several others to his plane.

The Legend of Charles "Speed" Holman

Holman's Laird was parked just east of the north-south runway, a real beauty. He had won the Gardner Trophy in 1930 with it. The Laird was a Speedwing (having a small taper on the top wing and quite a bit of dihedral on the bottom). It was coal black, must have had thirty coats of "dope" on it, and had rib stitches about a half-inch apart. Its 325-horsepower Wright engine was cowled with Laird's Speed Ring and spun a Hamilton fixed-pitch metal prop. The front cockpit was completely covered. The main gas tanks were in the top wing, but a small auxiliary tank was just ahead of the front cockpit. She was Matty Laird's ultimate biplane, just like the Negative Staggerwing was Walter Beech's, and Holman was the right guy to be flying her—"Speed" and Speedwing.

When we got out to the ship, he talked to the Rapid mechanic, Don Kelly of 4523 Grant Street, explaining that he only wanted the tanks half-full and no fuel in the auxiliary in front. He got his white coveralls out of the side compartment and gave me his suit coat. Holman was a big man, standing six foot four inches. He gave Kelly the inertia starter crank and told him not to put it back because, he said, "I'm going to do some flying, and I don't want it flying around." He was talking to the Chamber members, thanking them for the party the previous night, saying it was a great air show, and he hoped they would do it again next year. Wayne Selby and Harry Burkley were present, and there may have been others.

Holman pulled on the coveralls, took his Irwin chute out of the seat, snapped all the straps, and climbed into the cockpit. He pulled on his helmet and fastened the belts (he had a shoulder harness, and I saw him fasten it). He said, "I'm going to give the crowd one hell of a thrill." Switch off . . . switch off. . . . Kelly cranked the starter until she was humming good. Contact . . . contact. . . . The Wright went blurp-blurp-wham and took off.

Holman ran her for quite a while, wanting her running good. I moved the car back. He waved, turned her around heading south, then taxied over to the runway. He stopped at the wire fence, and several police and others bent it down until the wire was flat on the ground, and held it there. He taxied on the smooth grass a good three hundred feet. I knew what he was going to do: With the wind out of the south just under ten miles per hour, he was going to take off north and wanted to be going like heck when he flew by the stands. Well, he did just that. When he went by us, she was bellowing like the beautiful banshee she was, and he was holding her down. At about Noyes's grandstand, he pulled the stick back, and she started for heaven. Around fifteen hundred feet over the north edge of the field, she started to burble, so he did a hammerhead and headed down, leveled off at about fifty feet, and shot by the stands faster than any airplane had ever flown in Nebraska and probably most of the forty-eight states in this great Union of ours.

I drove the Chevy over to the starter's stand, and Holman was coming back north when I got there. He had pulled her up higher than the first time and did two aileron rolls to the right, then wing-overed back down. He was going faster this time. I was standing up in the seat yelling, alongside twenty thousand other air show fans. Ewing, the announcer, was giving a play-by-play to the crowd. The air was electric. The Wright roared like a hundred whirlwinds. The Hamilton had sound waves bouncing off of each other, and the flying wires were humming in high and low "C." When he pulled up this time at the north end, he had so much speed that he went up almost two thousand feet, did a triple aileron roll to the left, did a wing-over, and came back down again. Next, he pulled her up in a tremendous loop-and-a-half, then rolled out diving. This time he was almost on the deck.

He hadn't written out a routine, so it was anybody's guess what he was going to do next. He flew a little farther north, then pulled up to about two thousand feet, did a hammerhead left, and started down, but not quite as steeply or fast as before. I knew he was going to do something different! Lots of pilots said the same thing. North of the Ad Building at about five hundred feet, he half-rolled her to the right and was flying inverted-right over the runway. She didn't waver. The wings were level, but he was

The wreck of "Speed" Holman's plane

diving, and she kept right on coming down—and hit the runway. There was no big boom or loud noise; there was no fire. The pair just crumpled, like wadding up a newspaper, and skidded up into a pile of dust and debris. "Speed" and Speedwing flew into the Wild Blue Yonder. An old Red-tailed Hawk dipped his wings high above Pershing School, and Lincoln Beachey probably whispered, "Well, he certainly was flying some!"

It was the most unbelievable thing I had ever seen in my young life. All twenty thousand of us were stunned and almost completely silent. The announcer told everyone to be calm, as the police and other officials ran towards the wreck. A fire truck and ambulance roared out. The Central High School Cadet Band, led by Cadet Capt. Dallas K. Leitch, struck up "The Stars And Stripes Forever." Most of them were standing on their chairs to see what was going on, except Tom Marshall (now a retired prominent Omaha attorney), who had fallen off his chair in the excitement and was flat on his back. Tex Rankin ran out to his Great Lakes, had it in the air in minutes, and was soon doing acrobatics fifteen hundred feet up. Ewing did a great job, and the huge crowd behaved beautifully.

Holman's body was put in the ambulance. A tow truck hooked onto the wreck; its pieces and trash were loaded into another truck, and the place was cleaned up in less than half an hour. I took Myers and Mueller over to the Rapid hangar. Mueller took Holman's suit coat. They called Northwest Airlines in Minneapolis, and one of Holman's assistants there, Jack Malone, took off immediately for Omaha to make arrangements. Acting Coroner Paul Steinwender began an investigation into the death.

For all practical purposes, the air races were over. The huge crowd remained very quiet and sober and could not seem to work up a great interest in any of the other flying. "Speed" and Speedwing was certainly a hard act to follow.

In the final event, the Junior Chamber of Commerce trophy in the fifteen-mile free-for-all race went to Art Chester in his Waco. His average speed was 141.89 miles per hour. It was conceded that, had he lived to participate, Holman would have won the race. So ended the 1931 Omaha air races.

The "Speed" Holman Epilogue

I believe my description of Holman's death is accurate—I was there and saw all I described. I wrote down what I had seen when I got home that night, figuring I might write about it someday.

There probably were twenty thousand ideas about the cause of the crash: Tex Rankin said that Holman's belt broke, and most of the pilots agreed with him. Bill Ong believed his engine was running rough. H.W. Peterson, Boeing manager, and many others said that Holman had miscalculated and purposely crashed to keep from hitting the grandstand.

Those who have done inverted flying know that keeping the stick forward to maintain altitude is absolutely necessary. Holman was so good at inverted flying that he knew this particularly well. Holman had a shoulder harness. I saw it, but I don't know how it was anchored. If his lap belt came loose, he definitely would have started falling out. Then he would have spread his knees, trying to hook onto a longeron or something, and he would have tried to hook his feet on something. Rudder control wouldn't have been his big worry, so he would have grabbed anything with his left hand. He wouldn't have let go of the stick, though, and, if he was holding himself in with the stick alone, then he would have augured in to the ground because he would have been pulling it back, which would have brought the nose down. If the shoulder harness had held or if he held himself in with hands, arms, legs and feet, I think he would have gotten the nose up because he was a big, strong man and would have gotten the stick forward.

Ed Morrow, *World-Herald* staff writer, took a great picture of Holman's ship in the air, and Lee Clark, long time *World-Herald* photographer, snapped the crash. Morrow also wrote a fine report for the *World-Herald,* about which he and I talked many times when we were neighbors on Western Avenue in the fifties.

Speed Holman was buried at a funeral service in the Minneapolis Masonic Temple. There were fifteen hundred people in attendance, and military guards of honor stood at either end of the flag-draped coffin. Thousands of people lined the route from the Temple to Acacia Cemetery located at Fort Snelling. Planes from the Minnesota National Guard, Northwest Airways and the St. Paul Aviation Club flew overhead. Troops standing at attention lined the streets. Governor Floyd B. Olson delivered the eulogy. Roses were thrown down from a plane while the body was lowered into the grave.

While working in Minneapolis during 1946 and 1947, I learned that Holman's widow lived on the west side of State Highway 55, south of the Mendota Road. I often drove the highway to Farmington, Northfield, Faribault, and Owatonna, Minnesota, and I finally stopped one day and knocked on the door. A lady, maybe fifty, answered. I told her who I was and said, "I saw 'Speed' Holman crash." She looked at me for quite a time and said, "I just don't want to talk about it." "Yes, Ma'am," I said and walked back to my car. She was still watching me as I drove north. I always will wonder if she wouldn't have liked to have heard a little about that day.

In spite of Holman's death, the 1931 Omaha Air Races were a tremendous success, largely due to the work of Junior Chamber of Commerce leaders Crawford Follmer and M. M. Myers. There had been almost 26,000 paid admissions, an estimated three-day total of thirty thousand viewers from the field, and at least ten thousand outside viewers. The total admissions receipts were exactly $37.18 more than the total expenditures, so underwriters didn't have to put up one cent. More than fifty participating flyers had learned first hand of Omaha's great hospitality and about our airport. There had been some outstanding flying in some of the finest airplanes in the country, and Dorothy Hester had set a world record. Even the tragedy of "Speed" Holman brought world-wide attention to Omaha. After the air races, Omaha flying went back to normal—four Boeing flights a day, one Northwest flight a week, an occasional cross-country flight and the locals.

Marion Nelson, who was working for Midwest Aviation Corporation, was given the job of disposing of the wreckage of the Speedwing after Holman's crash. J. E. Boudwin, aeronautics inspector for the Department of Commerce, had arranged with Midwest to have it done. They burned everything that would burn on the south side of the hangar, and what was left was hauled to a dump (Nelson never told where). At the time, Nelson kept pretty busy flying, fixing, teaching, and barnstorming, and he flew every plane that came along.

During the summer of 1931, I worked for George Seemann's mom at her Lake Okoboji summer cottage. Don Baxter of Sioux City, Iowa, owned a Curtiss Robin with an OX-6 engine (really an

Marion Nelson and Holman's engine

OX-5 with valve modifications) that he flew out of Okoboji Airport. He had been a World War I pilot. The summer theater is in the hangar now, and the field was where one hole of the golf course and the motel are now. Anyway, I sailed crew for Baxter in his big X-Class boat. He let me fly the Robin, and I got pretty good. I flew to Sioux City and back with him a number of times. Landing on that field at Okoboji with trees all around was sort of thrilling. Once when we went to Sioux City, he let me do all the flying. We landed at the old airport, which was in South Dakota—the same field in which Rene Simon landed his Blériot in 1911—and it had been Baxter's dad's automobile that Sarah Bernhardt sat in watching Simon fly. Anyway, I was flying well, and Baxter said, "Robert, my boy! Fly her around once yourself!" Then he got out, and I flew her around. My first solo flight came almost out of the clear blue sky. I had flown probably twenty-five or thirty hours with Bob Steele and maybe ten with Baxter, but never by myself. Move over Lindbergh!

Omaha's Flying Weatherman

The Weather Bureau started collecting upper-air data in Omaha on August 8, 1931. This was done by fastening an aerometerograph on the wing of an airplane. The instrument recorded barometric pressure, temperature and relative humidity, and the Weather Bureau contracted by bid with a pilot or company to fly the instrument up as high as they could get each day. The flights were made in the morning, and the data was released at the same time by all of the weather stations in the country. This data was used primarily for forecasting, but it was the start of the sophisticated Pan Am Weather-mation system of today.

John Starr, Jr. (right)

John Starr, Jr., of Tulsa, Oklahoma, was the first weather pilot in Omaha. He was financed by W.G. Skelly, owner of the Skelly Oil Company and president of Spartan Aircraft Company. Starr flew a Spartan open-cockpit biplane with a 240-horsepower Wright engine. He kept the biplane at Midwest Aviation, and Marion Nelson maintained it. It had a very impressive insignia on the fuselage saying "Weather Bureau Observation Ship." Starr's contract required a minimum altitude of 13,500 feet each day, for which he received $24.55 for each flight. For every fifteen hundred feet or major fraction thereof over that altitude, he got a bonus of ten percent. For every fifteen hundred feet below that altitude, he got docked ten percent. The lowest altitude he could reach and still get something was fifteen hundred feet, in which case he got $2.40. Starr was going to get rich, if he didn't freeze or kill himself.

Aviation wasn't Omaha's biggest industry in 1932 (the packing houses were), but we had a pretty darn good airport with three surfaced runways.

One day in the spring, George Seemann and I were driving out West Blondo Street, about 120th today, when we spotted an airplane in a field on the west side of the Big Papio Creek. We stopped and went over the fence to take a look. A fellow was trying

Omaha Municipal Airport in 1932

to prop it, and he was by himself. It was a '26 or '27 Waco with an OX-5. I said I'd sit in the cockpit for him. He was happy to have help. I climbed in, pulled the stick back—it didn't have brakes—ran the switch and throttle, and we got it started right away. It wasn't running well. He came around and said, "You want a ride?" The front seat was big enough for two. Seemann was rearing to go, but something told me no, so we didn't. He said O.K., climbed in and hit the throttle. She wasn't running well, but he got into the air. There were some huge cottonwood trees about on the half-mile fence line, and he wasn't going to clear them. At the last second, he made a turn to the left. I thought he was going to stall out, but he somehow landed in a pasture on the west side of the road (120th Street), which is full of houses today. We drove down to where he was. He said that he was going to work on the engine, then fly her out later. He said, "My name's Jerry. Any time you're out here, stop, and I'll give you a ride." I couldn't help asking, "Where did you learn how to fly?" He just laughed.

Jerry, whose last name was Smith, evidently got the OX-5 running again, but his luck ran out, and so did some of his friends'. The *Omaha Evening Bee-News,* May 2, 1932, front-page story read, "Engaged Couple Dies In Crash of Plane." Barney Burnham told me later that Gerald "Jerry" Smith had taken lessons from them, but they never soloed him. He knew Smith had bought an old Waco and was flying it from the pasture on West Blondo. Would-be flyers and flyers will never change, but they forget that flying is so unforgiving.

The Omaha Air Races

Verne Vance, President of the Junior Chamber of Commerce, and Marvin M. Myers, President of the Air Race Association, announced that the most important program on the national aeronautical calendar that spring would begin Friday, May 26, at 1 p.m. and would last four days. This event, the Second Annual Omaha Air Races, would climax with the start of the National Balloon Elimination Races on Memorial Day afternoon. Phil Henderson was again the manager. Everything pointed to a more successful race than the year before. The National Balloon Elimination Races would draw many nationally-prominent pilots and airplanes and more than thirty thousand spectators. Advanced ticket sales were considerably ahead of the last year, Myers reported. With last year's experience under the organizers' belts, things would run more easily.

The stands for fifteen thousand people were completed. The field was in excellent shape. Everything was ready, and pilots and planes began to arrive on Wednesday. And what a bunch of flyers:

Russell Boardman, fresh from his record-breaking New York to Turkey flight; Ben Howard; Art Davis; Johnny Livingston; Steve Wittman of Oshkosh, Wisconsin (the man who started the legendary Oshkosh Fly-in); Earl Ortman; Clyde and Earl Cessna; Art Chester; the Hollywood Hawks—Frank Clarke and Roy Wilson; Betty Lund—more than forty pilots, fifty-seven airplanes and five balloons, the Navy Hell Divers and the Army's 130th Pursuit Squadron. Omaha meteorologist M. V. Robbins predicted fair weather. Police captain Charles Payne had traffic under control.

On Friday a crowd of four thousand was on hand at 1:30 when Mayor Richard L. Metcalfe gave the welcome, and the Tech High Band rendered "The Star Spangled Banner." For the four ensuing hours, everyone was thrilled, chilled and entertained. The fun started with Dr. John Bock of Kansas City winning the Sportsman's Race with his yellow Monocoupe. Then the dead-stick landing was won by W. W. Krantz of Louisville, Kentucky, who landed seven feet from the line. Betty Lund's acrobatics were superb and ended with her flying inverted completely across the field. Next came the great fly-by of the seven Curtiss Hawks of the 130th Army Pursuit Squadron, followed by George Harte of Wichita winning the fifteen mile approved-type certificate (ATC) race with his Cessna at an average speed of 121.33 miles per hour. John Livingston won a Novelty Race of three laps (the fliers had to land between laps for a sandwich and pop). Harold Neuman of Moline, Illinois, then won the 275-cubic-inch engine free-for-all at 96.77 miles per hour in his Monocoupe before the three Navy Hell Divers flew tight-diving formations. Clem Sohn of Lansing, Michigan, manipulated his parachute in the wind to a landing 120 feet from the center of the circle, and Wallie Franklin of Ypsilanti, Michigan, leisurely flew his glider in loops, turns, and steep banks (after being towed up to three thousand feet) and at the end landed right in front of the stands. The hair-raising acrobatics of the Hollywood Hawks ended with a landing on the finish of a loop, all the while trailing smoke in the sky. It was a great day of flying, and much more was to come.

With school out and a working pass from Wickham, I saw the second and third day's races. Saturday's crowd of seven thousand were treated to a great opening. The twenty-five ships entered in the dead-stick event took off north on the runway at five-second intervals, climbed single file to fifteen hundred feet while circling the field. Then one by one, with engines off, they started their glides to the spot on the field. It was a great show won by Marcellus King of Fairmont, Minnesota, who stopped his Travel Air five feet six inches from the line.

Art Davis won the fifteen-mile ATC race in his taper-winged Waco at 139.04 mph. George Shealy of Atlanta was second, and Cliff Kysor of Ottumwa, Iowa, was third. Roger Don Rae of Lansing, Michigan, proved his title of the best parachute jumper in the country by landing right on top of the flag in the center of the circle. Betty Lund's acrobatics were better than ever—loops, rolls and four turn spins. The Army Pursuits flew by, stacked and wing-overed into some great acrobatics. Brock again won the Sportsman Pilot's Race, nosing out meet director Cliff Henderson in his bright red Waco. John Livingston won the special Waco Race, beating Art Davis and Tex LaGrone. The fans loved this race because the time was counted from a race-horse start. They took off north, couldn't turn back south until they were past the north edge of the field, then flew the course three times. Nobody flew around those pylons even close to the way Livingston did. The irony was that Livingston was flying Davis's wife's Waco.

Harold Neuman of Moline, Illinois, won the OX-5 Race at 98.94 miles per hour. It was great. Eleven ships were entered. They took off at fifteen-second intervals, and there was some wild flying. Neuman's was the best. He flew on the deck all of the time. He had to rise perceptibly every time around to clear the trees on the north end of the course.

The awaited feature race, the twenty-five mile free-for-all, was a thriller. The seven entries were considered among the fastest planes in the country, and the prize money ($800, $600, $400 and $200) was only $200 less per finish than the National Air Races at Cleveland.

Livingston was first away in his revamped Monocoupe. Robert Clampett in his Rider Special went around Livingston on the back stretch. On the next lap Ben Howard in his new racer, *Pete,* passed Livingston. Howard was catching up on Clampett, who thought his lead was safe and had slowed to ease the strain on his engine. Turning a pylon, he saw Howard below and almost even. He pushed the throttle forward to get all his engine would give. The rest of the race was a tremendous battle. Clampett was only fifty feet ahead and was doing over two hundred miles per hour at the finish. He averaged a speed of 176.06 miles per hour. Howard averaged 175.04. Livingston's average speed was 170.44.

Jimmy Faulkner then put the Autogiro through some unbelievable maneuvers, rising and descending almost vertically, hovering almost at a standstill, and turning 180 degrees while hardly moving. The people loved it. Remember, this wasn't a helicopter, although it might be considered its great-grandfather.

The Hollywood Hawks exceeded their death-defying stunts of Friday. They both were intent on outdoing each other, pulling out of their dives as close to the deck as possible without hitting. They had the crowd thinking of "Speed" Holman a year ago. One wag said that an insurance man with a policy on one of them would suffer a breakdown watching them.

The show ended with parachutists Roger Don Rae, Jimmy Wessling, Billy Hamilton and Dick Hunter jumping out of Burnham-Miller's six-place Travel Air, floating down almost together, and landing right in front of the grandstand.

Sunday the weather was beautiful but with gusty winds. The air races drew the largest crowd yet: The stands were full. Myers reported that 14,500 tickets had been sold, and it was estimated that ten thousand people ("freeloaders") surrounded the field on the outside.

The Army Pursuit planes split into two flights of three and went through some precision loops, rolls, dives and Immelmann turns that looked like early versions of today's Thunderbird routines. They were good. A little before noon, they flew west about a thousand feet above Farnam Street stacked in right echelon. At 40th Street, Capt. W. B. Wright (he had the right name), the flight commander, wing-overed left back east, and leveled off at about two hundred feet above Harney Street with the other five right on his tail. Close to 20th Street, he pulled up to about fifteen hundred feet. They formed and maintained a "Lufbery" around the Fontenelle Hotel, then finally broke out and headed for the airport. I saw them over downtown when I came out of church at Trinity Cathedral at 18th and Capital.

It was rumored that the great fly-by had been arranged at the big reception and party Saturday night at the Fontenelle. It also was said that some of the old Aero Club guys were involved and that Bernie Wickham was the chief judge of the races. Department of Commerce regulations require that planes fly over a city at a height at which they can glide to the airport. When the *World-Herald* reporter asked Captain Wright about that, the captain replied that his pilots could have climbed to an altitude sufficient to reach the airport if there had been an emergency. You bet! It was a great show. The Navy Hell Divers weren't outdone as they flew some in their brand-new Boeing F-4Bs over the airport.

The first event was the very popular dead-stick landing. Twenty-eight pilots were entered. They again took off in five-second intervals. They made quite a sight as they flew south over downtown, then back on downwind at one thousand feet, turned on base-leg, pulled the throttle and landed. W. W. Krantz of Louisville, Kentucky, stopped his little Aeronca eight feet seven inches from the line. Ray Schenck of Clarinda, Iowa, landed his Travel Air nine feet eleven inches away, and Harold Neuman of Moline, Illinois, was ten feet eight inches away in his Monocoupe. Darn good flying!

John Livingston again won the Waco Race in Davis's plane. Tex LaGrone pushed him all the way, but Livingston owned the pylons. Art Davis, who came in third, probably was thinking, "No more loaning Johnny my wife's plane."

Roger Don Rae again won the parachute jump with a landing only forty-eight feet from center in the gusty wind. Dick Hunter of Minneapolis gave himself and the crowd a few breathless moments. He bailed out at three thousand feet. His pilot chute didn't get the main chute out fast enough, and Hunter fell into its folds. It took him several seconds to roll out and for the chute to open completely. He landed eighty feet from the center, taking second place but visibly shaken and maybe looking for a new profession. Nobody else got closer than two hundred feet.

Art Davis won the fifteen-mile Pony Express Race, in which pilots had to land after each lap, eat a sandwich, drink a glass of milk and take off again. Davis said that he finally found something he could beat Livingston at, and that was eating. The crowd loved this race. It took darn good flying to land close to your judge, eat and get back in the air.

The twenty-five mile free-for-all turned out to be all that it had been built up to be. First, Russell Boardman, transatlantic flyer, arrived late Saturday in his brand-new Lowell Bayless Gee Bee Senior Sportster, powered by a five hundred-horsepower Pratt and Whitney Wasp engine. The airplane had been timed at 250 miles per hour on the straightaway. Robert Clampett, owner of the

Rider Special *Miss San Francisco,* winner of the day before's thrilling race, named young Earl Ortman to fly his ship that day. Ben Howard and his crew worked on his speedster, *Pete,* most of the night. Livingston, Davis, Liggett, Wittman, Martin and Harte all were super pilots flying some of the country's fastest racers.

The start was wild. The nine ships lined up at the far north end of the field on an east-west line and took-off south at the drop of starter Ray Brown's flag. The only requirement was that they had to fly around the home pylon to start the first lap. Boardman's big engine got him into the air first, and he had plenty of speed when he got to the pylon, so he went around it in a breeze. The other eight ships had about the same horsepower (200 to 275), so they really were jammed together between each other's prop wash with the gusty wind and not enough speed. It was hairy around the pylon. Ortman was second, Livingston third, Art Davis fourth, Howard fifth, and Wittman sixth. Liggett, Martin and Harte were running neck-to-neck. Boardman rounded the home pylon two hundred yards in the lead after the first lap, and he kept that lead throughout the race. Ortman had a good lead over Livingston. Art Davis was fourth and Benny Howard fifth. On the next lap, Howard had passed the others and rounded the home pylon almost even with Ortman. From that point on, Howard pulled ahead and at the finish led Ortman by about two hundred yards. Livingston came in fourth, Liggett fifth, Davis sixth, Wittman seventh, Martin eighth and Harte ninth. Boardman's big engine was too much for the others. You could tell he was holding her back, but he let her out on the last leg to the pylon. He was probably doing close to three hundred miles per hour when he pulled up. He climbed her to at least 2,500 feet before she slowed down, then did a wing-over to the east, came in, and landed. Average speeds were: Boardman, 176.69 miles per hour; Howard, 175.74; Ortman, 174.28; Livingston, 169.18; Liggett, 168.99; Davis, 166.33; Wittman, 161.80 and Martin, 135.49.

After that, the Hollywood Hawks flew their usual breathtaking acrobatics. Wilson and Clarke were so good that the audience began to take them for granted. They flew Travel Airs with 240-horsepower Wrights and wobble pumps, but I wonder what they would have done with a Pitts or a Christian-Eagle? Lund flew a great show, then she ran into a parked truck when she was taxiing back to the line, broke the prop and banged up a wing. Jimmy Faulkner brought his Autogiro down vertically from five hundred feet and landed right in front of the stands as easy as a cat jumping from a chair.

Freddie Miller of Clarinda, Iowa, put on an "Old-Time Wing Walking Show." Miller crawled around Ray Schenck's Waco a thousand feet in the air as if it were his living room. He hung by his feet from a rope ladder fastened to the landing gear, climbed back up, walked out on a wing, held on to the strut with one hand while waving the other to the crowd, walked back, and climbed over the cockpit and out to the other wing strut. Schenck had been climbing while this was going on. He was a good two thousand feet up, flying south right over the entrance of the field. Miller again held on with one hand, waved—and fell off! Ewing cried, "Oh no," over the public address system. The audience paled—most of them sat or stood with their mouths open and no voice. Finally, Miller pulled the rip cord on his backpack and landed in front of the Ad Building. When he was driven up to the starter's stand, the crowd gave him a tremendous ovation. I heard that Miller got fifty dollars for the act. The day ended with a four-group parachute jump, which the wind made a little wild, but they all made it down without mishap.

Later, the Grand Island Whisker Club, promoting the city's 75th Anniversary, made Boardman, Howard and Livingston honorary members. The bewhiskered pilots flew to Omaha in three planes, led by Dr. A. Arrasmith, Emil Wolbach, Tom Keyy, Vern Hainline, George Bartenbach and Stover Deats.

Everyone was looking forward to the final races and to the National Balloon Elimination Race, but it rained all Monday morning. It finally stopped about 1 p.m., but remained cold, and the wind was gusty. And the ceiling was about two thousand feet. The weather really put a damper on the crowd, which totaled less than three thousand. In spite of the weather, however, all the flying events were held.

In the dead-stick landing competition, G. R. Lockhart of Wichita, Kansas, and Bill Reedholm of Boxholm, Iowa, (both flying OX-5 Travel Airs) stopped so close to the line that the judges had to determine from what part of the airplane to measure. They decided from the center of the landing-gear axles. Lockhart won with two feet six inches to Reedholm's three feet six inches. Pretty good driving!

Harold Neuman won the ATC race in his Monocoupe. Art Chester was second and Bill Reedholm third. Both Neuman and Chester lapped Reedholm.

The ceiling then rose to about four thousand feet, so the Hawks, Wilson and Clarke, put on some incredibly good acrobatics. They took off together and did loops, rolls, hammer-heads, wingovers and flew across the field inverted. On their pass south into the wind, both looked like they were standing still—inverted, mind you. They landed together, out of a loop, not twenty-five feet apart on the one-hundred-foot-wide runway (the field was too muddy). Marion Nelson, Dutch Miller, Barney Burnham, Steve Krantz and all of the old birdmen said that Wilson and Clarke flew the best acrobatics they had ever seen. These "Hawks" were named right—they really did fly with the old Red-tailed Hawk.

The final race of the meet, the free-for-all, was a fitting climax. Three of the fastest racers in the United States were in it, and six of the best pilots were flying. Clampett and Boardman had each won, and this would be the "rubber game." It figured to be a real duel between them. Because the field was so muddy, it was decided to take off south on the runway in groups of three. Time would start when each ship passed the home pylon. Boardman, Livingston and Art Killips (flying Art Davis's Waco) drew the first row. Ortman (flying Clampett's *Miss San Francisco*), Howard and Liggett drew the second. It was exciting when starter Ray Brown dropped his flag. Six throttles went wide open, and they were off! Boardman was in the air after three hundred feet, and so was Livingston at almost the same time. Howard, Killips, Liggett and Ortman lifted last.

Just as soon as they had enough speed, they turned left around the pylon and started their first lap. They were really going by then! The first time they came out of the north, past the home pylon, Boardman was in the lead. Livingston (how he flew those pylons!), Howard, Ortman, Liggett and Killips were right behind. On the second lap, it was Boardman, Howard closing, Livingston, Ortman, Liggett (in the tiny mid-winged Cessna with the retractable gear and a 165-horsepower Warner engine) and Killips. On the third lap, Boardman and Howard led. It looked like Ortman was going to pass Livingston, but he couldn't turn the pylon like Livingston. Liggett and Killips were right behind. On the fourth lap, the people sitting in the top rows of the stands could see Ortman and Livingston dueling for third as they rounded the other two pylons.

They started the final lap the same Boardman, Howard, Livingston, Ortman, Liggett and Killips. The real thriller, though, came at the north pylon when Howard dove under Boardman, made the turn ahead of him, and roared by the finish ninety yards ahead. His average speed was 179.96 miles per hour, the fastest time ever flown on a twenty-five mile closed-course race. Boardman's average speed was 179.06 miles per hour. Livingston barely stayed ahead of Ortman at 173.33 to 173.01 miles per hour. Liggett was fifth at 172.21 miles per hour, and Killips sixth at 153.88 miles per hour.

> "I had the ship wide open," said Boardman, "and I think I'd have won if I hadn't made a bad turn on the last pylon. I lost speed there."

Howard said he had planned to "take" Boardman just where he did, rounding the last pylon.

> "I didn't realize Howard had passed me," Boardman said. "I was on the home stretch when I happened to think, ' I wonder where Howard is? ' I looked ahead of me and there he was."[6]

Ortman had a thrill on the back leg, when his goggles ripped off. He later commented,

> I guess I'm lucky to be here now. . . .
>
> Oil sprayed my face and then I whipped my wing through the trees before I knew what had happened. I think I prayed for a minute, then, that I could get my face wiped off before the next pylon turn.
>
> After that I just ducked under the cockpit edge on the turns and tried to shield my face with my hand on the straightaways.[7]

Howard was elated. It was the first race he had won with his tiny white special-built racer and thus heralded the start of his very successful career as a designer and builder of his later famous DGA (Darn Good Airplane). Great race, great flying.

Boardman had asked and received permission to enter the competition for the "Speed" Holman Trophy. After he had refueled

---

[6]*Omaha World-Herald,* May 31, 1932.

[7]*Ibid.*

the Gee Bee, he flew. His take-off was reminiscent of Holman's: He took off south and held her down. The Wasp and prop were howling. Just in front of the stands, the birdman appeared to pull her straight up and disappeared into the clouds, which were at a good 3,500 feet. Then he came down and across the field. He said later that his speed was 350 miles per hour. Boardman pulled her up, did a triple roll and a hammer-head out, and dipped back down again. He flew a beautiful twenty minutes. Most of the crowd seemed relieved when he landed and stopped in the middle of the stands, and they gave him a great hand. The judges then conferred (wisely, no one ever knew who the judges were) and announced that Boardman was the winner. When Lida Whitmore, Queen of Ak-Sar-Ben, presented him the trophy, that ended the airplane flying for the 1932 Omaha Air Races.

The National Balloon Elimination Race was next. The winner was to represent the United States at the International Races in Switzerland in August. The positions in the race were drawn by the pilots at a banquet at the Fontenelle Hotel Sunday night. The lineup was, in numerical order: Chevrolet entry Tracy Southworth of Detroit; *City of Omaha,* E. J. Hill of Detroit; Junior Chamber of Commerce entry Pete Larson of Detroit; *Army Number One,* Capt. W. J. Flood of Washington, D.C.; *Army Number Two,* Lt. W. J. Paul of Rantoul, Illinois; and lastly *Goodyear VII* and its pilot Roland B. Blair.

The race officials were: referee major, W. E. Kepner of Wright Field, Dayton, Ohio; chief judge, Col. J. A. Pagelow of Scott Field, Illinois; assistant judges, Cliff Henderson, Ted H. Maenner and George B. Thummel; starter, Gould Dietz (who had been involved with every balloon or airplane event in Omaha since the start in 1900); and official timer, William Enyart of the National Aeronautic Association, Washington, D.C.

The object of the balloon race was to fly the furthest distance possible from Omaha. The gas bags were limited to 35,000 cubic-feet capacity, and the current world record for this capacity was 961 miles. Because of a gusty south wind, rain, and a low ceiling, the bags were filled with hydrogen at the north end of the field. This got underway about 3 p.m. The start of the race, which would have been a placid event in calm weather, was rendered tense and exciting by the wind. I know because George Seemann, Jim Musselman, Bennie Rimmerman and I were there, helping hold the balloons down. There were over a hundred people holding each balloon, and every now and then it got wild hanging on. The four of us knew how combustible hydrogen is because we were in Dr. Herbert A. Senter's chemistry class at Central High. The officials were nervous too. Fortunately, it was too windy and rainy, and the volunteers didn't smoke.

By the time the race started, nearly all the crowd had gone home. Finally, Major Kepner gave the order, and we released the Chevrolet balloon at 6:48 p.m. Man, did the two guys in it get a ride! The balloon bumped across the ground, rose a little, went through a fence, finally rose over the trees and the river, and disappeared in about ten minutes. *Army Number One* was next. When it was released at 7:15 p.m., it leaped into the air like it knew what it was doing and was into the clouds in a few minutes. *Army Number Two* dragged for a ways, then bounced, rose over the trees and disappeared. Although the Junior Chamber balloon didn't look full, it was released at 8:07 p.m. It dragged and bounced. They frantically threw out ballast sandbags, and it barely cleared the trees and disappeared. *Goodyear VII* got off in good shape at 8:25 p.m. and was soon out of sight. The *City of Omaha* was last because a tear had to be repaired, so it was released in the light of the runway flood lights at 8:46 p.m. and flew out of sight immediately. We were soaked, cold, exhausted, and covered with mud from head to foot, but we got those balloons into the air. You know, I still have never ridden in one.

I know it's hard to remember nowadays with the instant coverage provided through the eye of the television monster, but the only way we could track the balloons' progress was through the newspaper. Oh, there were a few telephone calls reporting sightings, but both the *Bee* and *World-Herald* did a great job.

How did the balloons fare? The Larsons landed at Fort Calhoun and practically walked back, but most of them went a long way on the wild south winds. *Army Number Two,* carrying Lts. W. J. Paul and John Bishop, landed north of Calgary, Canada, on Wednesday, June 1, having travelled 1,107 miles from Omaha. This set a new record and earned them the right to represent the United States in Switzerland.

As for the 1932 Omaha Air Races on the whole, while the show attracted more and better flyers and airplanes than it had the year before, financially it was a failure. The meet had broken even in 1931, but G. Crawford Follmer, secretary-treasurer of the Air Race Association, announced that 1932 receipts were $8,000 below expenses. The loss was borne by the businessmen who underwrote the show. According to Follmer, the poor weather and the large uncontrolled ground near the airport from which people watched the show without paying admission were the major causes of the shortfall. As I have said before, the sky belongs to everybody: All you have to do is look up. But I guess Gould Dietz's 1910 dead beats will always be around.

The Rest of 1932

In 1932 Boeing Air Transport formed United Airlines and started putting "United" on their airplanes. In addition, that spring the airline started service north to Sioux City, Iowa, as well as to Sioux Falls, Huron and Aberdeen, South Dakota. They advertised a flight from Omaha to Sioux City in fifty minutes for $5.50.

The summer of 1932 was tough for Midwest farmers because the market for farm produce hit bottom. In an effort to push up prices, dairy farmers banded together and refused to sell their milk to cities. Instead, they dumped it and stopped trucks trying to make deliveries. It got rough, and Sioux City was out of milk.

The photo shows Charles Gardner, manager of the Omaha Fontenelle Hotel and the same Gardner who had been ticket sales manager for the 1921 Omaha Aero Congress, consigning a can of cream to Clyde Sharrar. The pilot of the Boeing 40 B-4 is E. Hamilton Lee, long-time airmail

Lee, long-time airmail pilot who was just starting his thirty-year career with United.

Clyde Sharrar (a good neighbor of mine on Western Avenue) had just arrived in Omaha as Boeing's (United) local manager, ticket agent, salesman, baggage handler and ground crew. He held the job for thirty-eight years and retired in 1968. At the time, Eugene Eppley (after whom Omaha's Eppley Airfield is named) owned the Warrior Hotel in Sioux City. Sharrar made a deal with the hotel magnate to deliver cream by air to Sioux City for their fancy dining room.

Airplanes were about the only things flying in 1932. It was about the bottom of the Depression, and Franklin Delano Roosevelt was elected President. I worked for Neta (Mrs. George) Seemann at Okoboji in August and flew twice before school started. Once, Baxter took me with him to Des Moines and back and let me fly the Robin. The other time, I paid Stanley Fuller five dollars for a lesson in his Travel Air. Fuller ran a flying service at Milford, Iowa, and had his strip behind his house on the north side of town. When he found out I could fly, he let me shoot a bunch of landings, but no solo. I helped him one Sunday when he was hopping passengers on scenic lake rides for five and ten dollars. He let me fly at the end of the day, and we went all around Okoboji and Spirit Lakes. It was great!

CHAPTER NINE:
TOUGH TIMES (1933)

The year 1933 is generally considered the bottom of the Depression. It was also the year I graduated from Central High School and got a summer job working for the Omaha School District. My pay was thirty-five cents an hour. I thought everything was pretty good at the time; I guess I was too young to know we were bad off.

Times were tough around the world, and we were poised on the verge of momentous change. Adolph Hitler and his Nazis took over Germany; Benito Mussolini and his Fascists were running Italy; and Japan was involved in Manchuria; Stalin was purging Russia. France had its head in the sand behind their line. The sun still never set on an aging British Empire, and the United States was starting out on FDR's New Deal.

The Germans, Italians and Japanese were aware of the importance of airplanes. In 1933 our elected politicians cared about as much for the airplane as they cared for a one-night Charley in the county jail. As proof of that, Italo Balbo, Mussolini's Minister of Air, led twenty-four Savoia-Marchetti SM.55X flying boats from Italy to the Chicago "Century of Progress" World's Fair. They landed on Chicago's Lake Michigan front on July 15. The Italians then flew to New York before returning to Italy on August 12 and completing their world-shaking transoceanic flight. I didn't know much about world events, but I saw the Italian planes on Lake Michigan and saw them leave for New York. My mom, Cliff Schroeder, Quinn Earhart and I had driven to Chicago in our 1928 Huppmobile to see the fair and stayed with my folks' long-time friends, the John N. Joneses.

The fair was super! We danced to Guy Lombardo at the Edgewater Beach Hotel. We cheered Babe Ruth, Lou Gehrig, Tony Lazzeri, and Bill Dickey (with Joe McCarthy managing) in Comiskey Park. We saw Sally Rand and her fans. And we watched the Italian seaplanes take off, standing about where the north end of Miegs Field is today.

By then, I had saved a little money and enrolled at Omaha University that fall. My tuition was twenty dollars a semester. My books cost more than that. It was there I met William H. Durand. Durand and I sat across from each other in John Kurtz's and John

W. Jackson's Mechanical Drawing classes. Durand is without a doubt the best draftsman I ever saw. He has been flying gliders and airplanes that he built since 1934, was a Professor of Engineering at Omaha University (and established their Aviation Department), and taught Theory of Flight and Aircraft to most of the pilots around here in the thirties and forties. Today, he flies from his "Sky Ranch" field west of the Omaha Weather Station on North 72nd Street in his latest plane, the beautiful *Durand Mark V.*

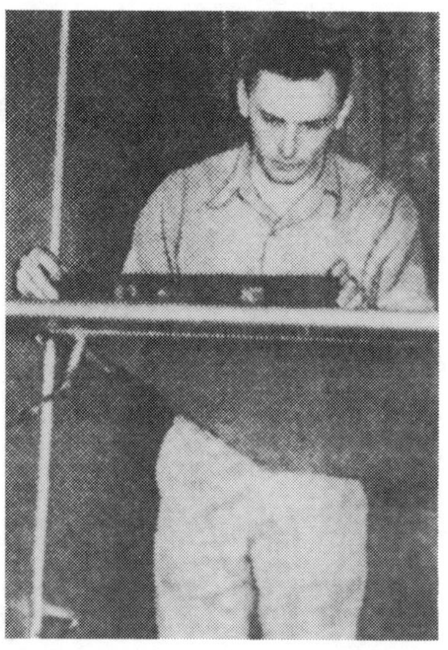

Bill Durand

By this time, Marion Nelson was doing pretty well. In 1933 he had made enough money to put the down payment on a brand-new Stinson Reliant four-place, powered by a 240-horsepower Wright. At about $4,000, it was almost the classiest airplane on the field. Nelson had the $1000 down-payment, but he had to do a lot of flying to make the $100-a-month payments. He was doing a great charter business and had several flying students. Probably his best charter customer was the Northern Natural Gas Company. That's how he met Ray L. Harrison, who became his silent partner (the same Harrison who owned the Staggerwing Beech that I flew). He taught Harrison how to fly in the Stinson. He also taught K. N. Fancher and Myrven Mead how to fly. They both worked for Northern.

United Airlines also was developing. In November, the airline put into service the best airliner built yet, the Boeing 247, the first twin-engined all-metal plane that would carry ten passengers at nearly two hundred miles per hour.

There were no Omaha air races in 1933. The underwriters said no, and neither the Junior Chamber nor anyone else tried to promote one. They did plan an Aviation Exposition to be held in the City Auditorium in November, but the promoters went broke a week before it was to open. Times and flying were tough.

Still, on December 17, thirty years after the Wright brothers' first flight, Omaha's new Municipal Airport Administration Building

Municipal Airport, Omaha, Nebraska

was officially dedicated. The program was sponsored by the Omaha Airport Association and the Chamber of Commerce Aviation Committee, of which Paul D. Selby and Lawrence Enzminger were chairmen. Speakers were Mayor Roy Towl, City Commissioner Harry Knudsen and A. L. Rushton (a pilot and Chamber member). Following their remarks, the ribbons across the doorway to the flying field were cut by Marline Burnham, daughter of Barney Burnham, the well-known area flight operator. A group of Omaha pilots, led by Ray "Wolf" Powell, put on a great airshow after that. The other pilots were Johnnie Kudrna, L. D. "Dutch" Miller, F. Robert Shields, Jimmie Goggins and Marion Nelson (with Bob Adwers as his passenger) in a brand-new Alexander-Eaglerock. My old North Field flying teacher, Bob Steele, was now sales manager for Alexander in Denver, and he let Marion Nelson fly the ship in the show. It was the first time I had met Nelson, and after the show he let me shoot a bunch of landings and take-offs. I did pretty well.

At this time, there were four major airlines whose principal source of revenue was their mail contracts—air passenger travel had not taken over yet. They were American, Eastern, TWA and United, the latter of which was the most innovative and profitable in the country. Its mainline route, east to west through the middle of the country, was operated better than any other airline's. The airlines didn't have much sway with the politicians in Washington, however, and north-south air service, especially in Omaha, was a hit-or-miss proposition.

The year 1934 also saw some aviation changes. Rapid Airlines evolved into Hanford Airlines, which later evolved into Mid-Continent Airlines, Inc., which later became Branniff. In addition, Donald Douglas had just rebuilt his DC-1 into the DC-2, which looked to be the best passenger and mail-hauling airplane ever built. To prove its worth, Jack Frye (TWA's president) and Eddie Rickenbacker (Eastern's president) flew TWA's DC-2, fully loaded with mail, from Burbank, California, to Newark, New Jersey. The flight took them thirteen hours four minutes, including two refueling stops. This was a new record.

Meanwhile, New Deal proponents in Washington were blaming the Depression, rightly or wrongly, on the monopoly of big business, and this soon had a heck of an impact on airplanes, flyers and aviation. Alabama Senator Hugo Black (later a U.S. Supreme Court justice) launched an investigation of the granting of airmail contracts. The airmail operators probably weren't all lily-white, but they were doing a pretty good job in a brand-new industry. In January, the illustrious Senator convinced FDR to cancel all of the private carriers' airmail contracts. Roosevelt then called in Gen. Benjamin Foulois, Commander of the Army Air Corps (the same Foulois that in 1909 had flown *Signal Corps Dirigible Number One* at Fort Omaha), and ordered him to fly the airmail. On February 19, the Army started flying the mail.

Man, did that order cause a heck of a hullabaloo. The airlines were stunned. The famous humorist Will Rogers raised ned in his daily column, which ran in all the Hearst papers. Lindbergh outspokenly condemned the action. Newspapers from coast to coast expressed great opposition.

Moreover, the poor Air Corps took a real beating. Their pursuit planes weren't built to haul mail, and their bombers had to be modified. Their radio equipment was almost nil. Their instrument flying was almost non-existent. Worst of all, most of the pilots had very little mountain- or high-altitude-flying experience, and they had never flown mail or passengers.

The first six weeks of Army Air Corps airmail delivery saw a series of disasters. Twelve Army pilots and crewmen were killed, and more than a score of airplanes were wrecked. Finally, FDR and Postmaster General Jim Farley returned the airmail contracts to the private airline carriers on May 8.

Some good came out of the bad, however. The Air Corps was starting to fly the mail pretty well before it was decided to return the mail to the airlines. This was largely because some excellent pilots had begun working at it—Lt. Col. Henry "Hap" Arnold, Capt. Ira Eaker and Lt. Curtis LeMay, for example. The woeful lack of airplanes, radios, navigation systems, maintenance, training, and so forth, which the Air Corps suffered was brought to national attention, and demands for Congressional funding began to be heard.

In the long run, the airlines became stronger, too. TWA's President Jack Frye (a long-time pilot) and young millionaire Howard Hughes (a very good pilot) became TWA's principal owners. American's President was C. R. Smith, a pilot-mechanic with a successful business background. United was run by Pat Patterson, a banker generally considered to have written the book on running an airline.

Back at home in Nebraska, in 1934 Nelson Flying Service joined Burnham-Miller Flying Service, Krantz Aviation and Enzminger Aircraft as the operators of the Omaha Municipal Airport. There still were only three hangars on the field. Nelson, Burnham and Miller used the original Legion hangar, as did Rapid Airlines. Krantz and Enzminger were in Enzminger's. United operated out of their hangar.

Omaha aviation was on the upswing, so the young Turks at the Junior Chamber began promoting a fourth Omaha Air Race, to be held August 11-12. The Chamber gave their approval, and twenty underwriters were secured. Most of the same people who organized the '31 and '32 races went to work putting it together. The 1934 Omaha Air Race was about the last grassroots air race in the country.

That was because airplanes were being designed for speed, and the ultimate goal for all speed designers, builders and pilots was to win the Bendix or Thompson Trophy, or both. Almost all these planes were built in a small shop, garage or hangar, without big business backing or government subsidy, and almost any pilot who was going to risk life, limb and property was going to do it at the national races, where the honor, glory and money lay. These competitions were held each September at the National Air Races in Cleveland, Ohio. The Bendix was flown cross-country from Los Angeles to Cleveland. Almost all its winners went on to the Atlantic for the transcontinental record. The Thompson race was a fifty-mile closed-course race. Thus the Cleveland Air Race became the aviation "Indianapolis 500."

When the promoters of the 1934 Omaha Air Race put up five thousand dollars in prize money, they attracted some of the country's best planes and pilots. This was the only Omaha Air Race that I didn't see, starting with the Pulitzer. I had a good summer job working for the Inland Construction Company, building a new road from West Thumb to Old Faithful in Yellowstone Park, so what I know of the races comes from the *Bee, World-Herald* and from the reminiscences of Marion Nelson, Dutch Miller, Barney Burnham, Steve Krantz, Rudy Mueller, Johnny Gillin, Bill Bloom, Ced Hartman and Larry Smith (of Dewey Chevrolet).

Day one of the air show began with a "bang." Roger Don Rae of Byron, Wisconsin, was testing his Kling Special on Saturday morning. On one pass he was clocked at 240 miles per hour. His

OFFICIAL PROGRAM

Fourth Annual

OMAHA AIR RACES

OMAHA MUNICIPAL AIRPORT

AUGUST 11 and 12, 1934

SPONSORED BY THE OMAHA JUNIOR CHAMBER OF COMMERCE
DIRECTED BY THE OMAHA AIR RACE ASSOCIATION

This Program Furnished by Courtesy

of

CHEVROLET MOTOR CO.

and

CHEVROLET DEALERS

of

OMAHA AND COUNCIL BLUFFS

....IT IS FREE TO YOU....

Sanctioned by the National Aeronautic Association and Held Under the Rules of the F. A. I.

tiny speedster landed at about a hundred miles per hour, touched down on the runway at the north end, hopped a ditch at the south end, and finally stopped with a bent gear and prop and a torn wing. The airplane had no brakes because they added extra weight and there wasn't room for pedals. Rae and the owner of the ship, Rudy Kling of Lemont, Illinois, immediately started repairing the damage, so he could fly in the two-thousand-dollars-prize Eugene Eppley Trophy Race on Sunday. A new prop was flown up from Kansas City, and Harold Philips, welder for Rapid Airlines, promised to rebuild the landing gear before the next day. It was agreed by all that Rae wouldn't have to qualify in the time trials to run later on that day.

More than 3,500 people were in the stands at 1:30 p.m. when Mayor Towl gave the welcoming address. The referee, for the third time, was Dr. J. B. Brock, a sportsman pilot from Kansas City. Brock signaled the start of the parachute jumping. Troy Colbach of Long Beach, California, was first at three hundred feet from the center. Roger Don Rae, who "flew" a parachute as well as an airplane, was second. Next, Milo Burcham of Muncie, Indiana, flying a Boeing B-12 pursuit plane, flew some hair-raising acrobatics. Muncie ended his flight by picking a rag off a line suspended between two fishing poles with his landing gear. Art Goebel, winner of the Dole flight from San Francisco to Hawaii, thrilled the crowd by skywriting "Dole" in half-mile-high letters. He then did some fancy acrobatics, leaving a smoke trail. Remember, this was way before the Thunderbirds and the impossible things flyers do nowadays.

Next, Wayne "Mile High" Wagener of Kansas City made a spectacular parachute jump from Burnham-Miller's Cessna, flown by Johnny Kudrna, Omaha's flying weatherman. He went out at twelve thousand feet and marked the course of his fall with a trail of flour for forty-two seconds before he opened his chute at about fifteen hundred feet. The crowd held its breath and loved it! Heck, not all airplanes could get up to twelve thousand feet in 1934, so imagine jumping from one. "Mile High" said that he had a twenty-five pound sack of flour and intended to empty it. He said that he would start higher the next time.

The real crowd pleaser was the B.V.D. Race for Omaha pilots only. The prize money was thirty, twenty and ten dollars. The stunt was to fly around the course once, land, climb out of the plane,

364

sprint over to the ground man, divest yourself of your pants, get back into your plane, fly the course again, land, put your pants back on, and fly the course once more. There were nine contestants, and it was wild. The *World-Herald* noted,

> The way some of those Omaha pilots side-slipped their ships in for quick landings, must have put the stunt flyers to shame. At the end of the first round, the field was covered with ships, and they were facing in all directions.
>
> The race was won by "Barney" Burnham in the weather ship, a Cessna, which is the fastest local plane. His partner, L. D. "Dutch" Miller, apparently is the fastest donner and devester of clothes, because he was away first after each stop, but he was fourth in the race. Marion Nelson was second, partly because of the breath-taking manner in which he side-slipped to land, and Steve Krantz was third.[8]

I heard this story on several occasions while "hangar flying" when those wonderful guys would get together.

During the time-trials for the pole position in the next day's Eppley Trophy Race, fans saw some of the fastest trials ever flown to that date. Each trial was made from a "flying start," and they really hummed! Gordon Israel, from Clayton, Missouri, flew the five-mile course at an average speed of 206.42 miles per hour. Lee Miles of San Bernadino, California, averaged 196.416; Art Chester of Glenview, Illinois, 194.2.; Steve J. Wittman of Oshkosh, Wisconsin, 173.783; and Joe Jacobsen of Kansas City, flying Benny Howard's famous *Pete,* 156.163. These five, plus Roger Don Rae, were the cream of the crop of American speed flyers.

The second day opened with clear weather, but there was a pretty stiff south wind which bothered the parachute jumpers. Roger Don Rae jumped first and landed a good thousand feet north of the circle. Mickey Efferson of Wilmington, Delaware, was second, and Troy Colbach, third. Next on the program, Milo Burcham and Art Goebel flew some hair-raising acrobatics. Marcellus King, former student of "Dutch" Miller, now of Minneapolis, did his famous "falling leaf" in his Challenger-powered Travel-Air. It was beautiful—coming almost straight down, nose high, and swaying side to side; he just about stalled out. "Dutch" Miller flew the best

---

[8]*Omaha World-Herald,* August 12, 1934.

"falling leaf" I ever saw, but they told me King was better. It is a tough maneuver to fly, and I don't believe anyone does them anymore. Johnny Kudrna won the B.V.D. Race, with Jimmy Goggins second, Marion Nelson third and John Kenwood fourth.

The flying was great, and everybody was having a good time; the air was electric. Everyone was waiting for the Eppley Race. Participating birdmen would fly fifty miles, six times around the course, for a chance at two thousand dollars total cash prizes. Finally, Jim Ewing turned the microphone over to Capt. Robert Dake, the chief starter, who told the racers to start their engines for the Eppley Race. The six tiny, beautiful, man-made hornets taxied north on the runway, their mighty four- and six-cylinder in-line engines tuned like the Philharmonic sang for the crowd of fifteen thousand as they glided by. At the far north end, they formed two lines and headed south. Israel was first on the left side of the runway, with Miles and Chester; on the right side were Wittman, Jacobson and Rae. They took off in that order, flew the course once in that order, and then every man struck out for himself.

What a sight! They shot by the stands sounding like six million hornets. When they passed the north pylon they were holding their positions, but when Starter Dake dropped the checkered flag all heck broke loose! Israel and Miles moved out together, with Chester right on their tails. They flew the first lap together. Rae moved up to fourth, Wittman fifth and Jacobson sixth (*Pete* had the smallest engine, a ninety horsepower Menasco, and he was flying 180 miles per hour). They flew these positions until the eighth lap. Israel had eased up because of his lead—a big mistake because Miles passed him. Fortunately for Israel, his 280-horsepower Menasco gave him enough speed to pass Miles on the home pylon at the end of the ninth lap. It gave the crowd a big thrill and Gordon a heck of a scare.

The first four were travelling almost 250 miles per hour when they pulled up at the pylon at the finish. Israel flew the winner's lap, while the others landed. Israel's approach was perfect, until he touched down at the far north end. Coming in like a bullet, the bird jumped back into the air, slammed down, went up on her nose, settled back, flew back up on her nose again, then flipped over on her back. Luckily, there was no fire. The crowd was stunned to almost complete silence.

Archie Baley . . . was first to the wrecked ship.

"Are you hurt?" he asked.

"No, lift up the tail so I can get out," replied Israel. His younger brother, Ed Israel, arrived at that time and shouted a warning about the danger of fire. He and Baley boosted up the tail of the ship, while others got the flier out. . . .Israel remarked afterward that he thought he had made a good three-point landing.[9]

Israel supervised the two dozen men who carried the broken little wasp to the Boeing hangar, then he returned to the starter's stand to receive the Eppley Trophy and a tremendous ovation from the much-relieved crowd. Israel's speed for the ten laps was 197.3 miles per hour; Miles, 195.24; Chester, 193.94; Rae, 192.34; Wittman, 176.34; and Jacobson, 168.63. The winner's total winnings were a little over one thousand dollars, and the damage to his ship was about three thousand dollars. The flyer didn't think the engine was damaged, and he was looking forward to the National at Cleveland the next month. What nerves of steel! No wonder I'm proud to be a part of the flying fraternity and fly with the old Red-tailed Hawk!

Israel (right) receives his trophy from Eppley

Thus the fourth Omaha Air Races came to an end. Treasurer Crawford Follmer announced that the total receipts of just under $13,000 would cover almost all of the expenses. It was thought that Eugene Eppley made up the difference.

Eppley's love of aviation goes back to his flying in the early days. In 1911 at Lima, Ohio, where he owned his first hotel, Eppley

---

[9]*Omaha World-Herald,* August 13, 1934.

and Cal Rogers owned an early Wright and flew it at county fairs. Rogers had been the first person to fly across North America; the trip took him three months and thirty-seven crashes. Rogers was killed in a crash, but it didn't destroy the Wright. Eppley had it shipped back to Lima and rebuilt. He then employed Andy Drew, a former St. Louis newspaper man, to fly it. They barnstormed around Ohio and Indiana until Drew spun it in one day. At that point, Eppley had had enough, called Orville Wright in Dayton, and canceled his order for a special-built barnstormer. When Eppley presented his trophy to Gordon Israel, he said that he had given up flying a long time ago, and that he'd ride with the young tigers from now on.

The flying kept up a furious pace. The next Sunday, August 19, six United States Army Martin B-10 bombers, equipped to carry enough bombs to destroy all of downtown Omaha, swept over Omaha at 12:15 p.m., swerved gracefully, and landed without a bump at Municipal Airport. Six of ten twin-motored, yellow-and-black B-10s on a flight to Alaska under the command of Lt. Col. Henry "Hap" Arnold, they were nearing the last lap of their trip. The birds refueled here, much to the delight of a large crowd at the airport, before leaving for Patterson Field, Dayton, Ohio.

Two days later, August 21, well-known local pilot Johnny Kudrna (Omaha's weather flyer) and a girlfriend, Irene Wostrel, were killed instantly at about 5 a.m. when Kudrna's ship crashed into a cornfield on the Iowa side of the river, two miles north of the airport. Kudrna had been flying the weather plane for Burnham-Miller for about two months and was one of the best pilots on the field. He was flying their new four-place Cessna. The ceiling was about a thousand feet when he took off around 4 a.m. The weather barograph recovered from the wreckage showed that he had reached 1,728 feet in seven minutes. Kudrna, as usual, had serviced the ship himself, and there was no one at the hangar when he took-off.

The flight usually took a little over an hour, so no one became worried until about 7 a.m., when he still hadn't come down or called. Marion Nelson and Barney Burnham took off looking for him. Nelson spotted the wreckage when he was coming back to the field shortly after 8 a.m. They got into their cars, drove to the site, and got there at about 10 a.m. The only bridge across the Missouri was the Douglas Street Bridge, and there weren't many roads, so it took time.

The Cessna was a real wreck. Marion said it looked like it went straight in. They found Kudrna's body about twenty or twenty-five feet from the plane. Then they unexpectedly found the woman's body about 150 feet from the Cessna. The Department of Commerce Inspector Harold Montee thought the crash had been caused by engine failure, and the sheriff seemed to be satisfied. No investigation was ever made, but when the guys talked about it later on there were knowing looks. Kudrna had a reputation as a lady's man, and he always loved to hot dog, as do most pilots.

Another unfortunate event occurred soon afterwards. Around 11 p.m., Friday, August 31, Rapid Airlines Flight Two from Kansas City to Omaha took off from St. Joseph, Missouri, in a thunderstorm. It was a fatal error in pilot Cleo M. Bontrager's judgement. Less than a half-hour later the Tri-motored, ten-passenger Stinson hit trees on the top of a hill just northeast of Oregon, Missouri, crashed into the bank of a highway, and burned. It was within a few hundred feet of the present Oregon Interchange of Interstate 29 and U.S. Highway 59. Bontrager and his four passengers were killed instantly. It was the first crash of an airliner bound for Omaha. The passengers were W. A. Truelsen (Manager of the *Omaha Journal Stockman*), Maude Shiffmacker (wife of Omaha cattle buyer Harry Shiffmacker), Frank Mahan of Chicago (who was going to Omaha to visit his folks, Mr. and Mrs. C. M. Mahan), and Dallas K. Leitch (a student at Dartmouth University who had been to a Beta Theta Pi fraternity meeting in Kansas City and was returning to his mother's home at the Conant Hotel in Omaha). This is the same Dallas Leitch who three years before led the Central High School Cadet Band that played at the air races and helped calm the ground with a stirring rendition of "The Stars And Stripes Forever" when "Speed" Holman crashed. No one played for the crash that Dallas died in, but maybe a Red-tailed Hawk watched from his perch on a storm-blown tree limb. I knew Leitch. He was the first contemporary of mine killed in an airplane.

The 1934 Air Races was the last big aerial flying event promoted or put on by Omahans to show flying and aviation to the people. Omaha had started these exhibitions way back with King Murphy in 1900. Since then, Omahans had seen Charles Hamilton in 1906; the Baysdorfers in 1907; Curtiss, Mars and Ely in 1910; Rene Simon and Rene Barrier in 1911; Lincoln Beachey in 1914; the Aero Congress and Pulitzer Trophy Race in 1921; and the Air Races in 1931, 1932 and 1934. There have been a few shows and exhibitions since 1934, and the annual STRATCOM Open House is a spectacular display of the equipment, organization and flying of the United States Air Force. But nothing can equal those earlier spectacles.

Although flying in Omaha didn't set the world on fire from 1935 to 1938, Omaha Municipal continued to grow. Marion Nelson's business grew as well. Burnham-Miller was one of the biggest operators and Steve Krantz was the other. On a national level, United started flying DC-3s in 1936, and Mid-Continent flew Lockheed 12s. The Department of Commerce passed the reins over commercial aviation to the new Civil Aeronautics Authority in 1936.

My flying really slowed to a standstill after 1934. All my efforts went towards getting through college. I graduated from the University of Colorado in September 1937 and went to work for General Motors as a cadet engineer. My career with GM ended in February 1938, and I went to work for Northern Natural Gas and returned to Omaha in April.

BOOK SIX:

The Airplane Reshapes the World
(1938 to the Wild Blue Yonder)

In the spring, I got serious about becoming a pilot and made a deal with Nelson Flying Service to get a CAA (Civilian Aeronautic Administration) commercial license and instructor's rating. Marion Nelson had just bought a new 65-horsepower Franklin J-3 Cub. He and Ray L. Harrison also owned a Stinson Gullwing SR-9. I was working for Northern Natural Gas Company but found time to fly before work in the morning and on weekends. I really hadn't done much flying for three years, since college had kept me busy. It came back fast, though, and I learned to drive that J-3 in a hurry. What a neat airplane! "Slim" Kenwood was my instructor. He turned me loose after three hours, and Nelson passed me for my private license in the required thirty-five hours.

Burnham-Miller had a new Waco F biplane with a 165-horsepower Warner, and I soon was flying it. Kenwood, Nelson, Ralph Brock, "Dutch" Miller and Billy Kimble taught me how to fly it. It wasn't long before I could snap, roll, and spin right along with them. The Waco F's brakes kept pilots on their toes. It had toe pedals on the rudders, and the pilot got pressure by pulling the throttle handle down. The handle was eight- or ten-inches long, and it was easy to crack the throttle when the pilot pulled it down. This resulted in some wild landing rolls. Anyway, I spent a lot of time at the airport, flew a number of different ships, and logged many hours of air time.

Then a great thing happened. Someone in Washington, D.C., decided our country should get serious about training pilots, so funds were appropriated for a CPTP (Civilian Pilot Training Program) to train 100,000 pilots. I don't know where the magic number 100,000 came from, but it was a start. This not only made it easy to learn how to fly, it assured flying-service operators all over the country financial success. Piper's business boomed, and its new sixty-five-horsepower J-3s were turned out by the thousands, alongside Aeroncas, Taylorcrafts, Luscombes, Stearmans, Ryans, Wacos, Cessnas and so on. The Omaha operators became busier than cats on a hot tin roof. Ground schools were held at Omaha University and many other places. There were primary and secondary courses.

It was great for me. I soon had my two hundred hours for a commercial license, passed the instructor's course, and went to

Kansas City, Missouri, for my CAA check rides. I arrived at the CAA hangar at Kansas City Municipal Airport, Kansas City, Missouri, at 9 a.m. after a five-hour drive from Omaha (there was no good way to drive to "K.C." in 1938). I waited for the usual hour, then an inspector named Baker came in and called, "Adders, let's go." I had my helmet and goggles, but they gave me a helmet with gossports. Instructors used to talk to the student pilots through gossports, a funnel stuck into a rubber tube that was hooked into earholes in the helmet. They checked out a chute to me, and we headed for the ship. It was a Waco UPF-7, and Burnham-Miller had a brand-new one in which I had flown my entire instructor's course. I loved that airplane, and I thought I could fly one as good as anybody. It was a beauty: black fuselage, CAA orange wings and tail, NACA (National Advisory Committee for Aeronautics) engine-cowling and pants on the wheels.

Baker told me to get in the front seat, take off south, fly a right-hand pattern, and shoot a 180-degree landing. There really wasn't very much traffic, but it was Kansas City. I was nervous, but darn it, I could fly. I taxied out S-turning, parked cross-T, revved up, checked the mags, cleared, got the green light from the tower, cleared, turned into the wind, locked the tail wheel, and gave it full throttle. She took off like a hawk. We flew a four hundred foot pattern at Omaha, so I figured that's what they did in Kansas City. It worked. I came downwind west of the field. He yelled to watch Fairfax traffic. I didn't see any. I cut the throttle, shoved the heat on, set a glide, turned on base, and got the green light. About the only planes with radio were the airlines, so the tower would give you a green light to land and a red to turn out. Pilots acknowledged by rocking the wings. I cleaned her out, turned on final, and greased her in alongside the hangar. He hit the throttle and yelled, "She's yours." This time we left the pattern and went northwest up the river valley and did S-turns, pylons, and forced landings. Then we went upstairs and did stalls, spins, chandelles and lazy-eights. Finally, I did loops, snaps and slow-rolls. Baker took the controls next. He did an Immelmann—a pretty good one. Then he told me my 360 power-turn wasn't too good. Heck, I had hit my prop wash, so he laid her over to show me what he wanted, kept tightening her up, and losing the nose. She started to burble. He kicked top rudder, and we spun over the top. He yelled, "What would a sensible man do?" I cut the throttle, booted full opposite rudder, and dumped the stick in a full NACA recovery. "Dutch" Miller had done the same thing with me once, and I did it with most of my more than three hundred students. Baker said, "Take me home," so I followed the Missouri back to Kansas City Municipal Airport.

I followed him into the ready-room and into the office. I thought that I must have done O.K. He sat down at his desk and said, "I'm passing you on your commercial, but your lazy-eights, power-turns and pylon-eights need more work for your instructor's. Go back to Omaha and work on them a little more." I told him that I had the day off from work and that I had to go back to Omaha. He got up and left, so I sat there for what seemed like an hour. He finally came back and said, "Trumbauer's coming in after lunch. I'll see if he will ride you. Can you wait?" "You'd better believe I'll wait." Baker left, and I waited.

I asked the office girls who "Trumbauer" was. They said he was a great guy, good pilot, a super musician, and had played in Paul Whiteman's band. I couldn't believe it!

If you're a jazz or big band fan, you'll remember Frankie Trumbauer, one of the immortal saxophone reed men. He started with Jean Goldkette, graduated to the great Paul Whiteman and played almost his entire career with the "King of Jazz." The Paul Whiteman Band played for the 1936 Junior Prom at the University of Colorado. He was playing at the Cosmopolitan Hotel, Denver, Colorado, and we got him for the prom at the bargain price of $1,500. What a prom! Frankie Trumbauer was the lead sax.

I didn't leave that office for fear of missing the man. He came in about 1:30 p.m. "Where's the man that wants to fly like a bird?" I jumped up and said, "Here." We shook hands, he pronounced "Adwers" right and said, "Let's go fly." Folks, we flew that Waco for an hour and a half, without a doubt one of my most memorable flying experiences. He could fly—had the smoothest and lightest touch on the controls I ever saw. We did everything, every one of the instructor course requirements. Whenever he thought I could improve he demonstrated, then gave her to me. Finally he said, "You can fly, son. Let's go upstairs and have some fun."

He was the first one to show me a snap at the top of a chandelle. I showed him my vertical cloverleaf. Line up on a road into the wind, dive, and do a loop, come out of the loop, and do a Cuban-eight. At the end of the eight, half-roll, then Split-S (watch your speed!), pull out, and do an Immelmann. I was hot and lucky.

She went like I knew what I was doing. He loved it, and we must have done ten before he was satisfied. We were flying about where Kansas City International Airport is today when he said we had better go home before they started looking for us. He signed my card, took me up to the Ad Building, and bought me a hamburger. I told him about the prom, and he remembered. I never saw him again, but he got me up there with the old Red-tailed Hawk. Thanks, Frankie Trumbauer. I know you and the Hawk have had some great flights.

When I got back to Omaha, I was sort of walking on air: I was now a commercial pilot and flight instructor. I could hardly wait to get out to the field. I really was a pilot now. Everyone sure was surprised that I got my ticket the first time. It seemed that it was an unwritten law that you never passed the first time, and I almost didn't. I know that it was Trumbauer who passed me, but Steele, Baxter, Kenwood, Nelson, Miller, Brock, Kimble and Harrison sure helped me. Flying was my avocation, and, although I made my living working for Northern Natural Gas, I got in a lot of flying on weekends.

The year 1938 was a fateful one on this earth. The Spanish Civil War (that most of us didn't know or wouldn't admit was going on) was approaching an end. Francos's rebels were winning, largely due to help they were getting from Germany and Italy. Russia was helping the government. None of the three were concerned much with the Spaniards. They were using the war to test their war-making equipment, mainly their air forces, and they found out what a great tactical weapon airplanes are. Meanwhile, England and France were increasingly fearful of Hitler and Mussolini, and the Soviet Union and Japan were involved in things in parts of the world most Americans thought existed only in *National Geographic.* So, we stuck our heads in the sand and thought those problems were minor and two oceans away. The Depression was a much bigger problem for us.

Finally, the country woke up. On September 28, 1938, President Roosevelt mandated the production of ten thousand long-range airplanes a year. He also appointed Gen. Henry H. "Hap" Arnold (the same Arnold who flew the mail and landed a Martin B-10 in Omaha in 1934) to replace the retiring Gen. Benjamin D. Foulois as director of the U.S. Army Air Corps.

At that time, the best pursuit or fighter plane we had was the Boeing P-26. A good airplane, it had been placed in service in 1933, and in 1938 we had less than two hundred of them. We had no long-range bombers, less than a hundred Martin B-10s and a hundred B-18s (an offshoot of the Douglas DC-3 which never amounted to a hill of beans). The Navy flying service was pretty well-equipped with carrier-based fighters but had few bombers.

Thank goodness for Foulois: His long career flying with the old Red-tailed Hawk inspired him to start the long-range heavy-bomber program which made the United States the world leader. In 1934 the Air Corps sent a proposal for a new bomber to all airplane manufacturers. The requirements were: top speed 250 miles per hour, cruising speed 220 miles per hour at ten thousand feet, and an ability to carry a useful load for ten hours. It was to reach operational altitude with load in five minutes and be able to maintain seven thousand feet with one engine out. Wow! Martin and Douglas received funding and started designing two-engine planes. Boeing received initial funding of $275,000, and they designed and built a four-engine long-range bomber.

The Boeing Model 299 first was rolled out for public display on July 17, 1935, and a reporter named it the "Flying Fortress." In August it was flown to Wright Field, Dayton, Ohio, and averaged 233 miles per hour. Phenomenal! Everything else we had looked sick in evaluation tests. But, Boeing wanted $197,000 for one Model 299, while Douglas only wanted $99,000 for a B-18. The 299 had everyone excited; the only drawback seemed to be its terrible price. Then on October 30, 1935, the Model 299 crashed after taking off on its last evaluation flight: Incredibly, the control locks were still locked. Douglas got an order for 133 B-18s.

Still, Boeing knew they had a darn good airplane and were able to build a new one right away. On January 17, 1936, they received a government order for thirteen B-17s at a cost of $3,823,807.

The first twelve B-17s were delivered by August 1937. They became the Air Corps's elite bombing unit—the Second Bombardment Group, based at Langley Field, Virginia. They were paraded all over the country, in war maneuvers on the East Coast, intercepted the Italian steamship *Rex* 750 miles at sea (to demonstrate long-range capability), and flew on a goodwill tour of South America. Ten modified B-17Bs were ordered late in 1937;

twenty-nine more were ordered in 1938; and delivery started in the spring of 1939. All in all, 12,731 B-17 Flying Fortresses were delivered to the Air Corps, the last one a B-17G in July 1945.

In August 1939 I delivered Northern Natural Gas's corporate records to 60 Wall Street when the original holding companies were divesting. I got to see the "Big Apple" and the World's Fair (which was held where La Guardia Airport is today). I also saw the first nine B-17Bs fly over as they made their record-breaking nonstop flight from Seattle to Mitchel Field, Long Island.

Republic was building Seversky's SEV-S2, a darn good fighter whose biggest bid for fame was winning the Bendix Trophy Races in 1937, '38 and '39. The SEV-S2 was rugged and tough, a good flyer and could cruise at 250 miles per hour. The Army finally bought fifty of them for delivery starting in 1939. They designated it the Seversky P-35. It evolved into the very successful Republic P-47 Thunderbolt fighter. Curtiss-Wright Corporation, Bell Aircraft and Lockheed (all using the new Allison "in-line" liquid-cooled engine) started designing and building Curtiss P-40s, Bell P-39s and Lockheed P-38s. The North American P-51 Mustang came a little later. Grumman Aircraft, Douglas, Chance-Voight, Martin and Consolidated Aircraft were working on planes for the Navy.

More important things were going on at home in Nebraska, though. Nebraska beat Missouri in Memorial Stadium using the airways when Omaha's Lloyd Grimm caught a pass from Lincoln's Bus Knight. It led to a score. Lloyd and Bus didn't know it at the time, but they were soon to become flying cadets.

CHAPTER TWO:
FUTURE PILOTS (1939 AND BEYOND)

Our tactical air forces (both Army and Navy) consisted of less than five hundred airplanes, and we probably didn't have that many tactical pilots. General Arnold set about doing something about it. Under his direction, our air forces had grown to 2,500,000 men and over 75,000 airplanes by 1945.

In spring 1939, General Arnold met with Charles Lindbergh, who had returned to the United States after living in England and France for several years. Arnold recalled that Lindbergh gave him the most accurate picture of the Luftwaffe, its leaders, its airplanes and equipment, its training methods, tactical plans and its shortcomings, that he ever received. After that, the Air Corps expansion program began, and the pilot training program was stepped up from three hundred to twelve hundred per year.

On September 1, 1939, World War II started when Germany invaded Poland. England and France declared war on Germany and, later, on its partner Italy, and World War II exploded. We were lucky. Three factors kept us out of the war for quite a while: We were a long way from the action, FDR rammed "Lend-Lease" to England through Congress, and, finally, the Japanese weren't ready to take a shot at us yet. Perhaps the Good Lord also was on our side. In this chapter, I'm going to tell some little-known stories about the war experiences of some Omaha, Douglas County and Nebraska guys.

Just before the war, three friends of mine from Omaha, entered the Aviation Cadet Program. They became flying cadets and graduated from the "West Point of the Air" at Randolph Field, San Antonio, Texas. I later spent fourteen months there as a pilot instructor.

Bennett Johnson's folks and my folks were lifelong friends, so we knew each other from the start. Dad and Frank Johnson worked for the Bennett Company back in the Baysdorfer days. Johnson also was secretary-treasurer of the Douglas Truck Co. whose plant was located at 30th and Sprague Streets (where Charles Baysdorfer flew the airship). Benny Johnson went to North High and Omaha University. He entered the Cadet program in the fall of 1939 and graduated as a fighter pilot from Kelly Field, Texas, in the Class of 40-J. In the summer of 1941, Johnson flew a brand-new Curtiss P-40 from March Field, California, to Omaha on a training mission. We all were at Burnham-Miller's hangar when he arrived. He shot a beauty of a fly-by, landed, taxied to the hangar, and parked. What an airplane! It was our best fighter at the time. I sat in it while he told me all about her. It was the first tactical airplane I had ever seen up close. It made us feel good after hearing how great the German planes were.

Johnson's squadron was on the aircraft carrier USS *Langley* somewhere north of Australia on December 10, 1941, when they were torpedoed by a Japanese submarine. Johnson and some of the pilots were picked up by a destroyer escort. The destroyer was torpedoed, and all hands were lost. Johnson never got a chance to fly against the Japanese, but I know he is with the old Red-tailed Hawk.

Benjamin Rimmerman and I were in the same class and played football together at Central High. Rimmerman and Jim Musselman were each other's best friends. Both Rimmerman and Musselman helped hold the balloons at the 1934 Air Races. Rimmerman went through the Cadet program and graduated as a fighter pilot in the Class of 40-E, and I saw him once when he was home on leave.

Ben Rimmerman

Rimmerman commanded the squadron of P-51s that flew the first fighter escort for B-17s all the way to Berlin in 1944. He was good and was credited with shooting down six-and-a-half Germans, which made him an ace. He also made full colonel. I'm almost sure that Rimmerman was the only ace from Omaha in World War II. He joined the World War I ace who called Omaha home, Eddie Rickenbacker. When Paris was liberated, Rimmerman and a friend flew a P-51 (it was pretty easy to make a one- into a two-seater) over there for a weekend of R & R (rest and relaxation). On the way back the weather was lousy. While making an instrument approach, something went wrong. They crashed short of the field. I'll bet Rimmerman showed the old Red-tailed Hawk a new way to do a full-gainer.

Fred Wupper is a native of Fremont, but he has lived in Omaha for years. Wupper entered the Cadet program after graduating from Northwestern University and got his wings in the Class of 41-H. By 1941 we were putting together our airborne armies, and we needed glider pilots.

In 1941 Wupper, his commanding officer and five mechanics were sent to Spencer, Iowa, to start up a civilian contract pre-glider school. The pilots-to-be were in the Army, and were given ten to fifteen hours in a J-3 (Piper) or Aeronca or Taylorcraft that the Army had bought from civilian flying-service operators. The trainees were given eight hours dual training, then they soloed. They never left the pattern and made every landing dead-stick, day and night. Those that survived were sent on to pilot army gliders, and the stories of their flying at Normandy are heroic and hair-raising. I saw a demonstration of three Waco gliders at Camp Hood, Texas, once, and there is no way I would have gotten in one. Wupper says he still shudders when he thinks of those days at Spencer and living at the Dickinson County Fairgrounds.

Wupper later had the best duty of any pilot in the Air Corps. In 1943 he was the Operations Officer at a primary school at Fort Stockton, Texas. They had just received Fairchild PT-19s when they got an order from the Training Command to fly two of them to Hollywood to be in a movie. And what a movie it was—*A Guy Named Joe,* starring Spencer Tracy, Irene Dunne and Van Johnson. Wupper and his commanding officer were on temporary duty for six weeks. Wupper did all the flying for Dunne, who was playing a Women's Auxiliary Ferrying Squadron (WAFS) pilot. They had the run of the sets and were treated royally. Tough duty! Incidentally, it's on television every now and then. The next time it comes on, take a look: Old Wupper is driving the PT-19. But, alas, the duty ended, and Wupper was in a Boeing B-29 Superfortress headed for Saipan when the big bomb was dropped.

It seems most people wanted to stay out of the war in Europe, and a lot of them, including a majority of the newspapers, weren't sure about FDR. Congress passed the so-called Neutrality Act, and the president signed it in November. No matter what, the die was cast, and it was only a matter of time.

382

*Cartoonist Ed Fischer paid an apt tribute
to Lindbergh in this cartoon after
Lindbergh's death in 1974*

Unfortunately, Charles Lindbergh got caught in the crossfire between those who favored American intervention and those who didn't. "Lucky Lindy" became a speaker for the anti-interventionist "America First Committee." In my opinion, he didn't deserve the beating he took from the press, which dated back to his flight in 1927, or the public attacks launched at him by many people, including FDR. He was so leery of the public that in 1940 he had to sneak into the Smithsonian Institution to see his beloved *Spirit of St. Louis.* He wrote in his diary:

> I have not seen the *Spirit of St. Louis* for several years, and I wanted to know how it is being cared for. . . . A dozen or so people were sitting on the benches on each side of the front entrance, watching all passers-by. . . . However, by blowing my nose at the right moment, I got by them all. . . . I stood looking at it [the plane] for nearly an hour, I think losing all count of time. Finally, I noticed two girls looking at me—they were not certain—in a moment they would ask. . . . I did not want to talk to people. I left.[1]

---

[1]Charles A. Lindbergh, "Friday, March 1, 1940," *The Wartime Journals of Charles A. Lindbergh* (New York: Harcourt Brace Jovanovich, Inc., 1970), 319-320.

After the attack on Pearl Harbor, Lindbergh served with distinction building Consolidated B-24 Liberators at the Ford plant in Detroit and toured the South Pacific showing our pilots fuel management. He was also credited with shooting down a Mitsubishi Zero-Sen.

Again, more important things were flying through the air in the Midlands—footballs! Nile C. Kinnick, Jr., an Omaha Benson High graduate and a friend of mine, dropped-kicked the extra point, and Iowa beat Notre Dame seven to six. In the summer of 1938, a bunch of us used to work out three or four nights a week at Benson High's practice field. What a bunch of football players we were—George Seemann, Hubert "Hub" Monsky and Lloyd Grimm (all of whom played at Nebraska), along with Kinnick and others who had played in high school. Kinnick, Seemann and I usually ended up on the same side, and we were hard to beat. At that time Kinnick didn't know that he was to be the fourth Heisman Trophy winner and would be voted the Athlete of the Year in 1939, beating out Joe DiMaggio and Joe Louis. Kinnick later told me that, when he got the phone call telling him he had won the Heisman Trophy, he didn't know for sure what it was.

Seemann (left) and Schwartzkopf after the win over Pittsburgh

Elsewhere in the state, the first oil well was drilled in Richardson County, just west of Falls City, Nebraska. Even more important, the Huskers beat mighty Pittsburgh fourteen to thirteen, and captains George Seemann and Samuel Schwartzkopf, later mayor of Lincoln, got their pictures on the front page of the *World-Herald*. The other front page picture was of Beth Hawley. Seemann became a pilot in the Air Corps, and Beth Hawley's husband-to-be, Bob Reichstad, was killed flying copilot in a B-17 crash at Kelly Field, Texas.

By 1940 all three Omaha Muni operators had Civilian Pilot Training Programs, which became their bread and butter. I flew quite often with Harrison on company business. On most weekends, I flew for Nelson or Burnham-Miller or Krantz, filling in for the regular instructors. I think I flew every plane on the field, so I was getting a lot of good flying time and making a good living at Northern Natural Gas. Life was pretty darn good.

I took my mom up for her first airplane ride. As great an aviation enthusiast as she was, she had never flown. Steve Krantz let me fly her in his brand-new Luscombe 8-A, a great two-place, side-by-side cabin job. She was really excited. We flew over the Woodman of the World Building, Douglas County Courthouse and Central High School. Then we flew out west above Dodge Street to Omaha University, then north to their house on Western Avenue. She was having a ball, but I noticed that she hung on to the strut on the high side whenever I banked or turned. "Relax Mom," I yelled. "Ride with her. She's just like a bird. Get hold of the stick (the dual was in) and fly her." She replied, "You mind your own business. I'm not one bit scared. I'm holding on in case the floor falls out."

In the spring, all the Omaha Muni operators and many other flyers participated in the twentieth anniversary of the Air Mail Service, receiving their instructions from airport superintendent Rudy Mueller. This event was promoted by the Chamber of Commerce. The airmail pilots who flew out of Omaha were Krantz, John Morrison, W. J. Devere, M. L. Powell, Ernie Ruckl, Ray L. Harrison, John Frazey, Marion Nelson, Barney Burnham, Mark Walker and Curtis Tucker. I flew with Harrison in his Stinson, and we picked up about thirty-five pounds of mail at Wayne, Nebraska. All the pilots furnished their planes, time, gas and oil, and even paid for the big dinner banquet at the Rome Hotel that night. It was a great occasion. Jack Knight was there, and I got to meet him. He was a United captain and was going to fly the same Cheyenne-to-Chicago flight he first flew in 1921, this time in a DC-3 full of night-flying instruments.

The war in Europe was going full-bore. Hitler invaded France in May. The huge French army of eight or nine million was still

Omaha airmail pilots, (left to right): Morrison,
Devere, Powell, Ruckl, Mueller (back to
camera), Harrison, Krantz, Frazey,
Nelson, and Burnham

sitting on the Maginot Line. The Germans went around them and took Paris and all of France in the summer. Somehow England survived the Battle of Britain: The Luftwaffe bombed the heck out of things, but the "Four Hundred" in their Supermarine Spitfires and Hawker Hurricanes made them pay dearly, and the people didn't flinch. The British fought the Italian Navy in the Mediterranean and the Italian Army in North Africa. All of this was taking a heavy toll and spreading Great Britain pretty thin, but the British hung on.

Meanwhile, the U.S. was picking up speed. Airplanes were built, so were ships, guns, trucks, tanks and other weapons. Even though we were "neutral," we were supporting England. Then in October, men between eighteen and forty signed up for the first draft. I signed up at my voting place, George Brandeis's garage at 60th and Dodge. My order number was 125, so my drafting was imperative. They sent me my status questionnaire that week. I appeared before the draft board in Waterloo's *Gazette* newspaper office the next week. I was married and had one son, so they classified me 3-A.

Most fellows who took ROTC and were graduating from college went right to active duty. One of my best friends, the late John E. Childe, graduated from Iowa State University in 1938 and was called to active duty in 1939. National Guard units got busy; Reserve units did, too. Every man under forty, whether he knew it or not, was in the big machine.

The November election was important nationally with FDR running for a third term, and locally Omaha voted on airport bonds worth $500,000. The Junior Chamber of Commerce sponsored an air show on Sunday, November 3. I was on the committee. There was a crowd of twenty thousand at the airport when the show opened with a fly-by of twelve ships. I was flying one of Nelson's J-3s in the small plane group. The pilots with big planes were behind us. After buzzing the field, we got into single file and dropped a five-pound sack of flour on a tractor parked in front of the crowd. Some had a heck of a time hitting the ground. There was some wild flying, and the people loved it.

The wind was out of the north. Everything started in front of the Ad Building. Kenwood won the small-plane race. It started with all of us taking off together, then flying east to the beacon at Rainbow Point, about three miles straight east of the Ad Building. We all flew about fifty feet high. We really tore up the place with those J-3s, Taylorcrafts and Aeroncas. Nobody could fly a J-3 like Kenwood.

Barney Burnham won the balloon-busting contest. Nelson was second. Jack E. Burnham, Barney's son, put on a good acrobatic-show with the Waco UPF-7. Miller put on his "Old Grandma Act." Later he really did give a Mrs. Morrison her first

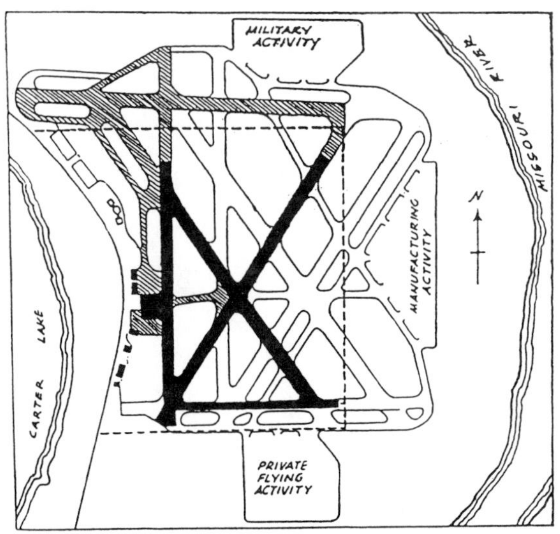

Plan for Omaha airport, after bond approval

flight at the age of eighty-six. All in all, everyone enjoyed the flying on one of those great late fall days. On election day the voters overwhelmingly approved the bonds.

CHAPTER FOUR:
THE WAR COMES TO US (1941)

The year started out wrong. On New Year's Day the mighty Cornhuskers, in their first and only appearance in the Rose Bowl, were defeated by the aerial bombs and ball-handling magic of Stanford's Frankie Albert. We in Omaha hung on sports announcer Bill Stearn's every word as he told us what was going on with his "radio word-pictures." We had to wait until the next day for the wire-photo pictures.

On March 3, 1941, Glenn L. Martin turned the first spadeful of dirt at Fort Crook, and construction of the Martin Bomber Plant started. It was the biggest undertaking Nebraska ever had seen, and it was for building airplanes. Did they ever build them—1,585 Martin B-26 Marauders, 531 Boeing B-29 Superfortresses and modifications on a thousand other Martin B-26 Marauders and North American B-25 Mitchells!

Things were busy at Omaha Municipal. All three operators had full Civilian Pilot Training Programs. Marion Nelson got approval to build a hangar just south of the original Legion hangar. When new bonds were approved, more land was acquired, and this brought the total to 761 acres. The three runways and taxi strips were completed, but other long-range plans were put on hold.

Great Britain was hanging on to the Mediterranean by their teeth. The Germans got their battleship *Bismarck* onto the high seas. It sunk the HMS *Hood*, and it was a real menace. In May, an RAF Catalina Dumbo, a lend-lease plane being flown by an American in the Royal Navy, found the *Bismarck*. Then two carrier-based Swordfish biplanes bombed her, then a destroyer torpedoed and sunk her. The Germans probably were embarrassed! That would be like a Missouri Class battleship getting bombed by a Stearman today.

Back at home, men were being drafted like crazy. I went up to the second floor of the Faidley Building on the northeast corner of 16th and Douglas Streets to see the Air Corps pilot-recruiting officer, Capt. Reed E. Davis. He assured me there should be no trouble finding a place for a commercial pilot with an instructor's rating. They tried to sign me up as a flying cadet, but that didn't thrill me much.

In June 1941, Hitler foolishly attacked the Soviet Union with 170 divisions of tanks, infantry, mechanized troops and 2,800 Luftwaffe airplanes. In collaboration with Vichy France, the Japanese took over Indochina in August. The United States imposed economic sanctions on Japan in response. Popular talk here was that our Navy could annihilate the Japanese Navy in seven days.

John Childe came home on leave in August. He was a major, having just graduated from the Command and Staff School at Fort Leavenworth, Kansas, and he was headed for Fort Benning, Georgia, to serve under Gen. George Patton. Childe told me that the U.S. Army couldn't have held up one little end of the fighting in Russia because we had less than ten infantry, mechanized, armored or what-have-you divisions, and the truth was we had less than one thousand tactical aircraft. This scared the heck out of me, and for the first time I realized that we'd better get into the act.

Childe spoke with the air officer of the Seventh Service Command (which was headquartered at 20th and Woolworth Avenue) for me, and I was told that I would be commissioned a first lieutenant. In August 1941 I was sworn in at the Post Office at 16th and Dodge Streets and sent to Kelly Field, San Antonio, Texas. In my four-plus years as a pilot instructor, some wild things happened.

Just before I left for Kelly Field, I had a long visit with Capt. Jim Musselman, M.D., my old friend from Lothrop School, Central High School and Omaha University. Musselman had helped hold the balloons at the 1934 air races. He had just finished his surgical residency at the University of Michigan and had been called to active duty with orders to go to the Philippines. When we met, he was on his way to Manila. I next saw him in May 1945. He had arrived in Manila the day before Thanksgiving, 1941, and was captured by the Japanese. The Omahan then survived the Bataan Death March and more than four years in the infamous Cabanatwan Prison Camp. Musselman served as the Chairman of the Department of

Surgery at the University of Nebraska College of Medicine and was on the Omaha School Board for many years. We know he had a tremendous influence on our son Jim, a doctor now. Musselman passed away in November 1990. Fly with the Hawk, Jim.

I was sent to Randolph Field in September 1941, the day after I passed my check ride. I instructed six cadets in the Class of 41K. When the class graduated, I got a seven-day leave, starting December 3. Two days before, I had the honor of flying a BT-9 in the final review for the retiring Commander of Randolph Field, Maj. Gen. Frank P. Lahm. General Lahm had been the Army's first pilot (soloed by Wilbur Wright in October 1908), had flown the American balloon in St. Louis in 1907 (when the Baysdorfers flew the *Comet),* and had commanded the Fort Omaha Balloon School in 1917. On leave, I caught a ride to Kansas City in a Douglas C-47, rode the Missouri Pacific to Omaha, and arrived home late on the fourth of December 1941.

I feel very strongly about December 7, 1941. Pearl Harbor Day, Sunday, December 7, 1941, was the day the Japanese made their unprovoked sneak attack on the United States of America, and they thought they could beat us. In alliance with the Nazis and Fascists, the Japanese were an incredible threat, and they hoped to rule the world their style. The Japanese bombing of Pearl Harbor brought the United States into the war, and after that it took our side four long years and millions of sacrifices to stop them. I firmly believe that our winning World War II did make this a better world. Remember Pearl Harbor!

I was cutting wood that Sunday with my brother Jack at his father-in-law's cabin at Todd's Lake, about three miles north of Waterloo, Nebraska, about eight miles from where I now live, and only a couple of miles from where Charles Baysdorfer made his flight in 1910. It was cold, so our wives were in the car with the heater on. About noon, they turned on the radio and heard the first news of the attack. They honked the horn and yelled for us to come. It was unbelievable, sitting there listening to the sketchy, incomplete and confusing reports, but we knew the Japanese had bombed Pearl Harbor. I called my squadron at Randolph Field and was told to get back, expediently! The attack was a complete success because no American had any idea it was going to happen. Looking back, I wonder if somebody didn't know it was coming and let it happen.

Two Omaha World-Herald headlines from December 8, 1941

Before I returned to Randolph, I visited the guys at the airport. Nelson's new hangar was almost completed, and all three were busy with what was now called War Training Service (WTS). The Martin Bomber Plant, the largest manufacturing plant ever built in Nebraska, was completed on December 10—the installation of the assembly-line equipment was started nine months after the first spadeful of dirt had been turned, and the first B-26 was delivered on August 10, 1942.

In 1942 the Japanese were riding high: They just about had the Pacific and the Orient in their pocket. It was very bleak for us. Thank goodness it's a big ocean! Hitler had bitten off a big chunk in Russia, which gave us a little time. We needed something good to happen!

In April old-time pilot Col. Jimmy Doolittle (Army-, Schneider-, Thompson- and Bendix-trophy winner, first pilot to land on instruments, and member of the Board at Mutual of Omaha) and seventy-nine other men flying B-25s bombed Tokyo.

Lt. Richard O. Joyce of Lincoln, Nebraska, was flying number nine. Joyce had been flying a B-25 on anti-submarine patrol at Columbia, South Carolina, when he volunteered for Colonel Doolittle's super-secret bombing project. They trained for a month at Eglin Field, Florida, under the direction of a Navy carrier pilot. The flight deck of the carrier *Hornet* was painted onto a runway, and the pilots flew a jillion practice takeoffs on it. The project was so secret that no one knew where they were, and they camouflaged the runway when they weren't flying.

Their B-25s had been modified to hold fuel enough for 1,975 miles while carrying a crew of five, one five-hundred-pound demolition bomb, 125 incendiary bombs, and about nine hundred rounds of .50 caliber and .30 caliber ammunition. Doolittle had picked the right airplane: I know the B-25 with its two 1,250-horsepower Pratt and Whitneys was a great flyer because I instructed a class in them.

In late March they flew to Sacramento, California, for new special props (costing $1,500 each), then to San Francisco to be loaded on the *Hornet.* Dick Joyce was twenty-two and a hot-shot pilot, so he had to hot dog and fly under the Golden Gate Bridge. He told me, "That's what it was there for, so I flew under it."

There the sixteen B-25s were loaded on the flight deck. The task force, under the command of Admiral William F. "Bull" Halsey of the *Hornet* (code name *Shangri-La*), then headed out for Japan. The B-25s took up four hundred feet of the eight-hundred-foot deck.

They hoped to get within four hundred miles of Tokyo before launching, but a Japanese patrol boat spotted them when they were eight hundred miles out. They sunk the boat, but it was feared they

394

had radioed Tokyo, so the mission started the morning of April 18. Dick Joyce told me the following account in 1953, when several of us had a great "hangar flying" session in the old Cornhusker Hotel:

The sea was rough. That, the excitement, worry and the rolling made most of us a little queasy. With the *Hornet* at maximum speed, we fired up the engines. Colonel Doolittle went first and got off well. So did two, three, four, five, six, seven, and eight. I was in number nine slot. When he started to roll, I prepared for take-off and put my left wheel on the white line (only fifteen feet from the edge of the deck). When the signal man dropped the flag, we started on our way. It happened so fast that, even with all our practicing, I was amazed how great she got into the air.

Joyce with a Japanese paper falsely reporting the Tokyo Raid results

The overcast was at about one thousand feet. We climbed through it right away, got our heading from the *Hornet*, and were on our way to Tokyo. It took us about five hours to get to the coast of Japan. My target was the Japanese Special Steel Company somewhere in South Tokyo. The anti-aircraft guns fired at us, but didn't do any harm. Some fighters flew above us, but we never were attacked. Remember, we were flying on the deck as close to the water as possible, and we surprised the heck out of the Japanese. We dropped our bombs at least close to our target and then got the heck out, heading southwest to China.

We were lucky. Thirteen of the sixteen planes made it to China, and we all bailed out just before running out of gas. That was the worst part for me. When I jumped out of the bomb bay it was pitch dark, and when my chute opened all I could think was that darn airplane is going to hit me.

It seemed like it was circling from the sound of the engines. Fortunately I landed in China with very few bumps and bruises, and after about a week later the Chinese got me to Heng Yang, so I could get back into the war.

Dick Joyce flew "the Hump" (that is, over the Himalayas) into China, returned to the states, and was discharged a light colonel. His Distinguished Flying Cross and his flying in the April 1942 raid on Tokyo places him in a pretty distinguished group. Joyce had a long and successful business career in Lincoln, Nebraska, served as Chairman of the Airport Authority, and flew into the Wild Blue Yonder in 1983. Fly with the Hawk, Dick!

Fifteen of the planes involved in the raid on Tokyo crashed and were destroyed. Thirteen made it to China; two crashed in the China Sea; and one landed in Russia where the crew was interned. Two of the crewmembers were reported missing, but later confirmed dead. One was Cpl. Donald Fitzmaurice, also from Lincoln, Nebraska. Joyce told me he only slightly knew Fitzmaurice, and I have no information about him. One man was killed in the crash of his plane. Of eight who were captured by the Japanese, all were tortured; four were executed as war criminals (two by beheading); and four were released after Japan surrendered. Many of the Chinese who rescued our pilots and hid them away were executed as well.

The Japanese would not answer the international agencies' requests for information about the captured men and refused to let neutral diplomats see them. They just said they had the terrible American war criminals who bombed hospitals and innocent school children. Finally, almost a year to the day later on April 22, 1943, some meager but ominous information was released.

In May we finally destroyed a large part of the Japanese navy—twenty-one ships and innumerable planes—in a gigantic battle in the Coral Sea north of Australia. Land-based bombers, the Navy's fighters, dive- and torpedo-bombers were heavily involved in the fighting. It was the beginning of the end for the Japanese Navy. Two friends of mine, Peter and David Lage (who were born and raised in Benson) were seamen on the USS *Lexington,* which was sunk in the battle. The ship was attacked by Japanese planes when its own planes were out on a mission, and after a fierce battle the *Lexington's* fuel supply exploded and the ship had to be abandoned. So with only the clothes on their backs, billfolds, and dog tags, the boys made the seventy-five foot jump from the deck to the water, and then climbed into the rafts to await rescue.

Then in June, one of the greatest naval battles in history took place near Midway Island. The ships never got closer to each other than a day away, but the airplanes had at each other. Our land-based B-17s and B-26s flew strikes on the deck. I have never confirmed that any Omaha pilots participated in the action, but a Martin B-26 that had just been modified in the Omaha plant to carry torpedoes was credited with sinking one of the enemy carriers. The Japanese Navy was just about finished, and this was only six months after Pearl Harbor.

The day after the Battle of Midway ended, Maj. Gen. Clarence L. Tinker (who had been the Air Officer in Omaha in 1922, led a fly-by at the dedication of Offutt Field in 1924, and commanded the flight of fifty-six fighters to Omaha in 1931) led a flight of six B-24s on their way to bomb the Japanese who had occupied Wake Island. Why General Tinker left a perfectly good desk in Hawaii for such a wild-goose mission seems to have baffled historians, but I know why he did it. Tinker was fifty-six years old, he was one-eighth Osage Indian, had been in the Army since 1912, was an "Old Bold Pilot," and had flown every airplane the Army had, but he had never flown a combat mission. The general knew that this would be one of his last chances to get into the act, so (as commander of the Seventh Air Force) he ordered the mission.

First Bomber Assembled at Fort Crook Takes to Air Amid Martin Workers' Cheers

It was a long haul from Hickham Field, Hawaii, to Wake Island, and about an hour out of Midway they ran into a horrendous cumulo-nimbus development. The air got wild; they saw the general's ship flying erratically; then it disappeared into the overcast. They never saw him again. The weather became so bad that they aborted the mission and made it back to Hickham. Tinker Air Force Base in Oklahoma City is named after him.

Back in Omaha, on August 11 the first Martin B-26 assembled in the Fort Crook plant roared into the air to the cheers of thousands of workers who had built her.

Meanwhile, the F model B-17 was rolling off of the line in Seattle. Sally (Mrs. Harry) Farnham, our good friend here in Elkhorn, helped build them. She was so small she could climb inside the wing to buck rivets. Both of us like to think that she helped stick together one of the planes I flew.

On August 27 in Victorville, California, Judson L. Hansen graduated from Advanced Flying School and got his wings and his second lieutenant bars. We worked

Jud Hansen (right) and fellow Omahan Craig Willy at Victorville

together at Northern Natural Gas and have been long-time friends. Hansen flew B-25s and B-24s in the Southwest Pacific. He started flying in a Civilian Pilot Training (CPT) course at Krantz Aviation and Burnham-Miller at good old Omaha Municipal. During the war, Hansen flew from the U.S. to New Guinea in a B-25, which is a long ways at two hundred miles per hour. He flew a half-dozen missions in a B-25, then they sent him to Australia to fly a B-24. The B-24 had the longest range of any of our bombers, and the distances were great in the Southwest Pacific.

In 1944, Hansen flew his first mission in a B-24—he was flying the lead of a nine-ship group. The Japanese had a big navy base in a natural harbor on Halmahera Island, about twelve hundred miles from the U.S. field in New Guinea. The plan was to turn at a landmark north and west of the naval base and come in from behind. They were at ten thousand feet over a heavy overcast. When they were about five hours out, the navigator called and said to let down—he thought they were right over their landmark. They broke out of the overcast at eight thousand feet. Hansen turned the group 180 degrees and was preparing to make his turn on the target-run when all heck broke loose from anti-aircraft fire. The navigator had put them down right over the base. They quick-dropped their bombs and got back into the overcast before encountering any serious fighter action. Everyone got home safe; photo-reconnaissance confirmed they scored some great hits; but the poor old navigator sure caught heck!

On October 24, the B-17 in which Eddie Rickenbacker was making an inspection trip of the South Pacific, radioed that it was out of fuel and was ditching somewhere in the vicinity where Amelia Earhart disappeared in 1937. The Army and Navy admitted the prospects of finding Rickenbacker and the crew were gloomy. On November 13, a Navy Catalina Dumbo spotted a raft a few miles from Samoa. Eddie and two others were still alive in it. Three more of the crew were rescued from the island. Col. Hans Adamson (who was in the boat with Rickenbacker) said that Rickenbacker said they wouldn't die, and they didn't dare doubt him. They had been adrift for twenty days.

Back in Omaha, every now and then an Army or Navy plane would land at Muni, and the town would get to see what a real military plane looked like. One day a Consolidated B-24 Liberator that was being ferried from Willow Run, Detroit, Michigan, took off on Muni's northwest runway. It didn't get into the air, crashed through the fence, hit the road, and bounced into Carter Lake. No one was hurt, but the B-24 was destroyed. The Army hauled it out of the lake and took it to the Lincoln Air Base.

Another time, S. Sgt. W. I. Sampson II (born in Alliance, Nebraska, and the Army's first glider pilot) was flying a Waco troop-carrying glider. He cut loose from the C-47 towing it about three thousand feet above Muni and glided like a rock to a successful landing. The three guys in the crew walked away O.K. Maj. Reed

Davis, the mayor and several business leaders wisely declined a ride. Two weeks later one of the same kind of gliders crashed in St. Louis killing fourteen men including their mayor, two councilmen and several business leaders. I have a dim view of gliders. People should be given the Air Medal just for getting in one.

On the first anniversary of Pearl Harbor, the United States had survived and was starting to make some headway in the war. The Japanese Navy was in big trouble. The United States was getting ready to start the offensive in the Pacific. Our forces landed in North Africa. The British controlled the Mediterranean Sea. The Germans controlled most of Europe, but were bogged down in Russia. In Southeast Europe, some great underground freedom patriots were rising to action. The Italians wanted life to return to normal and were becoming a burden on their Nazi friends. The Allies weren't in the lead, but there was a faint light ahead. The American people were rationing food, growing Victory Gardens, and working side by side on production and assembly lines. At Williams Field, Arizona, Huskers Bus Knight, George Seeman and Wayne Blue were in advanced bomber training.

I logged almost fifteen hundred hours in 1942, and ninety percent of it was instructing—both single- and twin-engine and sometimes a tactical aircraft. I was sure learning how to fly.

In spite of the war, a neat thing happened to me one night. I was instructing an advanced class in the North American AT-6 "Texan." The cadets were flying one of their last cross-countries. I was the final check-in instructor at Eagle Pass, about eighty miles west of Randolph Field, San Antonio, Texas, so I rode back with the last cadet. Those beautiful and nasty Southwest Texas cumulo-nimbus developments were all around. We were at three thousand feet. I was hot and tired, so I lowered down the back seat as far as possible, loosened my belt for more comfort, took off my shoes, and told him to take me home and not to mess around. Every now and then there would be lightning running in those clouds, but the air was stable and good, so I went to sleep. I woke up with my cadet frantically yelling, "Sir, we're on fire." I had my headphones above my ears, so he also had made a steep bank to get my attention. He did, but the prop really got it. Sparks were shooting off of it like a Fourth of July pinwheel. I woke up fast and checked the instrument panel. Nothing was out of place. Then the entire leading edge of the wing started glowing almost like a flame. Thank goodness it dawned on me—St. Elmo's Fire. I had seen it once when it ran like a fire ball on a barbed wire fence on my grandma's farm when I was a kid. Between the AT-6 moving through the air and the tremendous static charges all around us, the electrons were going nuts. Then the instrument panel started glowing, and it stopped on the prop. Now, I knew it was St. Elmo's Fire, so I told my cadet, "Son, you were getting too close to Satan and St. Peter's assistant, St. Elmo, was heading you off." Cadet John Haas probably never forgot that exciting night in the Texas sky. I haven't.

By 1943 everybody in Nebraska had some connection to the war. A lot of military installations were located here because we're in the middle of the country. There were ordinance (bomb-loading) plants at Mead and Grand Island, ordinance storage at Hastings and Sidney, the bomber plant at Ft. Crook, and air bases at Lincoln, Fairmont, Bruning, Harvard, Scribner, Grand Island, Kearney, McCook, Ainsworth and Alliance. Nebraska's many food processors were going full-bore, and shops were making everything from shells to piston rings. Our farms produced corn, wheat, beans, cows and pigs. We had a bumper crop. Nebraska was humming—we were a tremendous part of the war effort. President Roosevelt pretty well summed it up in his State of the Union report to the Congress: The Axis powers knew that they had to win the war in 1942 or everything would be lost. Roosevelt didn't need to tell Congress that the Axis powers hadn't won the war in 1942.

In 1943 Lt. Arthur Storz, Jr., was flying one of his last tactical cross-country flights in a B-17. His course took him over Omaha, his hometown. The Storzs were prominent citizens and lived in a fine mansion near the old Blackstone Hotel. His fly-by was done with style and class. He started at the Missouri River, flew up Farnam Street at a little higher than the downtown buildings, then got down almost on the deck, flew by Mutual of Omaha, and pulled up over his parents' house at 37th and Farnam Streets. He then flew a pylon-eight around Saint Cecilia's Cathedral, back over his house, and east over downtown. Then he took off into the Wild Blue Yonder. He was probably disciplined for the flight. Years later Storz told me he was going to fly under the old Douglas Street Bridge but lost his nerve.

By now a lot of our pilots were in final tactical training. My good friend Ralph Dickerson, who had gone to North High when I went to Central and consequently became my high school football opponent, had become a pilot too. The last time I saw Dickerson was in 1943 when we were both home on leave. He had just finished tactical in B-25s and was shipping out for the South Pacific. Some Omahans were getting into combat action, and some were getting hurt. Two Central High friends bought it, too. Bob Moose, Omaha's leading fighter pilot at the time, was killed when he and his wingman collided in combat action. Moose, a P-40 pilot, had shot down four Japanese, three in four minutes of furious action over New Guinea. He was one kill away from being an ace. You flew with the Hawk, "Mooser." Jim Milliken was killed when the B-24 of which he was the bombardier crashed in the Atlantic while on submarine patrol off Florida. Milliken just disappeared—no radio signal, no sighting, no nothing.

In April the Missouri River flooded. In those days, before the dams, the river flooded twice in the spring, when the snow melted

Omaha Municipal underwater

in April and when it rained in June. April was always the worst. The operators all moved their planes to high ground. John Savidge took this picture of Muni from Marion Nelson's Cub.

On June 3, Nile C. Kinnick, Sr., and his wife received "the telegram" informing them that their son, Nile C. Kinnick, Jr., was killed and his body was not recovered. My good friend Kinnick was no different than any of the thousands that were killed, but, when my mom wrote me about it, I couldn't believe it.

It was a long time before any detailed information about what happened was released, and at best the information was pretty sketchy. The Navy only said that on June 2 on a training flight off of the USS *Lexington,* Kinnick's Grumman F4F Wildcat crashed into the sea and sank. The body and wreckage were not recovered. The first story I heard was that he crashed on landing and went over the side of the deck. Over the years a number of different stories have surfaced, and they all boiled down to something like this: He was on his final tactical training cruise on the new *Lexington* somewhere off of the coast of Venezuela. The weather was clear and the sea calm. His engine developed a bad oil leak, and he was ordered not to land on the carrier, but to ditch close to an escort ship that was four miles from the *Lexington.* An Ens. Bill Reiter was said to have been flying next to Kinnick and saw him in the water free of the plane. Eleven minutes later a rescue party arrived at the site and found only an oil slick and paint particles. It was speculated that his belt failed, he cracked his head on the instrument panel, became unconscious, and the vacuum of the sinking plane pulled him down. Hogwash! So many holes can be shot in that story that even the Navy wouldn't make it up. As I said before, airplanes and flying are awfully unforgiving. Whatever happened probably was a real disaster because Kinnick was a winner and a survivor. He is sure remembered

Nile C. Kinnick, Jr.

by all the Iowa Hawkeye fans when they fill Kinnick Stadium on beautiful fall afternoons. I'll bet he and the old Red-tailed Hawk are flying high.

The tide was turning as our military forces came together in 1943. The Marines took Tarawa; Rommel was beaten in North Africa; and we landed in Sicily and bombed Italy. The fighting in Italy was tough. The Italians had had enough and ousted Mussolini, who had been their dictator since 1922. The Nazis took over. All was not roses in Japan, and Germany was retreating from Russia.

Harry Burrell and Bob Storz, both West Omaha boys, flew B-17s that dropped the first bombs on Rome. Later, Storz flew a B-24 on the Ploesti oil field raid in Romania. It was very hairy, and over half our planes were lost. Storz got to the target and got shot up good, but he managed to limp to Malta. Then two good things happened to him: He got the Silver Star, and he was rotated home. Another Omahan, Chuck Malec, whose folks owned Peony Park, was the lead bombardier in a B-17 on the first U.S. raid on Germany at Hamburg.

In Omaha, the Martin Bomber Plant was going full-speed, turning out fifty B-26s a month. A total of 1,585 were built. Modifications were made on 893 B-25s. One of them crashed in the middle of the plant on takeoff and did considerable damage, but the accident only threw production off for part of one day. Sue Seemann, George Seemann's wife, modified planes at the plant. Using a code, Sue sent George the serial numbers of the B-25s they modified. George was in the South Pacific at the time, and a couple of the planes his wife worked on arrived at his squadron. He flew one of them on several missions.

Later, the production of B-26s was tapered off, and the plant was retooled to build the new B-29 super-bomber. The first Martin-built B-29 came off the assembly line on June 1. The plant was soon turning out fifty-five a month. 531 B-29s of the total production of 3,970 were built in Omaha. More than thirteen thousand people got really good at building the best bomber the world had ever seen. For a while, Omaha was the "Air Capital of the United States"—finally.

In December President Roosevelt met with Sir Winston Leonard Spencer Churchill and Stalin in Teheran, Iran. The nature of the meeting made it historic, of course, but also note it was the first time a U.S. President had ever made a trip in an airplane. Roosevelt flew twenty-five thousand miles in the great four-engined Douglas C-54, alias DC-4—and it was forty years almost to the day that the Wright brothers had flown at Kitty Hawk. The President slept on a mattress laid across two seats from which the backs had been removed. The rest of the party enjoyed the comfort of those wonderful bucket seats—a far cry from *Air Force One* almost fifty years later.

We were turning out pilots in unbelievable numbers. The training program was good, and the students were taking to

*Robert E. Adwers (center) with two other pilots
at Mustang Field, Oklahoma, December 1943*

airplanes like birds. In just under a year they went from a walking
soldier in classification to learning how to fly in primary training,
to instrument and low-frequency radio navigation in basic training,
to refinement in single- or twin-engines in advanced training, to
tactical training for combat or support. They had less than four
hundred hours total training time when they were pronounced ready
for combat. There had to be a million great stories. Here are a
couple.

Ben Womble and I flew together at Randolph and at a
primary school in Oklahoma. Womble was a big Arkansas boy, slow-
talking, easy-going and a good pilot. He was flying a cadet's final
check ride in a Fairchild PT-19, a low-winged, open-cockpit trainer.
Snap-rolls were a required maneuver. To perform one, a pilot
started out at cruising speed, brought the nose up, snapped the stick
back, booted the rudder and (wham!) rolled around the longitudinal
axis, hit opposite rudder, neutralized the stick, and came back right-
side-up. His cadet was having trouble, so Womble demonstrated one
for him. Womble must have thrown the aileron in as she rolled (we

all did to get a good roll), and he accidentally must have released the safety catch on his seat belt. Then he told the kid to shoot him another one. He did, and it was a dandy. Old Ben Womble went flying out of the PT-19. He told us later, "You know, I knew something was wrong when I saw that airplane flying away from me." He pulled the rip cord of his old round twenty-four-foot chute and lit like a ton of bricks in an Oklahoma cane field. Since he had the rip cord, he became a member of the Caterpillar Club.

It must have been Oklahoma because an unbelievable thing happened to me there, too. I was flying a final check ride in a PT-19. It was hot. I was wearing my summer flight coveralls, so I had put my billfold in one of the breast pockets and forgot to zip it shut. My cadet did a beauty of a half-roll for me and (whoosh!) out went my billfold and a Parker fountain pen that my folks had given me for Christmas. A few of us had a running snooker game going a couple nights a week in Showen's Pool Hall in El Reno, Oklahoma. When I came in one night about a week after I lost the billfold, Arch Showen said to me, "Lieutenant, I've got your billfold." He honestly did. A farmer had found it, figured out it belonged to one of the fliers, and took it to Showen's. It had five dollars in it, which, the farmer told Showen, he kept for the reward. That was a fair enough trade, but I wonder where the Parker went?

It was a long way off, but we could see the end of the war by 1944. American forces were taking islands in the Pacific. We had so many men and materials in England that people were afraid the island would sink. The RAF was bombing the heck out of Germany at night, and the U.S. Army Air Corps bombed by day.

Many Omaha and Nebraska boys were among those flying over Europe. My old friend Lee D. Seemann (whose father was the George Seemann who let me drive the red roadster at the 1931 Omaha Air Races) had taken his tactical training in B-17s at Kearney, Nebraska, where I was an instructor. On March 26, 1944, he wrote his mother to tell her about his twentieth mission. The target on his twentieth mission was Brunswick, about a hundred miles west of Berlin and four hundred miles from the 96th Bomb Group's base near Snetterton Heath, England. They had a P-47 fighter-escort over the English Channel and the Netherlands, but not over Germany. This was before belly-, wing- and tip-tanks. As Seemann explained to his mother and told me many years later:

We were at 23,000 feet. It was all blue sky and beautiful sunlight with a pretty heavy overcast below, but we knew the Luftwaffe was tracking us trying to figure what our target was. I was flying the lead ship, so we had the thrill of seeing the ME-109s first. There must have been fifty of them. They always used the same tactic: dive from above and in front of you (our firepower wasn't as great forward, but a 17 could put out a lot of lead and steel from those twelve 50s). They came through us like a pack of wolves, and they were good. The next ten minutes reminded me of a Hollywood version of an air battle: Planes were blowing up and burning. Tracers were going in every direction. Twenty millimeter shells were exploding too near and too often, and there was the usual blossom of parachutes. My plane only took a few minor hits, but only about half of our group made it to the target. We took three direct hits, dropped our bombs, and headed for home. Then we found the number three engine was losing oil and smoking badly. We couldn't get it under control, so we had to shut down number three and feathered the prop. After boosting the power, the other three engines kept us flying in formation. Suddenly, number four lost its oil pressure. I shut her down, but the prop wouldn't feather. A 17 won't maintain altitude on two engines, so we waved goodbye to the formation and dove for the clouds. We threw out everything that was loose—guns, ammunition, etc. But I was still losing altitude, and those two Wrights were bellering and red hot.

We were losing three hundred feet a minute. Over Holland we were really concerned about our low altitude because we were a sitting duck for anti-aircraft guns or a fighter. Suddenly, two P-47s pulled alongside of us and assured us good protection.

I thought for a while we might make it to some point in England, then I realized we would have to ditch. The British Navy ran the air-sea rescue, and they were good at it. I tuned in channel D on the radio, and that neat Limey voice said, "What can I do for you, Yank. Hold your button down for ten seconds. Good! We've got your fix and will be waiting for you." We could see a ship quite a ways ahead of us. Everyone except the copilot and I got in the radio room (ditching procedure). I held her off as long as possible. The tail dragged, and I thought we were going to set down great, then (wham!) the engines and nose hit. It knocked all of us out for a while.

When we came to, she was floating. We climbed out the window on my copilot's side and got on the wing. All the guys made it. We saw a tail section, rudder and vertical fin, elevator and horizontal stabilizer and tail gunner's compartment floating about a hundred feet ahead of us. I was still dazed and thought, "Jeez, they've shot down a bunch of us!" It was our tail! Somehow it broke off and landed ahead of us.

Another thing I vividly remember was the hissing of steam coming off those red-hot engines. We all popped our "Mae West" valves, and they all inflated, except the copilot's. Then we released and inflated one life raft. The other was stuck. Our main fear was that we would go down with her when she sank. We all had on winter flying gear, including boots. There were ten of us, so I ordered six in the raft and four in the water.

The water was extremely cold, but luckily the waves were only two or three feet high. The guys in the raft tied those of us in the water to it with their boot laces. While in the water, I remembered when Bob Adwers, my brother George and I swam from Des Moines Beach to Arnold's Park once, and what a great time we had. I also thought a lot about Okoboji. I couldn't see anything except sky, and there was a P-47 circling us. The guys waved, and he rocked his wings. We saw smoke from a ship; the navigator still had his binoculars, so he took a look. He yelled, "It's a German. I can see the swastika!" The Germans always were looking for air crews to interrogate, so we were really worried. The 47 then made three passes on the ship, started a heck of a fire on the deck, and the ship turned away. We never found out who the pilot was, but we sure thanked him. I really only vaguely remember things. I remember my left foot hurt like heck, so I figured my right one was frozen because it didn't hurt [Seemann's left foot actually had been broken in the crash.]

After a while we spotted an approaching British ship. I was seeing those round, ring-like spots before my eyes, but I remember that ship. It was pretty good-sized, and painted in big white letters on its side was "HMS 515." Soon they hung a net on the side, caught me, and pulled me up. Man, was I happy, but I couldn't talk! They took us below and thawed us out by rubbing the heck out of us with wool blankets. They almost killed me when they rubbed my left foot, but I was so happy to be alive that I didn't care. I'm happy to say the whole crew escaped with only minor injuries—my broken foot was the worst.

I was transferred to an American hospital near our base, and there must have been a hundred beds in the ward. I was just lying there looking at the ceiling when someone said, "Lee Seemann, what in heck are you doing here?" It was Chester A. Waters, Jr., from Omaha. We had gone to Central High School together. Waters was a U.S. Army orthopedic surgeon assigned to the hospital.

Lee Seemann returned to his group after his foot healed and flew seven more missions before he was sent home in September. He received the Silver Star, Flying Cross and a Presidential

Citation. The medal Seemann cherishes most, however, is the Gold Fish Medal. He was assigned to Mitchel Field, Long Island, New York, which was great duty. It was a sort of "war hero" base. The airmen there got some wonderful assignments. His commanding officer called him one day and said, "Lee, you're going to get the Gold Fish Medal for ditching and surviving, and Captain Eddie Rickenbacker is going to present it to you." Seemann later recalled,

> All I could think of was when Bob Adwers and George used to read *Argosy* and *Aces,* the World War I flying magazines, and Captain Eddie Rickenbacker, our biggest ace, led the famous Hat-in-Ring squadron. He was as famous as Baron von Richthofen.

In October 1944 at the New York Aero Club, Omaha's Lee Seemann was given the Gold Fish Medal by Capt. Eddie Rickenbacker, whose Omaha ties dated back to 1909. Rickenbacker had received the first Gold Fish for ditching and surviving in the Pacific in a B-17. Today, Seemann lives in the old Waters's family home in the Fairacres division in Omaha.

Brothers George (left) and Lee Seemann with George's wife Sue, while on leave in 1944

Meanwhile in Hawaii, Lee Seemann's brother, Capt. George Seemann, had been sent to Hickham Field to fly a special B-25 that had a 75-mm cannon in the nose. The copilot's seat had been taken out, and the instrument panel and nose had been modified. The breech was quite far back, so the engineer loaded and removed the casing. The gunner aimed through a ringed sight, and the gun had a bungee cord recoil mechanism. United States forces would bomb and shell for days, but the Japanese would still be there when the infantry landed, so the cannon had been developed to shoot into the caves and trenches. Seemann told me, "It was a thrill when you fired it. The 25 shuddered, and you thought you had hit a wall for a second. I popped my nose on the damn sight the first time I fired it, so I threw it out and aimed the nose, and I got pretty good on the targets." George Seemann led the first flight that fired the cannon at the Japanese, on the little island of Maloelap. They came in right on the deck and fired straight at the caves. They could see the shell hit, but they didn't see any spectacular explosions. The Omaha boy later commented, "I don't know how effective we were, but I'll bet we scared the hell out of them."

The cannon episode didn't last too long because someone developed the skip-bombing technique, and the gunners really got good at it with the B-25s. Seemann flew a bunch of skip missions and low-level parachute bomb runs on Japanese airstrips. He got the Distinguished Flying Cross and a Purple Heart. We wrote each other, and he wrote that he got the Distinguished Flying Cross for opening two cans of beans under fire and the Purple Heart because he cut his finger opening the can. Later I found out they were for sinking a Japanese submarine caught on the surface. They used their last bomb in an "on the deck" run. The old Red-tailed Hawk was with the Seemann brothers.

Many other Omahans flew during the war, too. Owen Knudson, Omaha's long-time Superintendent of Schools, flew twenty missions in a B-17. Carl A. Ousley flew B-17s and B-24s in North Africa, Italy and England, and wound up a colonel with a Silver Star. Burkley Harding flew the lead B-24 on the bombing of Rabul. Hubert "Hub" Monsky flew C-54s, B-17s and B-24s. Grant D. Caywood flew a B-24 as squadron commander in the Fifteenth Air Force in Italy. He once wrote to his parents, "We unanimously agree the American Beauty is not a rose—it's a P-38." Don Moucka, Gus Baysdorfer's son-in-law, was a B-17 crew chief in the 91st Bomb Group stationed north of London. Maynard Swartz flew a C-47 and was fired upon by our own anti-aircraft while flying paratroops into Italy. Dick Lohse was a pilot instructor and flew many hours in a Curtiss AT-9, which all of us thought was about as hot as any airplane we had. Ernie Jones instructed at a half-dozen aviation schools. John Markel, Sr., flew B-17s in the Eighth Air Force. Dan Loring, Sr., drove B-25s, and former Omaha mayor Robert Cunningham was a Navy pilot. The late John Ashford flew a B-26 built in Omaha and named *Ak-Sar-Ben*. The late Harold Boelker was radioman on a B-17 and once dropped an old bicycle on Germany through the plane's bomb bay doors. Lloyd Grimm flew "the Hump" in India, was recalled in the Korean conflict, and stayed in the Air Force Reserve afterwards.

Although they weren't birdmen, Omahans John Childe and Dick Slaybaugh had notable experiences. As a "bird colonel," Childe served with General George Patton all the way from North Africa through the occuaption of Bavaria. He was the point for the Third Army during the Battle of the Bulge. Dick Slaybaugh, born and raised in Omaha's Dundee neighborhood, was an infantryman and landed on Omaha Beach in the third wave of the famous June 6, 1944, "D-Day" invasion. After the fighting ended, he and his unit pulled out their "K" rations, and a nineteen-year-old, Omaha-born young man sat on Omaha Beach, eating food prepared in Omaha by a company named Omaha Cold Storage—right smack in the middle of the greatest invasion force ever assembled. That old Red-tailed Hawk must have been somewhere around.

By the fall of 1944, we had trained over 100,000 pilots and built nearly fifty thousand planes. As a result, flight training programs was greatly reduced, even closed in many areas. I was on temporary duty at a number of bases while the programs were being phased out. In addition, many airplanes, mainly trainers, were declared surplus and were made available to the public through the Defense Plant Corporation. Back in Omaha, Marion Nelson was designated by the Defense Plant Corporation as a distributor, so many of the surplus airplanes were stored and serviced at Muni.

One day, a number of planes were flown to Omaha from Oklahoma City; then the pilots were going to return to Oklahoma City in a Lockheed P-VI. The P-VI took off on the northwest runway, climbed at a terribly steep altitude, stalled, and crashed before it got to the end of the runway. It burst into flames, and all seventeen people aboard were killed. It is the worst crash ever at Omaha, and it happened only a year after the B-24 crash. I knew two of the men on the plane. They were civilian instructors at Mustang Field in El Reno, Oklahoma, when I was the engineering officer there early in 1943.

The war in Europe was approaching the end. We were gearing up to attack Japan, and it was going to be one heck of a job.

What a year! Roosevelt died; Harry S. Truman became President; Germany surrendered; and on April 18, the Japanese bombed Omaha near 50th Street and Underwood Avenue. It was kept so hush-hush that nobody really knew about it until August when Japan surrendered, except my dad, who was opening the window in his bedroom at 5630 Western Avenue a little after midnight and saw the flash and heard the explosion. He and

> Mrs. Paul Pederson, 5004 Nicholas Street, thought it was a car backfiring. Mrs. Howard P. Smith, 5009 Western Avenue, thought the glare was the reflection of a car's headlights. Mrs. L. C. Hindman, 5021 Burt Street, said the light penetrated the drawn shades of her room. [Mrs. Hindman was a neighbor of mine much later and told me the story.] Miss Helen Sturgis, 5101 Izard Street, said the light appeared to be about the size of a football.[1]

Mary Holyoke (mother of Dr. Edward Holyoke) and her sister Leta Holdrige lived at 51st and Nicholas Streets, and both of them also saw it. Mary Holyoke reported that it looked like a ring of fire in the sky above Dundee.

Many other residents ran out in their nightclothes to see the light in the sky. It wasn't very high overhead and appeared a bright yellow color. The light turned a brilliant red and, swinging back and forth, floated slowly to the southeast. As it headed towards Dodge Street and downtown, the light disappeared. They didn't know it, but all of these folks had just seen a Japanese bomb balloon. It had been launched from Yokohama in southern Japan five days before and had ridden the Divine Wind 28,000 feet up at eighty miles per hour. The balloon had been set to fly to Ottumwa, Iowa, but evidently decided to thrill the Dundee folks on the way. It was found east of Ottumwa the next day.

Back overseas, Dick Longsdorff, an old friend from grade school days, was not so lucky. In 1945 he "bought it" when he ran out of fuel returning to Saipan from a mission in his B-29.

---

[1] *Omaha World-Herald,* August 15, 1945.

414

Also in April, another very hush-hush project started in Omaha at the bomber plant, "Operation Silver Plate." Five very special B-29s were built. They had many special changes and additions. My good friend Paul Floyd (an Okie who worked in the Martin Bomber Plant from 1941 till it closed in 1945) was busy around the clock scheduling and procuring parts not standard in B-29 assembly, and he said he really didn't know what the heck they were doing. Bill Durand, a professor at Omaha University at the time, was brought in to run the stress analysis on a special bomb-holding system. He didn't know what the bomb was, but he was given a size and weight so he could make his design. The special planes were built, then they were taken to the modification building where hardly anyone knew what was done to them. Later, five Air Corps crews (led by Col. Paul W. Tibbets, Jr.) flew the planes off. Floyd said plant workers knew they were to carry some different kind of bomb load from the changes in the bays, but that's about all they could figure out. They guessed that they had to do with the Manhattan Project.

The Manhattan Project was the name given to the secret military project for the building of the atomic bomb. My good friend Leo Quinn, also from Omaha, was involved with it. He worked for Eastman Kodak at a huge plant in Inkscap, Tennessee. I remember writing him and asking what the heck he was doing in Inkscap, Tennessee, and what was so important for him to have such a high security clearance. Even after the war he insisted he didn't know what they were doing. I'm really proud of those people who worked on the bomb, kept their mouths shut, and maintained secrecy.

On August 6, 1945, the whole world found out what had been going on when the first atomic bomb was dropped on Hiroshima, Japan. It was dropped from the *Enola Gay,* which had been built in Omaha, and Tibbets was the pilot on that mission. On August 9 a second atomic bomb was dropped on Nagasaki, Japan. Those two terrible bombs ended World War II.

In 1990 Tibbets and two of his crew members were in Omaha for a SAC Museum program. The *Enola Gay* is being restored in the Smithsonian's Paul Garber facility in Maryland and will be displayed at an aerospace museum when it is finished.

Most of the recent developments in Nebraska's aviation have been pretty well documented, so I am going to hit the highlights rather than repeat that work here. After World War II, most pilots returned to civilian life. I went back to work for Northern Natural Gas and stayed with them until I retired in 1972, but I never quit

Omaha Municipal Airport in 1945

flying. I had a couple of Stearmans and a Navy N3N3. I flew whenever possible, but I never flew full-time after the war.

Omaha Municipal developed like crazy and soon became a big-time airport. The day of the old-time flying service was over. Marion Nelson became Northern Natural Gas's first pilot. Northern bought a Stinson Voyager in 1947. It was one of the first companies in the area to have a corporate plane. I once flew with Nelson to our Ogden, Iowa, Compressor Station, and we landed in a cow pasture right beside the plant on the northwest side of town. All we had to do was climb over the fence, and we were on the station property. I'd like to see someone do that in a Citation, Saberliner or Lear.

Northern then bought a Cessna 195, hired "Slim" Kenwood, Ernest Ruckle, Donald Forman, and Arthur Pape, and started an "air force." The company bought a D-18 Beech, then a Lockheed PV, then a Douglas A-20 Havoc, and finally jets. The company even had a Super Cub and a Push-pull Cessna. Union Pacific, Mutual of Omaha, Henningson Durham and Richardson (HDR), Con Agra and many others started to fly and still do.

Omaha became important to the United States military in 1947, when the headquarters of the Strategic Air Command was moved from Washington, D.C., to Offutt Field. Gen. Curtiss LeMay put the command together. Now in 1994 the new U.S. Strategic Command, STRATCOM, is here, under the command of Gen. Lee Butler.

One of the most important airplane-happenings in 1948 was that the *Wright Flyer* came home. On December 17 at 10 a.m., exactly forty-five years after it flew at Kitty Hawk, the world's first aeroplane was given to the people of the United States by Milton Wright, a nephew of the Wright brothers. In an impressive ceremony attended by hundreds, it was unveiled, suspended high above the central entrance doors in the North Hall of the Smithsonian Institution's Arts and Industries Building. When the National Air and Space Museum was opened on July 1, 1976, it was moved to where it is today. The plaque reads:

The original Wright brothers aeroplane

The world's first

Power-driven, heavier-than-air machine in which man

Made free, controlled, and sustained flight

Invented and built by Wilbur and Orville Wright

Flown by them at Kitty Hawk, North Carolina

December 17, 1903

By original scientific research the Wright brothers

Discovered the principles of human flight

As inventors, builders, and flyers

They further developed the aeroplane, taught man to
Fly and opened the era of aviation[2]

The long road to the proper recognition of the Wright brothers' aeroplane is almost unbelievable. The Smithsonian, some of its directors, staff and boards were directly responsible for the forty-five-year delay. Samuel Pierpont Langley, secretary of the Smithsonian Institution and the nation's unofficial chief scientist, received a grant of $50,000 in 1898 from the War Department to develop a human-carrying flying machine. On October 7, 1903, his huge machine (called the *Great Aerodrome)* was launched from a barge anchored in the Potomac River. There was no flight: It slid down the launching-rail and fell into the water like a rock. Two months later, on December 8, its attempt at flight was even worse: The wings started coming off before it reached the end of the slide. For the next twenty years, the Smithsonian directors and others tried their best to prove that Langley's aeroplane was the first ever to fly, or the first that could fly. They even hired Glenn Curtiss to build a new *Aerodrome* to prove it would fly. He did, but any resemblance to Langley's original was purely accidental. Curtiss's participation in this affair added to the fight between him and the Wrights.

Tired of the bickering, Orville Wright sent the *Flyer* to England to the Science Museum in London in 1925. There it remained until 1948, when the Smithsonian admitted that its creators were the first ever to build and fly an aeroplane. When he sent the *Flyer* to England, Orville had stipulated that only by his order could it be returned to America. Unbeknownst to all, he informed the director of the London museum in writing on December 8, 1943, that he wanted the *Flyer* back. Orville died on January 30, 1948, and his unsigned will stipulated:

> I give and bequeath to the U.S. National Museum of Washington, D.C. for exhibition in the National Capital only, the Wright aeroplane (now in the Science Museum, London, England) which flew at Kitty Hawk, North Carolina, on the 17th of December, 1903.[3]

---

[2]Tom D. Crouch, "Capable of Flight: The Feud between the Wright Brothers & the Smithsonian," *Invention and Technology* (Spring 1987): 46.

[3]*Ibid.*

The executors of the Wright estate concurred, and so the *Wright Flyer*—the machine that started it all—was brought to Washington, where it hangs for all to see.

Many years later, in 1961, Omaha Municipal became Eppley Airfield. It was named after Eugene C. Eppley, who had always wanted to fly with the Red-tailed Hawk and who had claimed Omaha as his home since he first set foot in town in the early 1920s. His fortune has built everything from college buildings to an airport.

AUTHOR'S EPILOGUE

You know you are really lucky when you have been a part of something big. That's the way I feel about flying. On December 17, 1994, it had been ninety-one years since the Wrights flew the *Flyer* over the sand at Kitty Hawk. I have been flying for sixty-five years, two-thirds of the time airplanes have been here. That's why I wanted to get this record written down.

An airplane on the ground is like a fish out of water: It's awkward, unwieldy, fragile, clumsy, big and vulnerable. But when it starts its take-off roll, it becomes dynamic. When it starts to fly, it becomes unbelievably exciting! Once you have flown, you never forget the feeling or lose the desire to do it again!

I was privileged to be a part of "fun flying," barnstorming, wild flying, military flying, flying just for the heck of it, and commercial flying. I also had the honor of teaching over a hundred young men and women to fly, not to mention all the military pilots I instructed during World War II.

What an experience it has been! I got to fly the OX-5, Tail Draggers with no brakes and a tail skid, all of the Cubs (starting with the Taylor J-2), Alexander-Eaglerock, Negative-staggerwing Beech, Wacos, Fairchild 24, Travel Air, Luscombes, Howard DGA, the great gull-winged Stinson SR-9, the Stinson Voyager, Ten, Funk with a Model-B Ford engine, Arrow Sport, Curtiss Robin, Great Lakes, Stearmans, Cessnas, Bird and American Eagle, among others. I just got into the air once in a Heath Parasol when the Henderson motorcycle engine threw a rod and ended the flight. I flew Aeroncas (including the cabane-strut C-3), Taylorcrafts (one on floats off of Twin Lakes at Robbinsdale, Minnesota), a Culver Cadet with the hand-cranked retractable gear, all of the Air Corps Trainers in World War II, most of the tactical aircraft and the great B-17. I sat in the cockpit of a Spitfire, a ME-109 and a Zero (I flew them in my mind's eye). I even got to the jets, the first a Lockheed T-33. I've driven a little from the right side of a Straight Six (DC-6), a DC-7, a Turbo-prop Convair 400, a Boeing 727, and not long ago I got to fly copilot in a Cessna Citation. Last summer I spent over an hour in the cockpit of a B-1B while its great young crew told me all about her. I know I could drive her because it has a stick instead of a wheel. I even flew a two-place Switzer glider once at Miami, Oklahoma, where I was towed up by a Vultee BT-13. My only disappointment is that I have never flown a Jenny. What great stories Miller, Burnham, John and "Slim" Kenwood, Nielsen, Steele,

Orville E. Dickerhoof and all of those old guys used to tell about the famous Curtiss JN-4D, a very important airplane.

I have flown cross-country by guess and by golly; by section lines; by water towers; by cement plant smokestacks; by landing and asking, "Where in the heck are we;" by the Standard Oil Company road map; by the "iron compass" (railroad tracks); by pilotage; by dead-reckoning; by magnetic compass; by the gyro; by low-frequency radio (da-dit, dit-da); by one- and two-needle-width turns and ball; by cone-of-silence; and now by incredible computerized wizardry.

Orville and Wilbur Wright's flying machine has revolutionized the world. The airplane replaced the train and ship for distant passenger travel, and all mail and smaller things are flown everywhere. Anywhere on our earth is less than a day away. Anywhere in the contiguous states is under three hours from Omaha. We flew in space and walked on the moon.

The air up to forty thousand feet is full of airplanes. We blithely ride along at five hundred miles per hour only a few inches from a very hostile atmosphere of almost no oxygen and minus forty- or fifty-degrees below zero temperatures, worrying about getting to Denver or Los Angeles or Chicago or London in time for a meeting or dinner. Airplanes go faster than sound; they climb almost out of this world at fantastic speeds; they can put "G" pressures on pilots so fast that they can't survive even with special suits. Electronics guide, fly and navigate planes. I don't think planes really fly anymore—they "go." These systems are, for all practical purposes, foolproof, but I fear computers will be the pilots of the future. The O'Hares, Stapletons, Heathrows, and Kennedys are literally busy as beehives—there is no room for little or slow airplanes. Of course, as flying developed, the governments got into the act with their jillion rules, regulations, taxes and liability, too.

All of this is great, but I really believe that free-spirit flying is gone. The only thing that hasn't changed is pilots still fly airplanes with rudder, stick and throttle. I have been privileged.

Look Mom! See bird! I'm flying with the old Red-tailed Hawk!

It has really been grand!

Robert E. Adwers
Elkhorn, Nebraska
June 1994

BIBLIOGRAPHY

Books

Bahl, Gladys M. "The Forgotten Pilot: Errold G. Bahl, Early Instructor and Friend of Charles A. Lindbergh." San Bernardino, California, 1982. Unpublished photocopy.

Bilstein, Roger E. *Flight in America, 1900-1983: From the Wrights to the Astronauts.* Baltimore, Maryland: The Johns Hopkins University Press, 1984.

Boyne, Walter J. *The Smithsonian Book of Flight.* Washington, D.C.: The Smithsonian Institution, 1987.

Bryan, C. D. B. *The National Air and Space Museum.* New York, New York: Harry N. Abrams, Inc., 1979.

Carlisle, Robert L. *Tower, This is Andy.* Lincoln, Nebraska: Foundation Books, 1992.

Gilbert, James. *The Great Planes.* New York, New York: Grosset and Dunlap, Inc., 1970.

Hutchinson, Duane. *Savidge Brothers, Sandhills Aviators.* Lincoln, Nebraska: Foundation Books, 1982.

Josephy, Alvin M., Jr., ed. *The American Heritage History of Flight.* New York, New York: American Heritage Publishing Co., Inc., 1962.

Lindbergh, Charles A. *The Spirit of St. Louis.* New York, New York: Charles A. Scribner's Sons, 1953.

_____. *The Wartime Journals of Charles A. Lindbergh.* New York, New York: Harcourt Brace Jovanovich, Inc., 1970.

Mosley, Leonard. *Lindbergh: A Biography.* Garden City, New York: Doubleday & Company, Inc., 1976.

Oppel, Frank, ed. *Early Flight: From Balloons to Biplanes.* Secaucus, New Jersey: Castle, 1987.

Sterling, Robert J. *Legend and Legacy (Boeing).* New York, New York: St. Martin's Press, 1992.

Wescott, Lynanne and Degen, Paula. *Wind and Sand: The Story of the Wright Brothers at Kitty Hawk.* New York, New York: Harry N. Abrams, Inc., 1983.

Magazines and Journals

Crouch, Tom D. "Capable of Flight: The Feud Between the Wright Brothers & the Smithsonian." *American Heritage of Invention & Technology,* 2 (Spring 1987): 34-46.

Fetters, A. H. "Omaha, Air Capital of the Country by '40?" *Omaha Chamber of Commerce Journal,* March 24, 1928.

Newspapers

Des Moines Register and Leader (Des Moines, Iowa): *1911* June 5.

Detroit Morning News (Detroit, Michigan): *1911* July 1.

Elmira Advertiser (Elmira, New York): *1911* July 22.

Fremont Tribune (Fremont, Nebraska): *1913* August 7.

Galesburg Evening Mail (Galesburg, Illinois): *1911* June 14, June 16.

Humboldt Standard (Humboldt, Nebraska): *1920* July 2.

Lincoln Evening News (Lincoln, Nebraska): *1910* September 5, September 6.

Logansport Daily Tribune (Logansport, Indiana): *1911* June 23.

Logansport Journal (Logansport, Indiana): *1911* June 27.

Nemaha County Herald (Auburn, Nebraska): *1922* June 22.

New York Times (New York, New York): *1910* October 27.

Norton Telegram (Norton, Kansas): *1911* September 6.

Omaha Bee (Omaha, Nebraska): *1903* August 16, December 19; *1904* January 7; *1909* June 28, August 8; *1910* July 17, July 28, July 29; *1911* May 9, May 10, May 12, May 14; *1914* September 11, September 17, September 20, September 27; *1919* April 6; *1920* September 13, September 26; *1921* May 5, July 24, September 18, October 4, November 4, November 8; *1924* August 24; *1927* August 30; *1928* May 20, October 10.

Omaha Bee-News (Omaha, Nebraska): *1928* May 18; *1931* May 17, May 18, November 1; *1932* May 2.

Omaha Daily Bee (Omaha, Nebraska): *1903* December 19; *1910* July 16; *1915* July 1.

Omaha Daily News (Omaha, Nebraska): *1910* June 27; *1911* July 28, September 2; *1912* June 28; *1914* October 6; *1919* March 16; *1920* June 30.

Omaha World-Herald (Omaha, Nebraska): *1904* August 17; *1907* January 7, October 4, October 22; *1910* July 24, July 25, July 26, July 28, October 30, November 18, November 27; *1911* April 8, May 10, May 11, May 15,

September 13; *1912* May 30, October 27; *1913* August 14; *1914* October 6; *1919* December 5; *1921* May 15, July 24, October 26, November 4, November 5, November 6, November 7, November 8; *1924* September 18; *1925* May 5; 1928 May 13; *1928* May 27; *1929* July 21, July 28, October 6; *1931* May 13, May 15, May 16, May 18; *1932* May 31, August 24; *1934* August 12, August 13; *1938* May 19; *1939* June 21, November 19; *1940 November 7; 1941* December 8; *1942* August 27; *1943* June 4; *1944* October 15; *1945* August 15; *1961* April 16; *1967* September 17; *1981* March 1.

Papillion Times (Papillion, Nebraska): *1912* July 4.

Rochester Democrat and Chronicle (Rochester, New York): *1911* July 8.

Sioux City Journal (Sioux City, Iowa): *1911* May 24.

Terre Haute Tribune (Terre Haute, Indiana): *1911* June 16, June 17, June 19.

Personal Communication

Bahl, Errold Gordon, to Charles E. Marburger, San Bernardino, California, December 28, 1992.
Baysdorfer, Gerald. Interview, Omaha, Nebraska, April 1990.
Bishop, Lydia. Interview, Norton, Kansas, February 1989.
Boettger, Anna Otte. Interviews, Irvington, Nebraska, June 1988 to present.
Durand, William. Conversations, Omaha, Nebraska, September 1933 to present.
Foisey, Judy Baysdorfer. Interviews, Omaha, Nebraska, August 1990 to present.
Frost, George. Interview, Elkhorn, Nebraska, June 1988.
Gow, Flora May "Brownie" Rimmerman. Interviews, Omaha, Nebraska, 1970 to 1993.
Halsey, Harold. Conversations, Omaha, Nebraska, 1960 to present.
Holmgren, Frances. Interviews, Omaha, Nebraska, December 1988 to present.
Joyce, Richard O. Interview, Lincoln, Nebraska, June 1953.
Marburger, Charles E. Interviews, Humboldt, Nebraska, 1988 to present.
Meyer, William. Elkhorn, Nebraska, May 1989.
Moeller, Hans. Interview, Fremont, Nebraska, July 1988.
Moucka, Betty Baysdorfer (Mrs. Donald). Interviews, Omaha, Nebraska, February 1988 to present.
Nelson, Marion. Conversations, Omaha, Nebraska, 1938 to 1993; Modesto, California, 1993 to present.
Noyes, Mrs. Halsey. Interviews, Waterloo, Nebraska, 1987 to present.
Thanis, Polly, to Robert E. Adwers, Omaha, Nebraska, 1994.
Webber, Irene. Interview, Omaha, April 1989.

426

Wilson, Ralph.   Conversations, Waterloo and Elkhorn, Nebraska, 1980 to
     present.

Archives and Libraries

Auburn Public Library, Auburn, Nebraska.
Broken Bow Historical Society, Broken Bow, Nebraska.
Brunn Memorial Public Library, Humboldt, Nebraska.
Davenport Public Library, Davenport, Iowa.
Detroit Public Library, Detroit, Michigan.
Edith Abbott Memorial Library, Grand Island, Nebraska.
Galesburg Public Library, Galesburg, Illinois.
Historical Society of Douglas County, Omaha, Nebraska.
Kearney Public Library, Kearney, Nebraska.
Keene Memorial Library, Fremont, Nebraska.
Logansport-Cass County Public Library, Logansport, Indiana.
Nebraska State Historical Society, Lincoln, Nebraska.
Norfolk Public Library, Norfolk, Nebraska.
Norton Public Library, Norton, Kansas.
Ottumwa Public Library, Ottumwa, Iowa.
Public Library of Des Moines, Des Moines, Iowa.
Rochester Public Library, Rochester, New York.
St. Joseph Public Library, St. Joseph, Missouri.
Sioux City Public Library, Sioux City, Iowa
Steele Memorial Library, Elmira, New York.
Vigo County Public Library, Terra Haute, Indiana.
W. Dale Clark Library, Omaha, Nebraska.
Western Heritage Museum, Omaha, Nebraska.

438

Making History is a historical consulting firm in Omaha, Nebraska, that specializes in historical research, writing, lectures, exhibits and publishing, in addition to the restoration and reprinting of historic and heirloom photographs. If you are interested in preparing a special historical exhibit, if you need the perfect lecturer for an event or fundraiser, or if you have a non-fiction historical manuscript you wish to publish, write or call us at:

Making History

2415 N. 56th Street
Omaha, Nebraska 68104
402-551-0747

MAIL ORDER FORM

Please send the following to the address below (PRINT CLEARLY):

Name____

Organization____

Street/Box____

City__________________________State______________Zip Code__________

QUANTITY	TITLE	TOTAL
______	Jerry E. Clark, *Nebraska Diamonds: A Brief History of Baseball Major Leaguers from the Cornhusker State* (92pp; ISBN 0-9631699-0-4) $11.50	$______
______	Jerry E. Clark, *Anson to Zuber: Iowa Boys in the Major Leagues* (234pp; ISBN 0-9631699-1-2) $20.00	______
______	Jerry E. Clark and Martha Ellen Webb, *Alexander the Great: The Story of Grover Cleveland Alexander* (60pp; ISBN 0-9631699-2-0) $11.00	______
______	Robert E. Adwers, *Rudder, Stick and Throttle: Research and Reminiscences on Flying in Nebraska* (458pp; ISBN 0-9631699-4-7) $25.00	______
______	Dr. Allen Shepherd, Belvadine R. Lecher, Lloy Chamberlin, and Marguerite Radcliffe (eds.), *Man of Many Frontiers: The Diaries of "Billy the Bear" Iaeger* (553pp; ISBN 0-9631699-3-9) $25.00	______

Subtotal $______

Nebraska residents add 6.5% sales tax ______
(unless providing a tax number)

Add $3.00 for your first copy; $1.75 each additional (shipping) ______

Total balance enclosed (US funds only) $______
Please make check or money order payable to *Making History*.

Mail your completed order form and check or money order to:
Making History, 2415 N. 56th Street, Omaha, Nebraska 68104.

Organizational/wholesale discounts available.
Write or call 402-551-0747 for information.

MAIL ORDER FORM

Please send the following to the address below (PRINT CLEARLY):

Name____

Organization____

Street/Box____

City_____________________________State____________Zip Code________

QUANTITY	TITLE	TOTAL

______ Jerry E. Clark, *Nebraska Diamonds: A Brief History of Baseball Major Leaguers from the Cornhusker State* (92pp; ISBN 0-9631699-0-4) $11.50 $______

______ Jerry E. Clark, *Anson to Zuber: Iowa Boys in the Major Leagues* (234pp; ISBN 0-9631699-1-2) $20.00 ______

______ Jerry E. Clark and Martha Ellen Webb, *Alexander the Great: The Story of Grover Cleveland Alexander* (60pp; ISBN 0-9631699-2-0) $11.00 ______

______ Robert E. Adwers, *Rudder, Stick and Throttle: Research and Reminiscences on Flying in Nebraska* (458pp; ISBN 0-9631699-4-7) $25.00 ______

______ Dr. Allen Shepherd, Belvadine R. Lecher, Lloy Chamberlin, and Marguerite Radcliffe (eds.), *Man of Many Frontiers: The Diaries of "Billy the Bear" Iaeger* (553pp; ISBN 0-9631699-3-9) $25.00 ______

Subtotal $______

Nebraska residents add 6.5% sales tax ______
(unless providing a tax number)

Add $3.00 for your first copy; $1.75 each additional (shipping) ______

Total balance enclosed (US funds only) $______
Please make check or money order payable to *Making History*.

Mail your completed order form and check or money order to:
Making History, 2415 N. 56th Street, Omaha, Nebraska 68104.

Organizational/wholesale discounts available.
Write or call 402-551-0747 for information.